T0201819

An Introduction to Optimal Designs for Social and Biomedical Research

STATISTICS IN PRACTICE

Advisory Editor

Stephen Senn
University of Glasgow, UK

Founding Editor

Vic Barnett
Nottingham Trent University, UK

The texts in the series provide detailed coverage of statistical concepts, methods and worked case studies in specific fields of investigation and study.

With sound motivation and many worked practical examples, the books show in down-to-earth terms how to select and use an appropriate range of statistical techniques in a particular practical field. Readers are assumed to have a basic understanding of introductory statistics, enabling the authors to concentrate on those techniques of most importance in the discipline under discussion.

The books meet the need for statistical support required by professionals and research workers across a range of employment fields and research environments. Subject areas covered include medicine and pharmaceutics; industry, finance and commerce; public services; the earth and environmental sciences.

A complete list of titles in this series appears at the end of the volume.

An Introduction to Optimal Designs for Social and Biomedical Research

Martijn P. F. Berger

Department of Methodology and Statistics
Maastricht University, The Netherlands

Weng Kee Wong

Department of Biostatistics, School of Public Health,
University of California, Los Angeles, USA

A John Wiley and Sons, Ltd., Publication

This edition first published 2009
© 2009, John Wiley & Sons Ltd

Registered office
John Wiley & Sons Ltd, The Atrium, Southern Gate, Chichester, West Sussex, PO19 8SQ, United Kingdom

For details of our global editorial offices, for customer services and for information about how to apply for permission to reuse the copyright material in this book please see our website at www.wiley.com.

Library of Congress Cataloguing-in-Publication Data:

Berger, Martijn P. F.
  An introduction to optimal designs for social and biomedical research /
Martijn P. F. Berger, Weng Kee Wong.
    p. cm.
  Includes bibliographical references and index.
  ISBN 978-0-470-69450-3
1. Social sciences–Research. 2. Biology–Research. 3. Optimal designs
(Statistics) I. Wong, Weng Kee. II. Title.
  H62.B422 2009
  300.72–dc22

                                                          2009008347

A catalogue record for this book is available from the British Library.

ISBN 978-0-470-69450-3

Typeset in 10/12pt Times by Laserwords Private Limited, Chennai, India.
Printed and bound in the UK by TJ International, Padstow, Cornwall.

Contents

x   CONTENTS

Preface

One of the most crucial steps in the research process is at the design stage. In order to draw valid and reliable conclusions from a social, behavioural or biomedical study, the design has to meet a number of requirements. These requirements have been well documented and discussed in various textbooks in these fields, see, for example, Campbell and Stanley (1963), Cox (1958), Cook and Campbell (1979) and Cox and Reid (2000). One requirement that seems to lack emphasis and underappreciated by practitioners is the importance of a well-planned design. This is because a carefully designed study can provide valid and precise statistical inference at minimal cost given the constraints in the study.

Optimal design of experiment is a sub-field in statistics that provides the theory and construction of optimal designs in various settings. For many problems, algorithms are available to generate optimal or highly efficient designs. There are currently only a few monographs on this important topic and they are largely concerned with the mathematical treatment of the subject, including the analytical derivation of the optimal designs. Some examples are Fedorov (1972), Pázman (1993), Pukelsheim (1993) and Silvey (1980). The monographs by Atkinson and Donev (1992) and Atkinson, Donev and Tobias (2007) contain many examples drawn from engineering and pharmaceutical experiments.

This book takes a different route and addresses applied researchers from the social, behavioral and biomedical areas who have limited mathematical training and want to learn more about optimal design ideas and methods. Our primary aim is to present this important topic to our target audience in an expository manner. The focus is on design issues and methods to find more efficient designs for their studies using internet tools or otherwise. Using examples from the social and biomedical sciences, we demonstrate the methods to find efficient designs and how to compare merits of competing designs, including commonly used designs in practice. Throughout, we have tried to avoid matrix algebra as much as possible. For those readers who are familiar with matrix algebra, we have also included some matrix algebra near the end of the first eight chapters to facilitate discussion. However, they can be skipped without too much disruption. We were not able to avoid matrix algebra in Chapters 9–11, but kept its use to a minimum.

The book consists of 11 chapters. Each chapter focuses on the design issues for a commonly used statistical model. In Chapter 1, we give a short overview of the general statistical set-up, design methodology and ethical principles in conducting a study. We discuss the basic design elements and different types of research, including different forms of validity. We also discuss control mechanisms of unwanted variation in a study and provide a list of some of the ethical requirements expected in a scientific study. Finally, we describe examples taken from the social and biomedical research that will be used in later chapters.

Chapter 2 gives an introduction to design issues for a simple linear regression model. We describe the concept of simultaneous inference, present commonly used alphabetic optimality criteria and highlight the pros and cons of each type of optimal designs. Chapter 3 extends the concept of simultaneous inference to the multiple linear regression model. Because there are several parameters in the model, we introduce an optimality criterion for making inference only on a subset of the model parameters. We present some optimal designs for the polynomial regression models. Chapter 4 discusses design problems for the analysis of variance model, where the independent variables are typically qualitative variables. A two-factorial analysis of variance model is reformulated as a regression model with dummy variables and we show how the different dummy coding systems affect the outcomes. Chapter 5 discusses design issues for the logistic regression model with a binary response. In particular, we focus on designs for the logistic model with one or two qualitative and quantitative independent variables.

In Chapters 6 and 7, we entertain random parameters in the linear models. Chapter 6 introduces the multilevel model, along with designs for a cluster randomized and multi-center two-armed trial. When one of the arms is the control group, we compare efficacy of the treated group under different cost structures and provide designs with maximal power for testing the treatment effect. Chapter 7 discusses the design problem for random effect models in a longitudinal repeated measurement model. Here, we consider models with either a random intercept or a random slope with illustrations in the social and biomedical sciences. We also show that it is instructive to use graphs to display an optimal design for a longitudinal study as a function of the auto-correlation. Chapter 8 reviews design issues for a crossover study and discusses confounding and carry-over effects. We focus on the two-treatment crossover design and use a mixed-effects model to compare the efficiency of the 2×2 crossover design with competing designs while taking cost into account.

Chapter 9 introduces additional optimality criteria useful in practice for the linear models. Our discussion here includes Bayesian optimal designs, minimax optimal designs and multi-objective designs with illustrations. Chapter 10 addresses design issues for nonlinear models and starts off with the defining difference between linear and nonlinear models. We explain two core concepts in designing for a nonlinear model: the Fisher information matrix and local optimality. We discuss relative merits of using different design strategies for a nonlinear model and provide exemplary applications in the biological sciences.

Chapter 11 reviews popular sequential procedures and algorithms for obtaining an optimal design numerically. We describe the exchange algorithms and provide references for some of the more recent algorithms such as the generic and stochastic evolutionary algorithms. In addition, we provide a short list of computer programs currently available for finding optimal designs and introduce a new web site for generating a variety of optimal designs for selected models. This chapter concludes with demonstrations on the use of this freely accessible web site to generate tailor-made optimal designs after the user selects the model and inputs the design parameters.

Acknowledgements

We would like to thank the following individuals who provided valuable comments and assistance in the preparation of this book: In alphabetic order: Nick Broers, Gerard van Breukelen, Math Candel, K. C. Carriere, Heinz Holling, Valeria Lima Passos, Frans Tan, Fetene Bekele Tekle and graduate students who took the class Biostat 279 from the Department of Biostatistics at UCLA. We also thank anonymous reviewers from Wiley who made many suggestions that improved the book, Petry Kievit-Tyson for editorial comments on a portion of this book and Marga Doyle-Mommers for secretarial support. Of course, all remaining errors are ours. A large portion of the work by the first author was done while he was a fellow at the Netherlands Institute of Advances Studies in the Netherlands (2007–2008). His work was partly supported by a grant from the Netherlands Organization for Scientific Research (NWO, 400-05-702). The work of the second author was partly supported by a NIH grant award R01GM072876.

Martijn P. F. Berger
Department Methodology and Statistics
School of Public Health and Primary Care
Faculty of Health, Medicine and Life Sciences
Maastricht University
Maastricht, The Netherlands
martijn.berger@stat.unimaas.nl

Weng Kee Wong
Department of Biostatistics
School of Public Health
University of California at Los Angeles
Los Angeles, USA.
wkwong@ucla.edu

2009

1

Introduction to designs

1.1 Introduction

There are many statistical issues to consider in the design of an empirical study. Among these problems are the control of unwanted variation and the internal validity of a study. How can we be sure that a study is internally valid? In other words, how can we be sure that the treatment effect is attributed to the variables that are manipulated and not mainly influenced by unwanted variation? These questions and related problems are well documented and have been discussed in the social and biomedical literature by Cox (1958), Campbell and Stanley (1963), Cook and Campbell (1979), Cox and Reid (2000) and Shadish, Cook and Campbell (2002), among others. However, one important aspect that seems lacking in the discussion is the question whether the implemented design is the most efficient one for the objective or objectives of the study. We believe that researchers should always keep this question in mind when they design their studies. Doing so can invariably result in improved designs with higher statistical efficiency at minimal cost. Here and throughout, we use the term *statistical efficiency* or simply *efficiency* to indicate the accuracy of estimators of the model parameters in terms of the variances of the estimators.

Here is a simple illustration. Consider the 1987 Report of the Second Task Force on Blood Pressure Control in Children in the journal *Pediatrics* on the linear relation between systolic blood pressure and age. This conclusion was drawn from studies with samples of children with ages ranging from 12 to 18 years. Despite this informative finding, many subsequent studies continue to use inefficient designs to study the relationship between the two variables. Frequently, the design ignores existing information from the literature or simply comprises

An Introduction to Optimal Designs for Social and Biomedical Research   M. P. F. Berger & W. K. Wong
© 2009, John Wiley & Sons, Ltd

equal number of children from different age groups equally distributed over the age range. For instance, the design may sample equal number of children at 12, 14, 16 and 18 years old (*Design 1*), or the design may have equal number of children at 12, 15 and 18 years old (*Design 2*). These are called *uniform designs* because they have an equal number of observations from each age group that are equally spaced over the age of interest. Uniform designs are popular because they are intuitive and simple to implement. However, these designs can be rather inefficient in part because (i) they do not incorporate existing information on the two variables in the study and (ii) no rationale is provided for in the choice of the age groups included in the study. For instance, why were the age groups of 12, 15 and 18 selected in the second uniform design? Would another design with age groups of 13, 15 and 17 be equally acceptable?

These two questions often lead to a host of other design questions, for example: (i) Is *Design 1* better than *Design 2*? (ii) Is it a good idea to sample only two age groups in the study? (iii) What about sampling cost–can I additionally include cost considerations? (iv) What happens if sampling cost is proportional to the age of the children? Children older than 18 years are excluded from the sample, but I want to make inference on the relationship between blood pressure and children older than 18 as well. Is *Design 1* better than *Design 2* for this purpose? Can I have a design that is good for studying the relationship between systolic blood pressure and age of children up to 18 years old and also for adults as well? Is there a 'best' design for the study? These are complicated questions and sometimes there are no clear answers for a practical design problem that can have many more constraints and objectives.

Optimal design theory offers a useful foundation answering these and other design questions. The beautiful framework allows us to find the best design for a given problem using computer algorithms. For simpler problems, the theory also enables us to determine the design analytically. This is useful because with a formula for describing the design, we can study properties of the design a lot more easily. Although optimal design theory has been available for many years, social, behavioural and biomedical scientists seem to have little exposure to its theory and potential applications in their fields. The aim of this book is to promote interest among researchers from these fields in the use of optimal design theory and enable them to design more efficient and less expensive studies.

Since the work of Fisher (1925, 1935), statisticians have worked on optimal design problems. A small sample of useful references over the years are Cox (1958), Kiefer (1959), Kiefer and Wolfowitz (1959), Fedorov (1972), Box, Hunter and Hunter (1978), Silvey (1980) and Atkinson and Donev (1992). There was a surge in research in this area in the early 1960s after the seminal papers by Kiefer and Wolfowitz (1959, 1960) showing that a design can be simultaneously optimal under two very different and useful objectives in the study. The monograph by Kiefer (1985) is a voluminous collection of pioneering work in this area by the author and provides a good account of the chronological developments of optimal design theory. However, despite the ubiquity of design issues in all studies, the number of statisticians working actively in this important area has always been relatively small and continue to be so, compared to the larger statistical community working on data analysis issues. We believe there are several reasons for this.

Firstly, there has always been a strong focus on analytical solutions for optimal design problems, and the books published on optimal design theory require readers to have a good mathematical training to appreciate the theory. Although many social, behavioural and biomedical scientists receive training in statistical methods, this training does not include basic concepts of optimal design theory. Still another reason is that many statistics/biostatistics programmes at universities worldwide tend to give short shrift to design issues in their course curriculum. Another reason that optimal design ideas are underutilized in research is that its theory is generally perceived as too limiting. This is incorrect. For example, a popular criticism is that optimal design strategies are necessarily myopic. A common cited example is the optimal design for the homoscedastic simple linear model that requires an equal number of observations to be placed only at the two extreme ends of the design interval. This design cannot detect curvature and so why put all the eggs in one basket? Our response is that modern development in optimal design theory enables the researcher to construct more flexible designs that can capture the constraints and goals of the studies more realistically. In this particular example, one can construct an optimal design that balances the dual goals of estimating the two parameters in the linear model and at the same time estimates the curvilinear parameter with a user-specified level of efficiency. We will revisit this issue in Chapter 9.

However, there is increasing interest and realization that the theory can be more broadly applied to practical problems. Atkinson (1996) gave a compelling account of the usefulness of optimal designs and their potential applications to other fields. Some examples of use of optimal designs in social and behavioural research are McClelland (1997), Raudenbush (1997), Moerbeek, Van Breukelen and Berger (2000) and Berger (2005). The monograph by Berger and Wong (2005) is an attempt to show interesting applications of optimal design in different disciplines.

The use of optimal or highly efficient designs in social, behavioural and biomedical research has advantages. Fewer observations and therefore smaller sample sizes are required to find real effects, thus reducing the costs of the study. Our examples in this book show that optimal designs can reduce the number of observations from 20% to 40%, and even 50% in some cases when compared with the traditional or commonly used designs. This is especially beneficial in light of the ever-rising cost of conducting scientific studies. From an ethical viewpoint, a smaller sample is also highly desirable. For example, fewer patients may be required to undergo a controversial treatment or fewer animals need to be sacrificed in a toxicology study.

In this chapter, we review basic methodological concepts in the design of a study. We describe design terminology, types of different designs, requirements for a 'good' design and different kinds of validity issues that may arise in a study and how to control them.

1.2   Stages of the research process

In social and biomedical research, there are six stages in the research process:

1. *Formulation of the research problem.* Many research problems may be formulated as a relationship between a set of variables X and an outcome variable Y. The main task is to identify the set of variables X to include in the study and identify the outcome variable Y of interest. For example, a psychologist wants to study the effect of a systematic desensitization therapy (X) on phobic reactions (Y) of patients or a health scientist wants to study the relation between sources of job-related stress (X) and burnout rate (Y) of professionals in medicine.

2. *Choice of the research design.* A research design is used to structure the research and data collection. A design choice not only includes the selection of the number of independent variables, a distinction between qualitative versus quantitative variables, random versus fixed variables and a crossed or nested relation among variables but also the selection of the number of measurements, time points and subjects within groups. At this stage, we decide on the sources and amount of unwanted variation to control for in the design.

3. *Choice of statistical model.* For design purposes, the statistical model must be chosen before the data are collected. The model is a mathematical relationship posited in Stage 1 and describes the outcome variable Y as a function of the variables X and the error term.

4. *Data collection.* In this stage, the data are collected based on the design chosen in Stage 2.

5. *Analysis of data.* The data are analysed based on the statistical model chosen in Stage 3. Regression diagnostics are used to check model assumptions and whether the model provides an adequate fit.

6. *Conclusions.* Conclusions are carefully inferred from the data analysis in Stage 5 and they may lead to reformulation of the research problem or to additional research problems.

These six stages are visualized in Table 1.1. The bold horizontal line in Table 1.1 separates theoretical considerations from practical issues. Since the choice of a design and the choice of a statistical model are very closely related, Stage 2 and Stage 3 may sometimes be interchanged in practice. In Stage 4, data are collected according to the design chosen in Stage 2 and in Stage 5 the data are analysed based on the statistical model selected in Stage 3. The arrows in Table 1.1 represent the connection between the activities in Stages 2 and 4, and between the activities in Stages 3 and 5, respectively. The problem is that in order to be able to choose a 'good' design, information about the true model or the data generating process is needed and this information is not available in Stage 2.

Table 1.1   Main stages of the research process.

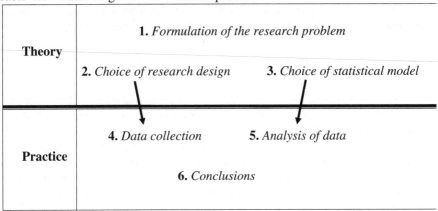

1.2.1   Choice of a 'good' design

This book focuses on the selection of a research design (Stage 2 of Table 1.1). The typical aim in social or biomedical research is to evaluate the effects of one or more independent (treatment) variables on an outcome variable. For example, we may want to design a study to ascertain whether a new teaching method is more effective than the current method in raising the average score in mathematics among high school students in the Los Angeles County. As such, it is important to have a design capable of estimating the treatment effect precisely and also has maximal power for testing the hypothesis of interest. In practice, an efficient ('good') design is chosen under a set of constraints that usually adds further complications to the design problem. These restrictions can be cost constraints, feasibility constraints and ethical constraints. An example of a feasibility constraint could be a study aimed at controlling the increasing obesity rates among children in the United States. A new method may be potentially effective in getting children to lose weight, but if the method requires children to be segregated in classrooms, this may be problematic to implement. Parents are likely to object and such a design may not be feasible.

Optimal design theory offers a systematic way for finding an optimal or a highly efficient design using all current information for the problem at hand. Once the statistical model is specified and the objective or objectives of the study are clearly stated, along with the constraints, there are proven optimization methods that will generate the optimal design. In many situations, these numerical procedures can be easily automated. This means that there are computer programs that can generate tailor-made optimal designs sequentially. Many of these algorithms are guaranteed to converge to the optimal design; more recent development includes applying these algorithms to find multiple-objective optimal designs and include cost constraints as well. We postpone discussion of these algorithms to Chapter 11.

In the next subsections, we describe the basic elements of research designs, different types of designs, a brief overview of the methodological and ethical requirements of a 'good' design and give references in the literature.

1.3   Research design

The research design describes how data are to be collected to test whether the posited relations among variables hold or not. It is a plan for collecting and utilizing data so that information is generated to test hypotheses. A good or poor research design may be characterized by the amount of information it generates and its power for testing the hypotheses. Specifically, the research design determines conditions under which the study is to be carried out. Conditions here refer to selection of the combination of levels of all the *independent variables*, including how the *units of analysis* are allocated to each of the conditions and how many *replications* are planned. The choice of the design is based on a hypothetical relation between *independent (predictor, explanatory) variables* and *dependent (response) variables* posited in Stage 1 where it is also assumed that variation in the independent variables leads to changes in the dependent variable (effect).

1.3.1   Choice of independent variables and levels

A researcher must address the questions of how many independent variables to use and how many values (levels) of these independent variables are needed to investigate the relation between the independent variables and the dependent variable. If a whole range of levels of an independent variable is of interest and the study only includes a few of these levels, then it is best to randomly select these few levels from that range. Such independent variables are classified as *random factors*. On the other hand, the so-called *fixed factors* are independent variables with a fixed number of levels, such that inferences are only limited to these fixed levels. For example, the factor Age can be either fixed or random. Age is a fixed factor if persons from a fixed set of age groups are included in the study and the researcher is only interested in drawing conclusions for these fixed age groups. The factor Age is random if the age groups are randomly drawn from the complete range of all ages in the population so that conclusions apply to the whole range of ages. Fixed and random factors must be clearly identified in the statistical model because they affect our tests concerning model parameters. The actual number of selected levels of a factor also influences the quality of our statistical inferences. For example, we show in the next few chapters that the variances of the estimated parameters and the power for finding real effects depend on the number of selected levels of the independent variables.

1.3.2   Units of analysis

The unit of analysis is the entity that is being analysed. *Units of analysis* are often referred to as *subjects*. Examples of the units of analysis are objects, organizations, general practices, patients, workers, pupils, and nurses. These units may be nested hierarchically within other units. For instance, a general practice has its own sample of patients. This sample of patients is nested within this general practice and not in another general practice. In the same way, nurses are nested within hospitals where they work, and pupils are nested within their own schools. Such a nested structure will lead to specific designs and the analysis of the data will have to take the hierarchical data structure into account.

1.3.3   Variables

A variable is defined as an attribute, property or characteristic of a unit. Examples are the colour of objects, size of organizations or general practices, blood pressure of patients, income of workers, ability of pupils, and workload of nurses. Different classifications of variables are distinguished in the scientific literature. To select both a research design and the proper analysis technique, the types of variables involved must be known. Variables are considered to be *qualitative* or *quantitative*.

A *qualitative variable* is a variable in which units are grouped into a set of mutually exclusive categories. Examples are gender, race, treatments, types of cancer, and departments in a hospital. Qualitative variables are also referred to as *grouping* or *categorical* variables. The differences are expressed in different groups or categories, for example, male and female or colon cancer and lung cancer. Stevens (1951) referred to these variables as *nominal* variables.

A *quantitative variable* is one in which units differ in degree. A quantitative variable is also referred to as a *numerical variable* and its measurement may be continuous or discrete. A measurement of such a quantitative variable may be on an *interval scale* or on a *ratio scale* (Stevens, 1951). An example of an interval-scaled variable is a patient's temperature. Interval variables are invariant under linear transformations. For example, temperature expressed in degrees Celsius may be transformed to the Fahrenheit scale without loss of information. Ratio-scaled variables have an absolute zero point that is meaningful. Examples of ratio-scaled variables are length, weight, response time, and blood pressure of patients.

Finally, an *ordinal-scaled variable* (Stevens, 1951), which merely assigns numbers to units to reflect the ordering of these units on a variable, cannot easily be classified as quantitative or qualitative. It is quantitative in the sense that the direction of any association involving the ordinal variable can be expressed. On the other hand it may be considered to be qualitative, because the intervals of the scaled measurement merely represent an ordering.

A quantitative or qualitative variable is usually symbolized by a capital letter X or Y. The corresponding realization of the measurement is symbolized by a lower case letter x or y, respectively. The direction of a hypothesized relation usually coincides with the distinction between the *independent* (*predictor*) variable and *dependent* (*outcome*) variable. If, for example, one is interested in the effect of dosages of a drug on the systolic blood pressure of patients, then the direction of the relation can be visualized as

$$X \text{ (dosage level)} \rightarrow Y \text{ (bloodpressure)}.$$

In practice an independent variable X can be measured differently in different studies. Dosages may be either represented by a rank ordering, that is, $x = 1$, $2, 3, 4$, and so on, or by the actual values of each dosage level, for example, $x = 6, 9, 15, 20$ mg/kg, and so on. In the first case, the independent variable dosage level X is measured on an ordinal scale, while in the latter case dosage level X is measured on a ratio scale.

1.3.4   Replication

The concept of *replication* is important in every study, because replications are needed to estimate the variance of the errors in a statistical model. Typically, replications are assumed to be independent; if they are not, the error variance will be underestimated and a Type I error results. We recall Type I error is the probability of rejecting the null hypothesis when it is true. In micro-array experiments, a further distinction between technical (within samples) replications and biological (between samples) replications is often made.

Replications should be distinguished from *repeated measurements*. Repeated measurements can also be obtained under the same conditions (same combinations of factor levels), but are not independent and often display a trend over a period of time. Such repeated measurements should be treated as a separate factor in the design. An example of a set of repeated measurements is the blood pressure readings at successive occasions from patients taking the cholesterol-lowering medication Zocor.

1.4   Types of research designs

Different research designs are used in social and biomedical research. They range from purely experimental designs to observational types of designs.

An enormous amount of literature is available for conducting *experimental designs* in a controlled environment. Examples are factorial designs, cross-over designs, blocked and Latin square designs. These designs are referred to as *experimental designs* because the researcher has full control over the design of the study. Full control means that the experimenter can manipulate the treatment (or independent) variables and the experimental units are randomly assigned to the conditions. Further details are in Winer, Brown and Michels (1991), Kirk (1995) and Montgomery (2000), among others.

In *observational designs* the researcher does not have full control over the variation of the variables. Examples of such designs without complete control in biomedical and epidemiologic research are (prospective) cohort designs, cross-sectional designs and case–control designs, see Rothman and Greenland (1998) for details. Examples of observational studies in social research are quasi-experimental designs, survey designs, and non-experimental designs. See Cook and Campbell (1979), Campbell and Stanley (1963) and Shadish, Cook, and Campbell (2002) for details on the problems connected with quasi-experimental designs. Quasi-experimental studies may contain a manipulation of the treatment variable, but they are carried out in existing environments with existing groups of subjects. In general, observational designs are more exposed to threats to internal validity mentioned in Section 1.5.2.

One could infer that observational designs have less control over all possible sources of unwanted variation than experimental designs and as such maintain less internal validity. On the other hand, observational studies are often chosen because the study cannot be easily carried out in a controlled environment. Concato, Shah and Horwitz (2000) and Benson and Hartz (2000) conducted meta-analyses to compare results from observational and experimental studies and concluded that well-designed observational studies may produce results that are more or less comparable to those obtained in purely experimental studies. However, it should always be kept in mind that observational studies are not experimental and therefore cannot guarantee that treatment effects are unbiased. Carefully designed randomized controlled studies are always preferable (Pocock and Elbourne, 2000).

1.5   Requirements for a 'good' design

The objectives of a 'good' design are to provide *interpretable*, *accurate* and *valid* conclusions. A number of methodological requirements for a good design have been described in the literature on the design of experiments and other studies, see, for example, Cox (1958, Section 1.2), Box and Draper (1987, Chapter 14) and Atkinson and Donev (1992, Chapter 6), among others. Although these requirements were originally only formulated for experimental designs, they also apply to other types of studies with less control over the independent variables.

A 'good' design should provide valid and reliable statistical inference. Campbell and Stanley (1963) were probably the first to distinguish internal validity from external validity and describe possible threats to both forms of validity. Cook and Campbell (1979) expanded this classification into four distinct types of validity, namely, statistical conclusion validity, construct validity, internal and external validity:

- *Statistical conclusion validity* is concerned with the validity of the conclusions based on the statistical methods employed. If the statistical analysis is inappropriate, the conclusions may not be valid.

- *Construct validity* is the correspondence between the measure and the construct that is being measured. Construct validity is usually investigated by empirical testing of hypothesized relations.

- *Internal validity* is concerned with the question whether the effect found in a study can be attributed to the variables that are manipulated.

- *External validity* is the generalizability of the conclusions. Can the conclusions of current study be generalized to other populations?

A 'good' design at the very minimum should provide control over the threats to *statistical conclusion validity* and *internal validity*. We focus on these two types of validity in the following subsection. For a detailed explanation of construct validity and external validity, we refer the readers to Cook and Campbell (1979).

1.5.1   Statistical conclusion validity

There are a number of threats to statistical conclusion validity and they can take on several forms. A good design should guard against these threats. Below is a short list of these threats and some advice on how one can minimize these threats.

1.5.1.1   Lack of statistical power for finding real effects

A 'good' design should have sufficient power for finding real effects. Low power may lead to the erroneous acceptance of the null hypothesis.

This problem may arise because of inappropriate choice of design, and parameters are poorly estimated. A carefully chosen design can improve the parameter estimates and power of the tests of interest. One way to minimize such a threat is to make sure that the ranges of the independent variables are properly selected. In practice, this means that each of the independent variables in the study should have large enough variance to fully capture the outcome variation as the value of the independent variable varies. For example, consider the outcome as the reaction time between application of the stimulus and a response to that stimulus. The reaction time of subjects decreases as young subjects grow older but then increases as older subjects become older. Consequently, if one limits the age range by selecting only young subjects in a study, one will not have enough information to know and test for the curvilinear relation between age and reaction time.

Another way to minimize the lack of power threat is to have a large enough sample size in the study. The cost of a study is always an important factor and so the researcher always has to find a balance between reigning the costs of the study and having enough power to detect treatment effects. Optimal design theory offers methods that can find optimal designs with sufficient power and efficiency at minimum cost.

A way to find out that there is lack of power for finding real effects is when we observe that the standard errors of the estimates are too large. One possible cause may be because of inaccurately measured variables. The error variance of the statistical model becomes large when the independent and dependent variables have large measurement errors. This may happen, for example, when the environment of the study is unstable or when the selected subjects have characteristics that cause the dependent variable to have relatively large variation. Such unreliable measurements can result in a high probability of incorrectly failing to reject the null hypothesis and loss of power for finding real effects.

Additional statistical issues on power and sample size calculation are discussed with examples in basic statistics monographs, such as Pocock (1983) and Chow, Shao and Wang (2007).

1.5.1.2 Violation of model assumptions

To prevent incorrect inferences, a design should allow the researcher to check model assumptions and to check for goodness of fit of the model. One way to infer that the standard errors are small is to check for model fit. This can be done before the actual study is performed in a so-called pilot study or this can be done by means of the data in the study itself.

A 'good' design should generate data that allow the researcher to perform a lack of fit test on the postulated model and enable the researcher to check specific aspects of the model. We emphasize that a design may be efficient for one statistical model but not for another. Consequently, it is important to keep in mind that goodness of a design assumes an underlying model. For example, if the relation between a predictor and a dependent variable can be adequately described by a linear model, then only two distinct measurement points for data collection are sufficient to estimate the model parameters. If, however, it is not certain whether the relation is linear, and one suspects it to be curvilinear, then more than two distinct measurement points are needed.

1.5.1.3 The model is unnecessarily complicated

In any study, the researcher should first make a concerted effort to decide on the appropriate level of complexity he or she is willing to entertain in the model and the design. The key is that the model should not be unnecessarily complicated and the design should be as simple as possible to meet the study objectives. In other words, the design should aim at striking a right balance between simplicity on one hand and statistical efficiency and practicality on the other. If a design is too simple, this may lead to invalid or weak conclusions, while a highly efficient design may disguise invalid conclusions. For example, consider a Latin square design with blocking and matching variables and controls for order effects at the same time. If the statistical model for the data analysis includes different covariates as well, then the interpretation of the results can become difficult. On the other hand, if we simplify the design by not controlling the effects of extraneous variables, we may end up with false conclusions.

1.5.1.4 Conclusions based on invalid extrapolation

The range of conditions under investigation is one of the requirements for a design to be externally valid. It is desirable that the study be undertaken in a broad range of conditions to increase the chance that the conclusions from the current study are valid for other ranges of values of the independent variable. A random effect design, where the levels of the independent variable (factor) are drawn randomly from a population of levels, has the advantage that the conclusions are generalizable to other population levels of the factor. See Cook and Campbell (1979) for a list of threats to external validity. Threats to statistical conclusion validity arise when conclusions are inferred outside the experimental conditions.

For example, in the study of the relationship between systolic blood pressure and age of children, any inference drawn from the study for people aged 25 or higher is obtained from extrapolation. We have only data for children up to 18 years of age and yet we attempt to infer blood pressure levels for people aged 25 or higher. This is risky business because while there may be evidence that a linear relationship holds for the two variables between 2 and 18, there is no reason to believe that the same linearity assumption applies to people in the higher age groups.

1.5.2   Internal validity

Internal validity is concerned with the question of whether the effect can be attributed to the variables that are manipulated. A 'good' design for a study should always control the threats to internal validity. In this subsection, we provide a short list of the main threats to internal validity and refer the reader to Campbell and Stanley (1963) and Cook and Campbell (1979) for a more in-depth discussion. The main threats are *history*, *maturation*, *testing*, *selection*, *regression towards the mean*, and *mortality*.

1.5.2.1   History

Changes in outcome variable may not be caused by the treatment variable but by an event that took place during the study. For example, in assessing the effectiveness of the biology lessons given by a teacher, higher grades of pupils on a biology test may not be attributable to the biology lessons (treatment) taught by the teacher, but by an on-going television home programme that was seen by these pupils during the same period of time. The event (television programme) represents a history threat and can be separated from the treatment effect by including a control group in the study, which does not receive the treatment (biology lessons).

1.5.2.2   Maturation

Change of outcome may be caused by changes that people undergo during the period of a study. This so-called maturation effect can also be controlled by inclusion of a separate control group of subjects with the same maturation effect.

1.5.2.3   *Testing*

Change in outcome scores may be induced by the effect of practice or memory. An example from education is that students being tested twice, may perform better the second time because they remember what kind of questions were asked the first time. Such effects are likely to arise in designs with repeated measurements and can be controlled by counterbalancing the order of presentation or by the inclusion of a suitable control group.

1.5.2.4 Selection

In observational studies, subjects are not randomly assigned to groups. Controlling extraneous variables is usually difficult to achieve. This problem is referred to as the *selection effect* and it may account for variation of the outcome variable that is not caused by the effect of the treatment that one is interested in. Some control can be obtained by matching and blocking procedures. The idea behind blocking is that units are grouped in such a way that all units within the same block have the same characteristics and are likely to respond in the same way. Matching is grouping units into pairs, so that in the absence of treatment effects the pairs of units will produce the same responses.

1.5.2.5 Regression towards the mean

Regression towards the mean is related to the selection effect. This threat to internal validity can best be explained as follows. Any athlete knows that it is extremely difficult to beat one's own record a second time. How can this be explained? Assume that any measurement y consists of two elements: the true measurement y_T and measurement error e, that is, $y = y_T + e$. The true measurement is assumed to remain the same, while the measurement error changes due to chance. The record of an athlete is a high score and such a high score can be caused by a high true score y_T, and/or by a high value of the error e. The second time that the athlete runs under similar conditions, the high true measurement y_T will remain high, while the high error will be expected to become smaller due to chance. Overall, the measurements y are likely to become smaller the second time. Control of this effect in a design can be established by random assignment to groups or by inclusion of a control group.

1.5.2.6 Mortality

Incidental loss of data or dropout and loss of subjects during the process of the study can differentially affect the dependent variable measurements over different conditions and treatments. The dropout pattern can influence the internal validity. In selecting a 'good' design, one has to take this threat into account. For example, in the design of a clinical trial, it should not be longer than it is necessary to observe a meaningful effect because a longer trial is almost certain to have more dropouts and missing data than a shorter trial.

1.5.3 Control of (unwanted) variation

It is usually difficult to identify extraneous variables that distort the specific effects under study. These extraneous variables are often referred to as *confounders* and may interact with the independent variables. In general these variables can be controlled by *inclusion*, *exclusion*, *statistical control* or by *randomization*.

1.5.3.1 Inclusion

Inclusion of an extraneous variable means that the variable is included in the study and that its specific effects on the dependent and other variables are taken into account. In fact the extraneous variable is added as an extra factor in an experimental design. For example, in a medical study, if one assumes that gender is a confounder and that it could distort the relation between dosage level and blood pressure, it can be 'controlled' by adding it as an extra factor in the design.

1.5.3.2 Exclusion

Exclusion or elimination means that the impact of the extraneous variable is eliminated from the design. In this way, the variable is held constant for all the units in the study. A potential distorting covariate may be gender. Holding a variable constant means that, for example, only male or female patients are included in a study. Of course, exclusion may affect the generalizability of the results negatively.

1.5.3.3 Statistical control

Control of extraneous variables is also possible via statistical manipulation. Examples of such a statistical control are inclusions of extraneous variables as covariates in a regression model or in an analysis of covariance. Schematically, the idea behind statistical control can be explained by Figure 1.1a and b.

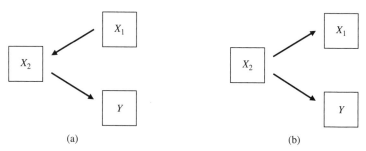

(a)   (b)

Figure 1.1 Two diagrams for the relation between variable X_1 and variable Y influenced by the effect of a third variable X_2.

The arrows in Figure 1.1a indicate that the variable X_1 affects X_2, which in turn affects Y. The relation between X_1 and Y is fully explained by the successive effects of X_1 on X_2 and of X_2 on Y. In Figure 1.1b, the variables X_1 and Y are both affected by X_2. One often refers to the relation between X_1 and Y in Figure 1.1b as *spurious*, because both have a common cause X_2. In both diagrams, the variable X_2 plays a role in the relation between X_1 and Y. Any correlation between X_1 and Y would disappear if the variable X_2 is excluded or held constant. Proper handling of the variable X_2 in a regression analysis or analysis of covariance model would also lead to the same results.

1.5.3.4   Randomization as method of control

Randomization was introduced by Fisher (1926) for agricultural experiments comparing the effect of different fertilizers on the growth of wheat. Randomization plays a role in scientific studies in two ways. Firstly, randomization ensures generalizability of the conclusions. This is because random samples are often drawn from a finite population to generalize conclusions, and in a simple random sampling procedure, theoretically every unit in the population is equally likely to be included in the sample. Secondly, randomization can be applied to control the effects of extraneous variables. For example, random assignment of subjects to a treatment group and a control group can be established by flipping a fair coin for each subject: heads means assigning the subject to the treatment group and tails means assigning the subject to the control group. This process guarantees a 50–50% chance that each patient is assigned to either group and helps to ensure that all extraneous variables are equally distributed between the two groups. Of course, such a process can only be applied in purely experimental studies.

Randomization, however, should not be automatically done in all studies. Certain combinations of factor levels may not be possible and in some situations random assignment may not be practical or feasible. For example, random assignment in quasi- and non-experimental studies is usually not possible and control must be established by other procedures, such as restricted randomization, matching and blocking. See, for example, Box (1990) for a discussion on randomization.

1.5.3.5   Matching, blocking, restricted randomization and balancing as methods of control

There are also more or less *ad hoc* procedures to control for unwanted variation. For example in a learning study, students with the same age can be grouped into three treatment groups. The variable age is then *matched* to ensure that in all three treatment groups there are students with exactly the same age. In general, *blocking* procedures isolate or partition out variation that is attributable to an extraneous variable, so that it does not influence treatment effects and estimates of error variance. Blocks of units are homogenous with respect to the extraneous variable. *Balancing* is a procedure to obtain groups of subjects with, for example, the same number of students in the same age category. *Restricted randomization* is used to control the randomization in small samples to achieve a balance in group sizes or a balance between groups on other characteristics. An extension of balancing is *counterbalancing*, which is used in designs to control for possible order effects by changing the order of presentation of treatments. Examples of such designs are the so-called cross-over design and the Latin square design. Cross-over designs are discussed in Chapter 8.

1.6 Ethical aspects of design choice

A 'good' research design should also meet a number of ethical requirements. Although ethical requirements may conflict with the methodological requirements, in an ideal situation ethical requirements can be reinforced by the methodological requirements of a study. A 'good' design in social, behavioural and biomedical studies should be both ethically and methodologically sound. A discussion on the benefits of both approaches is given by Palmer (2002).

Ethics codes for social, psychological and biomedical research have been developed via international consensus over the past 60 years. Starting with the Nuremberg Code (Nuremburg Code, 1947; Trials of War Criminals, 1949), research has been ethically guided by the various updates of the Declaration of Helsinki (2000).

Nowadays professional societies, such as the *American Medical* and *Psychological Associations*, have adopted these codes for the protection of the rights and interests of humans and animals (American Psychological Association, 2002). These codes specify what is required and what is forbidden and they are used by the various human subject research committees at universities and research institutes as guidelines in assessing research proposals. Apart from the commonly known principle of *informed consent*, these codes have quite firm requirements for the design of a study. Emanuel, Wendler and Grady (2000) proposed seven ethical requirements in designing clinical studies, which are also applicable to the design of studies in the social and biomedical field. Without discussing the ethical requirements in depth, we will briefly list the requirements proposed by Emanuel Wendler and Grady (2000).

Social and scientific value. The study design should enable evaluation of a treatment intervention to improve health and increase knowledge.

Scientific validity. A study should use accepted scientific methods, including statistical techniques to provide valid and reliable data. This requirement is closely related to the statistical conclusion validity of a study.

Fair subject selection. A study should not specifically select vulnerable individuals for risky research and reserve the socially strong subjects for the more beneficial research projects.

Favourable risk-benefit ratio. Potential risks to the subjects should be proportionate to the benefits to the subject and society.

Independent review. Reviewers of the proposed design of a study should be independent and not affiliated with the research project itself.

Informed consent. Subjects should be informed about the purpose of the research, its procedures, potential risks and benefits, so that the individual understands it and can make a voluntary decision whether to enrol, continue participation or stop.

Respect for potential and enrolled subjects. The research protocol should include respect for subjects by permitting withdrawal from research, by protecting privacy, by informing subjects of newly discovered risks or benefits and by maintaining the welfare of subjects.

It should be emphasized that the ethical requirements that focus on the well being of the subjects are important and can easily be overlooked or forgotten when a study is designed. Designers of research studies sometimes have the tendency to concentrate on the more methodological requirements of a good design than on ethical aspects of designing a study.

To summarize, a 'good' design for any type of study should ensure sufficient information and power for finding real effects in the region of interest, be robust against violations of model assumptions, enable adequate check of model fit, enable adequate check of model assumptions, allow for adequate control of extraneous variables, ensure simplicity of data patterns and computations, require minimum costs and sample size and enable valid extrapolation of the conclusions. In addition to these methodological requirements, ethical aspects should be taken into consideration as well, especially those that focus on the well being of the subjects and their informed consent.

1.7   Exact versus approximate designs

This book focuses on the use of optimal or highly efficient designs in social and biomedical research. Requirements for such optimal and highly efficient designs have been discussed in the previous sections. In this subsection, we make clear the distinction between two types of designs: *exact* and *approximate designs*. This distinction is important for understanding the material in the rest of this book and reasons for our choice to work with approximate designs instead of exact designs.

To understand the distinction between *exact* and *approximate designs*, let us consider a typical design problem for a dose–response study. The researcher has to decide in advance how to select from a given dose interval, *the number of dose levels* to use, the *dose levels*, and the *number of* subjects to assign to each of these dose levels. Suppose that the available resources allow us to take a fixed number of subjects, say N, in this study. An *exact design* tells us how many subjects to assign to each dose. If $N = 60$ subjects are available for the study, an exact design may require two different doses with 20 subjects given to the first dose level and 40 subjects given to the other second dose. Alternatively, if resources only allow $N = 57$ subjects, an exact design may allocate 22 subjects at a first dose level and 35 subjects at the second dose. Of course, these doses must all be selected from the dose interval of interest that was specified in advance.

Approximate design is another way to allocate a given number of N subjects. Such a design may specify that one-third of the subjects be given to a first dose level and two-third of the subjects be given to the other dose level. If $N = 60$, this results in a design that has 20 subjects at the first dose and 40 subjects at the second dose. The defining characteristic of an approximate design is that it is specified by the number of dose levels, the dose levels and the *proportion* of subjects or units assigned to each dose. Approximate designs may appear to be exact designs but there is a key difference. Unlike exact designs, approximate designs can be specified regardless of the value of N. For instance, if $N = 180$ in the same example, the allocation scheme is the same, that is, one-third to the first dose and two-third to the second dose, resulting in the design that has 60 subjects at the first dose and 120 at the second dose, or vice versa. So the approximate design coincides with the exact design in this specific instance.

What happens if $N = 100$? The above approximate design then requires $100/3 = 33.3333$ subjects at the first dose and 66.6667 subjects at the second dose. Silly? Well, this is precisely the explanation behind its name. Approximation is the name of the game. Sometimes these designs are also called *continuous designs* suggesting that the allocation scheme has a continuum connotation. In practice, an approximate design is rounded in some natural way so that it becomes an exact design before implementation. For the approximate design just discussed, the implemented (rounded) design can take one of the following forms:

1. 33 subjects at dose 1 and 67 subjects at dose 2 or

2. 34 subjects at dose 1 and 66 subjects at dose 2.

So the design that we use in practice may not be unique, but, nevertheless, should be very similar to any one of the candidate designs listed above.

Kiefer pioneered the approximate design approach and his extensive work in this area is well documented in Kiefer (1985). There was criticism of this approach initially, but it is now widely accepted as a practical way of finding optimal approximate designs. Kiefer gave three powerful reasons for working with approximate designs instead of exact designs.

The first reason is that optimal exact designs are very difficult to find and they depend on the specific value of N in addition to being dependent on the model and optimality criterion. This means that for each model and each criterion, we need to have literally an endless list of optimal designs for the practitioners because each different value of the sample N results in a different optimal design. In contrast, the optimal approximate design is independent of the value of N and consequently they are easier to describe and be listed. Second, there is rounding involved when we implement an approximate design. One can show that the implemented optimal approximate design is always close to the (unknown) optimal exact design and the difference between the exact optimal design and the implemented design vanishes if N gets large. The third reason is that optimal exact designs usually require complicated mathematical theory and many optimal exact designs still cannot be found for relatively simple problems. In contrast, optimal approximate designs are available either in analytical form or they are found using iterative methods via computer algorithms. In summary, there are compelling reasons to work with approximate designs in practice and much of the rest of this book will focus on approximate optimal designs.

1.8   Examples

In this section, we first illustrate the research process using a simple hypothetical radiation dosage example. We then present studies taken from the social, behavioural and biomedical literature to illustrate how different types of design issues can arise in practice. For each study, we describe the problem and design issues, but defer how one may improve the designs to a later chapter.

1.8.1   Radiation dosage example

Suppose that a radiologist is interested in the linear effect of radiation dosage (X) on tumour shrinkage (Y) and assumes that the relation between radiation dosage levels and tumour shrinkage can be adequately described by the simple linear regression model. The mathematical relationship can be written as $y_i = \beta_0 + \beta_1 x_i + \varepsilon_i$, where β_0 is the intercept parameter and β_1 is the slope parameter. The tumour shrinkage for the ith patient is y_i and the radiation dose is x_i. We assume each error term ε_i has mean 0 and constant variance, and all observations are independent. In clinical practice, the range of dosage levels must be restricted, because an overdose can harm the experimental units, which are patients in this case. On the other hand, too small a dosage is likely to be ineffective. We assume for illustrative purposes here that the radiologist can only use eight equally spaced dosage levels and that sufficient funding is available to include only $N = 16$ patients. If these dosage levels are indicated by the numbers 1, 2, 3, ..., 8, a simple and intuitive design is to allocate n patients to each dose levels. This implies we must have $n = N/8 = 2$ patients per dosage level. Table 1.2 schematically shows this design for the study. The numbers 1 through 8 indicate the dosage levels and these dosage levels are equally spaced between the minimum and maximum dosage levels.

Table 1.2   A balanced radiation dosage design.

Dosage levels							
1	2	3	4	5	6	7	8
n	n	n	n	n	n	n	n

Following Section 1.2, we have now covered Stages 1, 2 and 3 of the research process. We identified radiation dosage as the only variable X to study its effects on the outcome variable Y, which is tumour shrinkage. That fulfilled Stage 1 of the process. The posited linear regression model between the tumour shrinkage and the radiation dose along with the choice of the design in Table 1.2 fulfilled Stages 2 and 3. In Stage 4, we collect data to test the scientific hypothesis of interest. In this case, the null hypothesis is that there is no linear relation between dosage level and tumour shrinkage, that is, $H_0 : \beta_1 = 0$. An important design question is whether this design is able to estimate β_1 efficiently and whether there is sufficient power for testing the null hypothesis $H_0 : \beta_1 = 0$ to detect if the data support a linear relationship between tumour shrinkage and radiation dosage level.

The design issue for this problem is particularly pressing because we know that we have only a small sample of $N = 16$ patients and that measurement errors in radiation studies are usually quite large. This means it is absolutely crucial to choose the design carefully to minimize cost and maximize efficiency. What design would that be? Would it be more efficient to assign patients to a smaller number of dosage levels, such as dosage levels 1, 5 and 8? Or would it be more efficient to assign patients to the most extreme dosage levels 1 and 8? We will show in Chapter 2 that the design with one-half of the $N = 16$ patients assigned to dosage level 1 and the other one-half to dosage level 8 is the most efficient for estimating the slope parameter β_1. When the linear model assumption holds, this design that assigns equal proportions of patients to the extreme two dose levels is also optimal for several other purposes, but we defer further discussion to Chapter 2. Of course, an optimal design would still have to comply with ethical requirements. How efficient is the design in Table 1.2 compared to the optimal design for estimating β_1? It can be shown that this design is not efficient; in fact, its efficiency is only approximately 40%. We explain later on how this percentage is computed and what the efficiency means in practice.

1.8.2   Designs for the Poggendorff and Ponzo illusion experiments

Over the years, psychologists and cognitive scientists have been interested in the so-called Poggendoff and Ponzo illusions. These illusions are displayed in Figure 1.2. The Poggendorf illusion is that the ends of a straight line segment passing behind an obscuring rectangle seem to be offset when in fact they are aligned. The railroad or Ponzo illusion is that the upper horizontal line appears much longer than the lower horizontal line. In the Ponzo illusion, the subject is required to adjust the length of the lower horizontal line to match that of the upper horizontal line. In the Poggendorff illusion, the subject positions the

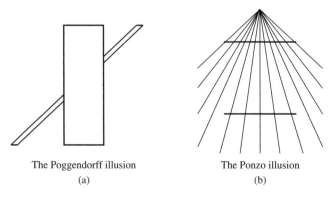

The Poggendorff illusion   The Ponzo illusion
(a)   (b)

Figure 1.2 Graphs of the Poggendorf and Ponzo illusions.

right-hand bar as a continuation of the bar on the left. The subject's error in these adjustments (in inches) is measured and recorded as the magnitude of the illusion. Bock (1975) reported two experiments conducted by Leibowitz and Gwozdicki (1967) and Leibowitz and Judisch (1967), which studied the magnitude of pictorial illusions as a function of age. The original design for each of the two studies was the same and had 16 children in each of the following age groups 5.0, 6.0, 7.0, 8.5, 10.5, 12.5, 14.5, 16.5 and 19.5. This design with a total of 144 children is summarized in Table 1.3. Bock (1975, Chapter 4) showed that the magnitude of the Poggendorff and Ponzo illusions as a function of age could be adequately described by a quadratic and cubic regression model, respectively.

Table 1.3   Design of Ponzo and Poggendorff studies.

	Age groups								
	1	2	3	4	5	6	7	8	9
Age	5	6	7	8.5	10.5	12.5	14.5	16.5	19.5
n	16	16	16	16	16	16	16	16	16

Figure 1.3 displays the fitted polynomials. It clearly shows the curvilinear relationship between age and the magnitude of the illusion. For the Poggendorff illusion, the magnitude of the illusion decreases as the age of the children increases and the magnitude increases again for older children (Figure 1.3a), whereas for the Ponzo illusion, the magnitude of the illusion first increases with the age of the children, then decreases and finally increases again (Figure 1.3b).

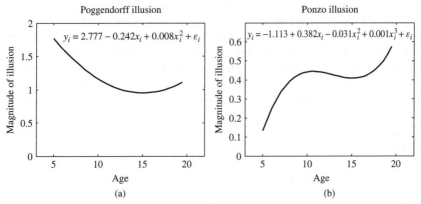

Figure 1.3 Magnitude of the Poggendorff and Ponzo illusions as function of age (data from Bock, 1975, Chapter 4).

The *design* for such studies requires selection of the age groups and the number of children in each of the age groups. One design question can be raised here is whether the nine age groups used in the original design are all needed to estimate the polynomial functions that describe these two magnitudes of illusions. More specifically, how many distinct age groups are needed to estimate the quadratic and cubic function for these two illusions as efficiently as possible? Another question is whether these 144 children all contribute the same amount of information to estimate the parameters in the two functions. More importantly, is it possible to choose the number of age groups and the number of children to include in each of these age groups to obtain the most precise estimate of the polynomial relationship at minimal cost? In Chapter 3, we provide some answers to these questions using optimal design theory.

1.8.3 Uncertainty about best fitting regression models

Suppose that an epidemiologist studies the relationship between body weight and the joint effects of height and age of nutritionally deficient children. This hypothetical example is described in Kleinbaum *et al.* (1998, Chapter 8, Example 8.1) using data from 12 nutritionally deficient children with ages lying between 6 and 12 years. The heights of the children were roughly between 50 and 60 cm, and their weights roughly ranged from 50 to 80 kg.

The design question for this example is as follows: Which design will give us the most efficient estimate for the relationship between weight and the joint effects of height and age? This question is difficult and really quite impossible to answer unless further assumptions are made. For instance, what is the posited relationship between weight and the two predictors, height and age? Is the epidemiologist aware of postulated models in the literature or able to perform a pilot study to obtain a preliminary estimate of this relationship? Clearly, the answer to the design question depends on the assumed statistical model. What are plausible models? In the absence of good prior information, one considers the simplest possible models, and hope that the data will tell us which model seems most appropriate later on. From the design perspective, one can determine an efficient design for each plausible model and use these designs to come up with a reasonable model. In practice, researchers usually begin by considering simple linear models. Here are some common regression models to consider for two predictors X_1 and X_2 and a single outcome Y. For the ith child, we denote the value of the dependent variable, body weight, by y_i and the values of the independent variables, height and age, by x_{1i} and x_{2i}, respectively:

$$\text{Model 1: } y_i = \beta_0 + \beta_1 x_{1i} + \beta_2 x_{2i} + \varepsilon_i. \tag{1.1}$$

This model considers only the main effects for age and height on weight. If, in addition, we entertain an interaction between age and height, the model becomes

$$\text{Model 2: } y_i = \beta_0 + \beta_1 x_{1i} + \beta_2 x_{2i} + \beta_3 x_{1i} x_{2i} + \varepsilon_i. \tag{1.2}$$

The epidemiologist may also be interested in a second order effect for height (X_1). In this case, the model is

$$\text{Model 3: } y_i = \beta_0 + \beta_1 x_{1i} + \beta_2 x_{2i} + \beta_3 x_{1i} x_{2i} + \beta_4 x_{1i}^2 + \varepsilon_i. \qquad (1.3)$$

If a second order effect for age (X_2) is also entertained, the model becomes

$$\text{Model 4: } y_i = \beta_0 + \beta_1 x_{1i} + \beta_2 x_{2i} + \beta_3 x_{1i} x_{2i} + \beta_4 x_{1i}^2 + \beta_5 x_{2i}^2 + \varepsilon_i. \qquad (1.4)$$

Model 4 contains linear and quadratic terms for both the independent variables and also an interaction between these two variables. To complete the list of possible models for two variables, Model 5 below contains a second order term for both independent variables, that is, it is the same as Model 4, but without the interaction effect:

$$\text{Model 5: } y_i = \beta_0 + \beta_1 x_{1i} + \beta_2 x_{2i} + \beta_4 x_{1i}^2 + \beta_5 x_{2i}^2 + \varepsilon_i. \qquad (1.5)$$

Details on the actual analysis of such regression models can be found in Kleinbaum *et al.* (1998, Chapter 8), among others.

From the design viewpoint, we know that the choice of an efficient design depends on the model specification. The design problem confronting the epidemiologist is that data needs to be collected to validate the assumed model, but it is impossible to specify how data should be collected without a design. For this study, the design problem consists of the question what combination of levels of the independent variables (height and age) will provide the most efficient estimators of the parameters and how many children should be selected for each of these combinations. This design problem is explained in further detail in Chapter 3.

1.8.4 Designs for a priori contrasts among composite faces

Galton (1878) and Stoddard (1886) were probably the first to compose portraits of photographic exposures of faces. Langlois and Roggeman (1990) provided empirical evidence that composite faces seem to be 'better looking' than the original individual pictures. In their study, Langlois and Roggeman found a strong (curvi)linear relation between attractiveness and the number of faces entering the mixed or 'averaged' composite face. Illustrations of such a morphing effect can be found on the web site: http://www.beautycheck.de/.

The experiment performed by Langlois and Roggeman (1990) consisted of five levels of composite faces that were obtained by 'averaging', respectively, 2, 4, 8, 16, and 32 faces. The attractiveness was rated by subjects on a five-point Likert scale. Figure 1.4 shows the mean ratings of the five levels of female composite faces using data from Table 1 in Langlois and Roggeman (1990). The vertical lines in Figure 1.4 indicate the standard deviations at the five composite levels. It is assumed that faces per composite level were rated by different samples of raters.

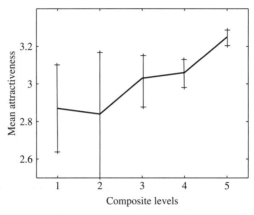

Figure 1.4 Average attractiveness ratings for female composite faces (data from Langlois and Roggeman, 1990).

Figure 1.4 shows that there is a relation between the composite level of the faces and their attractiveness. Overall, the trend shows that the faces were judged more attractive as more faces were entered. There also seems to be more consensus among raters about the attractiveness of the faces as the composite level increases. Figure 1.4 shows that the standard deviation tends to decrease as the composite level increases.

Although a clear trend is shown in the figure, it is not clear how the differences are manifested. Is this relation between attractiveness and composite level mainly a linear relation or is it also curvilinear? In the original design, the same number of raters was used for each composite level. In this study, all five composite levels are needed and so there is no question about the choice of number of levels to include. One can, however, take issue with the number of raters needed at each composite level. What is the optimal number of raters at each level to obtain the most efficient estimate for the linear or curvilinear effect of the composite level on the attractiveness? More details about this problem are discussed in Chapter 4.

1.8.5   Designs for calibration of item parameters in item response theory models

In educational research, a lot of resources are spent on the design of achievement tests, and various researchers have studied the problem of optimally designing achievement tests to decrease the costs of measurement. An example is the computerized GRE® General Test (Graduate Record Examinations), which measures verbal reasoning, quantitative reasoning, critical thinking, and analytical writing skills that have been acquired over a long period and that are not related to any specific field of study. The GRE test scores are frequently used as a criterion for admission to graduate studies in the United States.

Item response theory (IRT) models are usually used to estimate the characteristics of items in such a test. Van der Linden and Hambleton (1997) is a useful reference for IRT models. An often used model is the two-parameter logistic (2PL) model, which models the probability of a student with ability θ_j to correctly answer an item as a logistic function:

$$p_i(\theta_j) = \frac{\exp\left[a_i(\theta_j - b_i)\right]}{1 + \exp\left[a_i(\theta_j - b_i)\right]}. \tag{1.6}$$

The response probability $p_i(\theta_j)$ for item i is modelled as a function of the ability level θ_j in the interval $-\infty < \theta_j < \infty$. The item parameter b_i represents the location parameter with range $-\infty < b_i < \infty$ and the item discrimination (slope) parameter is represented by a_i with range $0 < a_i < \infty$. The exponential function is $\exp(x) = e^x \approx 2.7184^x$. It should be noted that this is actually a logistic model, but with a quantitative latent variable θ_j. The difference between the logistic model described in Chapter 5 and the IRT model is that the IRT model assumes that the ability levels of students are unknown and have to be estimated as well, whereas the logistic model assumes a manifest independent variable. Usually, marginal maximum likelihood estimation of the parameters is applied with a normal density function for the distribution of the θ_js (Van der Linden and Hambleton, 1997).

Figure 1.5a shows a typical set of nine response functions from the 2PL model. These response functions vary in location and slope. Figure 1.5b shows one of these nine functions for which the slope $a_i = 1$. The dotted lines show that the probability of correctly answering the item is $p_i(\theta_j) = 0.5$ for a person with ability $\theta_j = b_i$. It is seen that as the ability of the student increases, the probability of answering the item correctly also increases, and that the steepness of the probability function depends on the size of the slope parameter a_i.

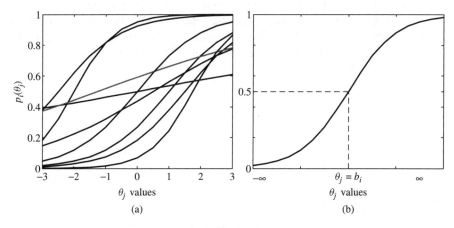

Figure 1.5 A set of nine typical response functions.

A computerized adaptive testing (CAT) setting is often used to estimate the ability levels of students and the item parameters as efficiently as possible. For example, the GRE General Test is implemented in CAT form where a computer program sequentially selects items for a student to answer. The CAT procedure selects items that are not too difficult and not too easy for the student. It uses the information received from answers of a student to previously administered items. In this way, the ability of the student can be efficiently estimated with fewer items than with a traditional paper and pencil test. However, to adopt such CAT procedures, a sufficient large item bank with calibrated items is needed. To prevent the item bank from being exhausted by the CAT procedure, that is, when the items become out of date or over exposed, these item banks usually contain a huge number of items and builders of item banks have to conduct costly sessions to calibrate all these items.

Different procedures to build such huge item banks have been proposed. In many cases, calibration of items takes place by administering the items to a large fixed sample of students. This may, however, result in large parts of the data with little or no information on the item parameters. More efficient sampling designs have been suggested by Berger (1992, 1994), among others. The characteristics of the items that provide more efficient estimators for the item parameters are explained in Chapter 5.

1.9 Summary

This introductory chapter reviews the basic elements of a research design and lists some key requirements of a 'good' design. These requirements can be method-ological, statistical or ethical in nature. The main focus of this book is to construct research designs that are as efficient as possible at minimal cost. Efficiency here can refer to the accuracy of the estimates for the model parameters or the power level of test or tests of interest. We distinguish two types of designs in the literature: exact and approximate (or continuous) designs. We also provide examples from the social and biomedical fields to illustrate different types of design problems we may encounter in practice and what design issues actually entail for each of the problems.

2

Designs for simple linear regression

2.1 Design problem for a linear model

Suppose that we are interested in the relationship between two quantitative variables, that is, we want to know how large the effect is of an independent (predictor) variable X on a dependent variable Y. For example, we could be interested in the relation between radiation doses X and the reduction of tumours Y in a sample of breast cancer patients. Assuming that there are no ethical objections, we could randomly assign a sample of patients to different dosage levels x_i and afterwards measure the tumour shrinkage y_i for each patient. Each individual subject i in the sample is associated with a pair of values (x_i, y_i).

The relationship between the random variables X and Y can be visualized by means of a scatter plot. The first problem that arises is how to determine the type of relationship between the two variables. What effect of X on Y is of interest? We will assume in this chapter that the relation between X and Y is linear and therefore adopt a linear model to describe this relationship. But there is no law that says that this relation must be linear. In fact, many relations among two variables are intrinsically nonlinear. The linear function, however, is not only the most simple function to understand, but is also often an accurate approximation of the true relation between two variables, especially over a narrow range of values for the X variable.

In the design problem for a linear model, we can distinguish two questions. The first question concerns the selection of values or levels of the independent variable. To investigate whether radiation dosage has an effect on tumour reduction, we have to first choose different radiation dosages to observe the

An Introduction to Optimal Designs for Social and Biomedical Research   M. P. F. Berger & W. K. Wong
© 2009, John Wiley & Sons, Ltd

tumour reduction. Of course, these radiation dosages have to be chosen from a pre-selected range that meets all feasibility and ethical requirements. The second question concerns the actual number of patients assigned to each radiation dosage. Specifically, how many patients do we assign to each of these radiation-dosage levels?

For the simple linear model, both design questions can be answered definitively if our primary interest is in estimating the intercept or the slope parameter in the model. Before we present the optimal design for estimating each of the single parameters, we caution that the optimal design is constructed based solely on theoretical considerations and so they may not always be feasible in practical settings, where all kinds of practical and ethical constraints need to be taken into account. Our constant guiding and over-riding theme in this book is that even if an optimal design is not implementable in practice, we can use the characteristics of the optimal design to come up with an implementable design that retains as much features of the optimal design as possible. In other words, we allow the design to stray from the optimum to meet practical constraints, but keep constantly in mind the resulting loss in efficiency relative to the optimal design. In this way, even if the resulting design is not optimal, the design is informatively chosen. We believe that adopting such a design strategy is likely to produce a design that meets both practical and theoretical considerations, and on average, should still outperform many of the frequently used research designs, where rationale for their use often seems to be lacking.

2.1.1   The design

A design is determined by the choice of the so-called design points or support points and by the number of subjects assigned to each of these design points. For example, suppose that we want to study the effect of radiation-dosage levels on tumour reduction and wish to include eight different dosage levels in our study. Let us denote these eight different design points by $d_1, d_2, d_3, \ldots, d_8$ and the number of patients to be assigned to each of these design points (dosage levels) by $n_1, n_2, n_3, \ldots, n_8$, respectively. If the total number of patients in the study is N, then all n's sum up to N. Different designs rise when we have different design points and/or different numbers of patients at each of the design points.

For a total sample of N patients, the independent variable X (radiation dosage) has N values $x_i (i = 1, \ldots, N)$. Among these N values there are $m = 8$ distinct values associated with the radiation-dosage levels. Consider the specific design in Table 2.1 where the design points d_j's are displayed together with the scores x_j's of the independent variable X. The x_j values are equal to the d_j values themselves or equal to some linear transformation of the d_j values. For this table, x_1 and x_2 are equal to d_1, and x_3 is equal to d_2, and so on. It is plainly visible that this design has an unequal number of patients assigned to each design point (dosage level). For example, only two patients are assigned to the first design point d_1, while design point d_5 has four patients.

Table 2.1   An example of a design for the radiation-dosage study.

			Design points, d_j				
1	2	3	4	5	6	7	8
x_1	x_3	x_4	x_6	x_9	x_{13}	x_{14}	x_{15}
x_2		x_5	x_7	x_{10}			x_{16}
			x_8	x_{11}			
				x_{12}			
$n_1 = 2$	$n_2 = 1$	$n_3 = 2$	$n_4 = 3$	$n_5 = 4$	$n_6 = 1$	$n_7 = 1$	$n_8 = 2$

Note: The N subjects are usually assigned randomly to the design points d_j, $j = 1, \ldots, m$.

2.1.1.1   Choice and scaling of design points

The values of the independent variable X have an upper and a lower limit x_{max} and x_{min}, respectively, that is, $x_{min} \leq x_i \leq x_{max}$. For the radiation-dosage design, one can imagine that there will be a lower limit for the radiation dosage at which there will be no effect on tumour reduction and an upper limit where the radiation dosage will cause unacceptable damage to the tissue. These upper and lower limits mean that all design points d_j have to be selected within these two limits and so are bounded between $d_{min} \leq d_j \leq d_{max}$. The set of all possible design points is called the *design region* or *design space* and is symbolized by $\Delta = \left[d_{min} \leq d_j \leq d_{max}\right]$. Although the boundaries of the design space $[d_{min}, d_{max}]$ usually coincide with the boundaries $[x_{min}, x_{max}]$ of the x_i values, this is not necessarily always the case. The boundaries of the design space can sometimes be restricted and have a more narrow range of values.

We frequently recode the design points using a linear transformation to facilitate interpretability and comparison of the results. A popular scale is to have the design points constrained between -1 and 1, that is, $-1 \leq d'_j \leq +1$. This can be accomplished by the linear transformation of the original design points $d_{min} \leq d_j \leq d_{max}$:

$$d'_j = \frac{d_j - \tilde{d}}{d_{max} - \tilde{d}}, \text{ where } \tilde{d} = d_{min} + \frac{d_{max} - d_{min}}{2}. \tag{2.1}$$

As an illustration, consider the dosage level example with eight different dosage levels, properly scaled at dosages 1, 2, 3, 4, 5, 6, 7 and 8, that is, we have $1 \leq d_j \leq 8$. These design points can be re-scaled to values within the interval $-1 \leq d'_j \leq +1$. To do this, we first verify that the centre of the dosage scale is $\tilde{d} = 1 + 3.5 = 4.5$, and upon substitution of $\tilde{d}$ into $d'_j = (d_j - 4.5)/(8 - 4.5)$, we have the transformed eight design points d'_j : -1, -0.7143, -0.4286, -0.1429, 0.1429, 0.4286, 0.7143 and 1.

Other transformations of the design points are also possible, but it should be kept in mind that some characteristics of a design will change after recoding and that in many cases the optimal designs may change after recoding. For example, when we change the design space, one may be unable to deduce the new design points for the new optimal design from the design points of the optimal design in the original design space. We will discuss this issue more fully in a later chapter.

2.1.1.2   Choice of number of observations

A study is usually set up by first determining the total sample size N. After selecting the appropriate values for the independent variable to study, we choose design points from the pre-selected design space Δ, that is, choose $d_j \in \Delta$, and the number of observations n_j from each design point such that $\sum_j n_j = N$. We can write such a design with m design points as

$$\xi_N = \begin{Bmatrix} d_1 & d_2 & d_3 & \dots & d_m \\ n_1 & n_2 & n_3 & \dots & n_m \end{Bmatrix}. \tag{2.2}$$

Each design point d_j has a corresponding weight n_j/N. This notation of a design stresses the fact that the n_j's are integers and that the design ξ_N is called a *discrete* or *exact design* with N observations; see Section 1.7, Chapter 1 or Atkinson and Donev (1992), among others.

In practice, all designs have to be discrete for implementation. This is because whole units are assigned to the different design points. However, working with exact designs is generally not an easy task and usually results in very difficult optimization problems. Even for relatively simple problems, the optimal exact design cannot be described in closed form. Therefore it is mathematically more convenient to work with the so-called *continuous* or *approximate designs*, generically defined and denoted by

$$\xi = \begin{Bmatrix} d_1 & d_2 & d_3 & \dots & d_m \\ w_1 & w_2 & w_3 & \dots & w_m \end{Bmatrix}. \tag{2.3}$$

Here, the continuous design also has m design points, but the key difference is that the design is formulated with continuous weights such that $\sum_j w_j = 1$, and $0 \leq w_j \leq 1$, for all j. The integral of this design measure ξ over the design space is equal to 1. The weights in the continuous design approximate the n_j/N's in the exact design. Another distinguishing feature between these types of designs is that continuous designs are defined only in terms of proportions, and so do not require the total sample size N to be specified in advance.

We emphasize that a continuous design does not necessarily have to be equal to an exact design even when the two designs have the same support points and the same sample size. It depends how the approximate design is rounded up to form an exact design. Typically, an exact design ξ_N is formed from a continuous design by multiplying its weights w_j's by N and by rounding the product to the nearest integer value. All these rounded integer values should then again sum up to N. Pukelsheim and Riedel (1992) provided approximation rules with optimal properties. When the sample size is large, both the exact and continuous designs should be close.

The continuous version of the design for the radiation-dosage example with eight different dosage levels on the design interval $\Delta = [1, 8]$ can be written as

$$\xi = \begin{Bmatrix} 1 & 2 & 3 & 4 & 5 & 6 & 7 & 8 \\ w_1 & w_2 & w_3 & w_4 & w_5 & w_6 & w_7 & w_8 \end{Bmatrix}. \tag{2.4}$$

If an equal proportion of patients is assigned to each dosage level, all the weights are equal to $w_j = 1/8$. On the other hand, if patients are only assigned equally to the upper and lower limits of the radiation-dosage scale, that is, $d_1 = 1$ and $d_8 = 8$, then the corresponding weights are $w_1 = 0.5$ and $w_8 = 0.5$, and all other weights are zero.

Suppose, as an illustration, we have a total sample size of $N = 90$ patients and we would like to assign them equally to each of the eight dosage levels. Then we have $N \times 1/8 = 11.25$ at each dose, and rounding off to the nearest integer, 11, gives the following design:

$$\xi_{90} = \begin{Bmatrix} 1 & 2 & 3 & 4 & 5 & 6 & 7 & 8 \\ 12/90 & 11/90 & 11/90 & 11/90 & 11/90 & 11/90 & 11/90 & 12/90 \end{Bmatrix}, \tag{2.5}$$

where the two weights at both ends of the scale are set at $12/90 = 0.1333$ to comply with the design condition $\sum_j n_j = N$. Of course, this rounding off to an exact design for implementation is not unique; for example, one could have the middle dosage levels 4 and 5 with 12 patients and all other dosage levels with 11 patients. If we wish to facilitate interpretation and comparison of different designs, the d_j values can be recoded between -1 and 1 by the transformation $d'_j = (d_j - \tilde{d})/(8 - \tilde{d})$, where $\tilde{d} = 1 + (8 - 1)/2$. The resulting design is

ξ_{90}

$$= \begin{Bmatrix} -1 & -0.7143 & -0.4286 & -0.1429 & 0.1429 & 0.4286 & 0.7143 & 1 \\ 12/90 & 11/90 & 11/90 & 11/90 & 11/90 & 11/90 & 11/90 & 12/90 \end{Bmatrix}. \tag{2.6}$$

2.1.2   The linear regression model

We assume that the relation between the variables X and Y can be described by a linear regression model. The simple linear regression model is given by

$$y_i = \beta_0 + \beta_1 x_i + \varepsilon_i, \tag{2.7}$$

where the intercept β_0 and the slope β_1 are unknown parameters and the ε_i's are random error terms assumed to be independent and normally distributed with zero mean and common variance σ_ε^2.

In Figure 2.1, a scatter plot together with the linear regression function of tumour reduction on dosage level is visualized. The estimated means are $\bar{x}$ and $\bar{y}$, respectively. The responses y_i and the values of the independent variable x_i together form the $N = 16$ pairs of artificial data (x_i, y_i), $i = 1, \ldots, 16$, which are displayed as points in the scatter plot.

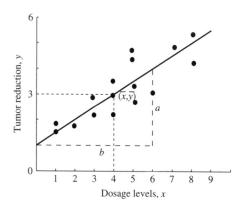

Figure 2.1 Scatter plot and linear regression of tumour reduction on dosage level.

The slope parameter β_1 in the regression equation in Figure 2.1 is the change in the mean tumour reduction that results from a change of one unit of dosage, that is, $\beta_1 = a/b$, where a and b are the lengths of the large dotted lines in Figure 2.1. The intercept β_0 represents the mean tumour reduction for a dosage level equal to zero, that is, for $x_i = 0$. The point $(\bar{x}, \bar{y})$ in Figure 2.1 represents the means of dosage levels and the tumour reduction, respectively.

2.1.3   Estimation of parameters and efficiency

For the linear regression model in Model (2.7), we can estimate the parameters β_0 and β_1 by the least squares method. The least squares estimators $\hat{\beta}_0$ and $\hat{\beta}_1$ are found by minimizing the sum of the squared errors $\varepsilon_i = y - \left(\hat{\beta}_0 + \hat{\beta}_1 x_i\right)$ in the model. In other words, least squares estimates are obtained by minimizing $\sum_i^N \varepsilon_i^2$. Using calculus, one can directly obtain the least squares estimators of β_0 and β_1 to be

$$\hat{\beta}_0 = \bar{y} - \hat{\beta}_1 \bar{x} \text{ and } \hat{\beta}_1 = \frac{\text{cov}(x, y)}{\text{var}(x)} = \frac{\text{SS}_{xy}}{\text{SS}_x}, \tag{2.8}$$

where the sum of squares between x and y or covariation is $\text{SS}_{xy} = \sum (x_i - \bar{x})(y_i - \bar{y})$ and the sum of squares of x is $\text{SS}_x = \sum (x_i - \bar{x})^2$. In addition, we have $\text{cov}(x, y) = \text{SS}_{xy}/N$ and the variance of the x's is $\text{var}(x) = \text{SS}_x/N$.

A useful property of the least squares estimators $\hat{\beta}_0$ and $\hat{\beta}_1$ is that they are unbiased. This means that their expectation is equal to the parameter value, that is,

$E(\hat{\beta}_0) = \beta_0$ and $E(\hat{\beta}_1) = \beta_1$. Under the normality assumption, these estimators also have minimal variances among all linear unbiased estimators of β_0 and β_1. Therefore, these estimators are *efficient*.

In general, the efficiency of an estimator increases as its variance becomes smaller. A larger variance of an estimator expresses more *uncertainty* about the estimator. In practice, there are three approaches to describe uncertainty about the estimators.

2.1.3.1   Three measures of uncertainty

The first approach measures uncertainty by the variance of the estimators. The variances of the two estimators $\hat{\beta}_0$ and $\hat{\beta}_1$ for the simple regression model are

$$\text{var}(\hat{\beta}_0) = \sigma_\varepsilon^2 \left(\frac{1}{N} + \frac{\bar{x}^2}{\text{SS}_x} \right) \text{ and var}(\hat{\beta}_1) = \frac{\sigma_\varepsilon^2}{\text{SS}_x} = \frac{\sigma_\varepsilon^2}{N \, \text{var}(x)}. \qquad (2.9)$$

These variances are used to test hypotheses on the single parameters. For example, the null hypothesis that there is no relation between dosage level X and tumour reduction Y is $H_0 : \beta_1 = 0$. The corresponding test statistic $t = \hat{\beta}_1 / \sqrt{\widehat{\text{var}}(\hat{\beta}_1)}$ is distributed as a t distribution with $(N - 2)$ degrees of freedom, where $\widehat{\text{var}}(\hat{\beta}_1)$ is an estimate of $\text{var}(\hat{\beta}_1)$. To compute $\widehat{\text{var}}(\hat{\beta}_1)$, we require a sample estimate of σ_ε^2. This estimate is frequently called the *mean squared error* and is equal to $\text{MS}_e = \sum_i^N (y_i - \hat{\beta}_0 - \hat{\beta}_1 x_i)^2 / (N - 2)$. Because the power of the test increases as $\text{var}(\hat{\beta}_1)$ decreases, we observe from the formula of $\text{var}(\hat{\beta}_1)$ that a more powerful test of this hypothesis or a more efficient estimator of β_1 can be obtained when

- the variance of the error terms in the model σ_ε^2 decreases and/or

- the variation SS_x or variance $\text{var}(x)$ increases and/or

- the sample size N increases.

It can be seen from the formulas that there is a trade-off among these three components. A simple way to increase power and efficiency is to increase the sample size N. Although this is often done in practice, it may lead to additional costs of collecting the data. Therefore, it may be more desirable to reduce measurement error by having more precise measurement of the variables to reduce the value of σ_ε^2. This second approach is probably more appealing in social and behavioral research where the error of measurement can sometimes be very large. These two approaches are commonly used in observational studies. Still another way to improve efficiency of the estimators in the simple linear model is to select the levels of the independent variable in such a way that the variance $\text{var}(x)$ increases. Clearly, this approach is feasible only in studies where the researcher has control over the sampling and selection process. The above discussion also clearly applies to the case where estimating or testing the intercept β_0 in the simple linear model is of interest.

A second measure for the uncertainty of the estimators $\hat{\beta}_0$ and $\hat{\beta}_1$ is the confidence interval for the parameters β_0 and β_1. The smaller the width of the confidence intervals, the more efficient the estimators are and the more power for the tests of hypotheses for these parameters. Confidence intervals are based on the variances of the parameter estimators. This can be seen in the following confidence interval formulas:

$$\hat{\beta}_1 - t_{\alpha/2, N-2}\sqrt{\frac{\mathrm{MS_e}}{\mathrm{SS}_x}} \leq \beta_1 \leq \hat{\beta}_1 + t_{\alpha/2, N-2}\sqrt{\frac{\mathrm{MS_e}}{\mathrm{SS}_x}} \qquad (2.10)$$

and

$$\hat{\beta}_0 - t_{\alpha/2, N-2}\sqrt{\mathrm{MS_e}\left(\frac{1}{N} + \frac{\bar{x}^2}{\mathrm{SS}_x}\right)} \leq \beta_0 \leq \hat{\beta}_0 + t_{\alpha/2, N-2}\sqrt{\mathrm{MS_e}\left(\frac{1}{N} + \frac{\bar{x}^2}{\mathrm{SS}_x}\right)}, \qquad (2.11)$$

where we recall that the mean squared error $\mathrm{MS_e}$ is the sample estimate of σ_ε^2 and $t_{\alpha/2, N-2}$ is the $100(1 - \alpha/2)$th percentile of a t distribution with $N - 2$ degrees of freedom. Percentiles of the t distribution are widely available either from statistical tables provided in textbooks or from a statistical programme. The above formula shows that as $\mathrm{MS_e}$ decreases or the sum of squares SS_x increases, the width of the confidence interval decreases and our estimators $\hat{\beta}_0$ and $\hat{\beta}_1$ become more accurate.

A third approach to express uncertainty of the parameter estimators is to consider the variance of the predicted value $\hat{y}_0$ for the response y_0 corresponding to an arbitrary value x_0, that is, $\hat{y}_0 = \hat{\beta}_0 + \hat{\beta}_1 x_0 = \bar{y} + \hat{\beta}_1(x_0 - \bar{x})$. The variance of the predicted value for the linear regression model is

$$\mathrm{var}(\hat{y}_0) = \sigma_\varepsilon^2 \left\{ \frac{1}{N} + \frac{(x_0 - \bar{x})^2}{\mathrm{SS}_x} \right\}. \qquad (2.12)$$

Here, again, the variance of the predicted value $\hat{y}_0$ is inversely related to both the sample size and the variation of the independent variable, SS_x, and is directly related to the error variance σ_ε^2. Kleinbaum et al. (1998) gave additional details on confidence intervals for regression analysis.

It is clear that to estimate the parameters β_0 and β_1, we need to have the response data y_i's (Equation 2.8). The variances of the parameter estimators can be computed without any response data, but in order to estimate the actual variance of these parameter estimators in Equation (2.9), prior information on the response data y_i's is needed to estimate the error variance σ_ε^2. The same holds for the confidence intervals and for the estimation of the variance of the predicted value $\hat{y}_0$. Given σ_ε^2, we only need information that is already available at the design stage of the study, namely, the sample size N, the levels of the independent variable X and the allocation of the observations to each level of the independent variable.

2.2   Designs for radiation-dosage example

Consider again the simple linear model for studying the effect of radiation dosage on the reduction of tumours in breast cancer patients. Assuming that there are no ethical objections to assigning $N = 16$ breast cancer patients to eight radiation-dosage levels, it is instructive to consider the four designs in Table 2.2.

Table 2.2   Four designs assigning breast cancer patients ($N = 16$) to eight different radiation-dosage regimes.

Design points, d_j	Design 1	Design 2	Design 3	Design 4
1	2	2	3	8
2	1	2	2	0
3	2	2	2	0
4	3	2	1	0
5	4	2	1	0
6	1	2	2	0
7	1	2	2	0
8	2	2	3	8
N	16	16	16	16
SS_x	70	84	108	196
$var(\hat{\beta}_1)$	0.0143	0.0119	0.0093	$0.0051(\times\sigma_\varepsilon^2)$
$var(\hat{\beta}_0)$	0.3441	0.3036	0.25	$0.1658(\times\sigma_\varepsilon^2)$

The discussion in the previous section and Equation (2.9) tell us that when both the sample size N and error variance σ_ε^2 are fixed and variance of the dosage levels $var(x)$ is maximized, the variance $var(\hat{\beta}_1)$ will be minimal. This is the same as saying $var(\hat{\beta}_1)$ is minimized when the sum of squares SS_x is maximized. Table 2.2 shows how the sum of squares SS_x and the variance $var(x)$ influences the variance $var(\hat{\beta}_1)$ using the four designs all with $N = 16$ patients.

Design 1 is the original design and all 16 patients are randomly assigned to the eight different dosage levels. The variance $var(x)$ for *Design* 1 is $var(x) = SS_x/N = 70/16 = 4.375$. *Design* 2 has two patients for each of the eight dosage levels and the variance is $var(x) = 84/16 = 5.25$. *Design* 3 is a design with relatively more patients assigned to the smaller and larger dosage groups than to the middle dosage groups. The corresponding variance is $var(x) = 108/16 = 6.75$. The last design, *Design* 4, equally assigns patients only at the extreme dosage levels, that is, at $d_1 = 1$ and $d_8 = 8$. The variance of the dosage levels is $var(x) = 196/16 = 12.25$. This design has the largest value of $var(x)$, compared among all the four designs.

For this example, *among the four designs, Design 4 is the optimal design for estimating the slope parameter because it has the largest value for* $var(x)$ *and this in turn implies the smallest value of* $var(\hat{\beta}_1)$. Table 2.2 shows that this optimal design consists of two distinct design points and assigns $N/2$ patients to dosage level 1 and $N/2$ patients to dosage level 8. It can be written as

$$\xi_{16}^* = \begin{Bmatrix} 1 & 8 \\ 8/16 & 8/16 \end{Bmatrix}. \tag{2.13}$$

The results in Table 2.2 also show that the original *Design* 1 is the worst design in terms of efficiency and that as more patients are assigned to the lower and higher dosage levels, as is the case in *Designs* 3 and 4, the efficiency increases. The *optimal design* in Table 2.2 guarantees that for a fixed sample size N and a fixed error variance, the power of the single test of the hypothesis $\beta_1 = 0$ is maximal.

It is straightforward to construct an interval estimate for β_1 in this example. If one uses the optimal design (i.e. *Design* 4 in Table 2.2), it can be shown that a 95% confidence interval for the slope parameter β_1 is

$$\hat{\beta}_1 - 2.145\sqrt{\frac{\text{MS}_e}{196}} \le \beta_1 \le \hat{\beta}_1 + 2.145\sqrt{\frac{\text{MS}_e}{196}}, \tag{2.14}$$

where $t_{\alpha/2,N-2} = 2.145$ is the 97.5th percentile of the t distribution with $N-2 = 14$ degrees of freedom. With $N = 16$ and a fixed error variance, the width of this interval is the smallest possible among all four designs. For instance, we observe that this width is about one-half the width of the corresponding confidence interval for β_1 constructed from *Design* 1. Note that a confidence interval for the intercept β_0 can also be similarly constructed and compared.

It is important to note that the optimal design found above was obtained among four designs. The optimality of the design clearly depends on the class of designs under comparison. An optimal design found from a smaller class of designs may be no longer optimal when more designs are included for comparison. In this particular example, we show later on that *Design* 4 is *universally optimal* for estimating the slope parameter. This is a very strong and desirable property and it means that *Design* 4 is optimal among ALL designs on the given design space. In other words, there are no other designs on the same design space that can produce a smaller variance for the estimated slope parameter than *Design* 4. When there is no ambiguity, the *universally optimal design* is often simply referred to as the *optimal design*.

2.3   Relative efficiency and sample size

Let us return to the commonly applied method to increase efficiency, namely, an increase in sample size. Given the fact that increase in sample size N can have the same effect on efficiency as an increase of the var(x), a relative efficiency measure can be formulated. This measure makes it possible to compute how many patients are needed to ensure that *Design* 1 has the same efficiency as the *optimal design* (*Design* 4). We can define the relative efficiency (RE) as the inverse ratio of the $\mathrm{var}(\hat{\beta}_1)$ of that design and the variance $\mathrm{var}(\hat{\beta}_1^*)$ of the *optimal design*:

$$\mathrm{RE} = \frac{\mathrm{var}(\hat{\beta}_1^*)}{\mathrm{var}(\hat{\beta}_1)} = \frac{N \ \mathrm{var}(x)}{N^* \ \mathrm{var}(x^*)}, \qquad (2.15)$$

with N^* and $\mathrm{var}(x^*)$ being the sample size and variance of x for the *optimal design*, respectively. It should be noted that the ratio of the two variances $\mathrm{var}(\hat{\beta}_1^*)$ and $\mathrm{var}(\hat{\beta}_1)$ in Equation (2.15) is actually a normalized efficiency measure for continuous designs. This RE is always less than or equal to one and assumes that $N^* = N$.

In the radiation-dosage example, we compare the efficiency of *Design* 1 relative to the *optimal design*: $\mathrm{RE} = 0.51/1.43 = 0.36$. To find out how many patients are needed in *Design* 1 to have the same efficiency as the *optimal design*, the numerator in Equation (2.15) can be set equal to the denominator, that is, $N \ \mathrm{var}(x) = N^*\mathrm{var}(x^*)$. In terms of sample size, we require about $N = N^*\mathrm{var}(x^*)/\mathrm{var}(x) = 16 \times 12.25/4.37 \approx 45$ patients in *Design* 1 to have the same efficiency as the *optimal design* for estimating the slope parameter. Alternatively, we need about $45/16 = 2.8$ times as many patients in *Design* 1 to obtain the minimum variance of the treatment effect estimator $\mathrm{var}(\hat{\beta}_1^*)$. It can be put differently by saying that based on a sample of $N^* = 16$ patients in the *optimal design*, about $(\mathrm{RE}^{-1} - 1)100\% = (0.36^{-1} - 1)100\% = 180\%$ more patients than for the *optimal design* will be needed. Of course, these calculations assume that increasing the sample size will not affect the variance $\mathrm{var}(x)$ and that the (normalized) relative efficiency adequately applies to exact designs for small samples.

2.4   Simultaneous inference

An optimal design that minimizes the variance of one parameter estimator may generally not minimize the variances of other parameter estimators in the model. The estimators of the different regression parameters are often correlated and it may therefore be more informative to identify uncertainty simultaneously for all the parameters in the model.

In the linear regression model the slope parameter β_1 is usually of main interest, but the intercept β_0 also needs to be estimated. We recall that when we have a $100(1 - \alpha)\%$ confidence interval of a single parameter, the coefficient $100(1 - \alpha)$ represents the percentage of times that the confidence interval contains the true parameter with repeated random samplings, and α is the pre-selected Type I error rate. If several parameters are to be estimated, such an interpretation for the confidence coefficient no longer holds. If the estimates of both parameters β_0 and β_1 are independent, and each has a confidence coefficient $(1 - \alpha)$, then the joint probability will be $(1 - \alpha)^2$, which is smaller than the original confidence coefficient $(1 - \alpha)$. The probability that at least one of the hypotheses on the two parameters is erroneously rejected will then be $\{1 - (1 - \alpha)^2\}$, which is of course larger than the original Type I error probability α. Since the individual parameter estimates are usually not independent, the confidence coefficient cannot always be precisely determined. A better way is to construct a simultaneous confidence interval or ellipse for both parameters β_0 and β_1. It can be shown that the quadratic form for β_0 and β_1

$$F = \frac{N(\hat{\beta}_0 - \beta_0)^2 + 2\sum x_i(\hat{\beta}_0 - \beta_0)(\hat{\beta}_1 - \beta_1) + \sum x_i^2(\hat{\beta}_1 - \beta_1)^2}{2MS_e}, \qquad (2.16)$$

has an F distribution with 2 and $N - 2$ degrees of freedom, and that

$$\text{Prob}\left(F \leq F_{\alpha,2,N-2}\right) = 1 - \alpha, \qquad (2.17)$$

where $F_{\alpha,2,N-2}$ is the $100(1 - \alpha)$th percentile of an F distribution with 2 and $N - 2$ degrees of freedom. The expression within the brackets in Equation (2.17) characterizes a confidence ellipse that contains the parameters β_0 and β_1 in $100(1 - \alpha)\%$ of the times that the study is replicated. The probability that this will happen simultaneously is $(1 - \alpha)$. Miller (1966) and Montgomery and Peck (1992) gave a general review of simultaneous inference and technical details.

Figure 2.2 shows four different ellipses for the two parameters β_0 and β_1. The intersection of the dotted axes in these ellipses represents the point estimators $\hat{\beta}_0$ and $\hat{\beta}_1$. The lengths of the axes in the ellipses are related to the variances of each of the estimators $\hat{\beta}_0$ and $\hat{\beta}_1$. Elongation of the ellipse along the axis of one parameter implies that that parameter is not as well estimated as the other parameter.

The covariance between these estimators $\text{cov}(\hat{\beta}_0, \hat{\beta}_1)$ determines the direction of the axes. A positive covariance $\text{cov}(\hat{\beta}_0, \hat{\beta}_1) > 0$ indicates that the two estimates $\hat{\beta}_0$ and $\hat{\beta}_1$ are positively correlated. This means if one estimate is large, the other also tends to be large, and conversely. If we have a negative covariance, that is, $\text{cov}(\hat{\beta}_0, \hat{\beta}_1) < 0$, this means that the two estimates are negatively correlated and the two estimates tend to move in opposite directions.

Figure 2.2 shows how the covariance between the two estimates influences the confidence ellipses. Figure 2.2a and b shows the case when both estimators are negatively correlated, that is, $\text{cov}(\hat{\beta}_0, \hat{\beta}_1) < 0$. Figure 2.2c and d displays the situation where the intercept and slope estimates $\hat{\beta}_0$ and $\hat{\beta}_1$ are uncorrelated.

The confidence ellipses in Figure 2.2 contain all information about the uncertainty of the two parameters β_0 and β_1. When more than two parameters are involved, ellipses can be extended to form ellipsoids in more than two dimensions. In general, the information about the uncertainty of such ellipsoids can be represented by the *volume* of the ellipsoid or by the *contour* of the ellipsoid. The *length of the axes* is also a measure of uncertainty. In the following section, we explain how we can base design optimality criteria on various characteristics of a confidence ellipsoid.

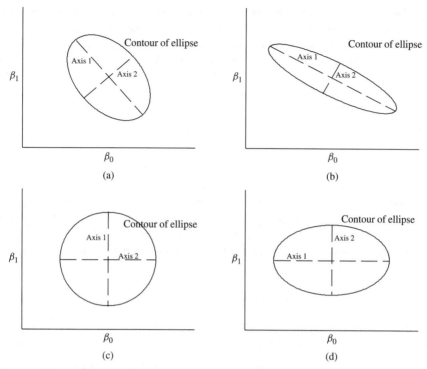

Figure 2.2 Confidence ellipses for two parameters β_0 and β_1 in the simple linear model.

2.5 Optimality criteria

Confidence ellipsoids for the parameters can be used to find an optimal design for the simultaneous estimation of these parameters. The problem is, however, what property of these ellipsoids can be used to define an optimal design? Let us denote the two parameter estimators $\hat{\beta}_0$ and $\hat{\beta}_1$ as a vector $\hat{\beta} = (\hat{\beta}_0, \hat{\beta}_1)'$, where the prime indicates that this vector $\hat{\beta}$ is a 2×1 column vector and $\hat{\beta}_1$ is placed underneath $\hat{\beta}_0$. The variance–covariance matrix of the parameter estimators $\hat{\beta}_0$ and $\hat{\beta}_1$ is given by

$$\text{Cov}(\hat{\beta}) = \frac{\sigma_{\varepsilon}^2}{N\text{SS}_x} \begin{bmatrix} \sum x_i^2 & -\sum x_i \\ -\sum x_i & N \end{bmatrix}. \tag{2.18}$$

This matrix, by definition, is symmetric and so the off-diagonal elements are necessarily equal. The variances of $\hat{\beta}_0$ and $\hat{\beta}_1$ are on the main diagonal of this 2×2 matrix, and the covariance $\text{cov}(\hat{\beta}_0, \hat{\beta}_1)$ is the off-diagonal elements. From this matrix, we note for example, the variance of the estimator $\hat{\beta}_1$ is $\text{var}(\hat{\beta}_1) = \sigma_{\varepsilon}^2/\text{SS}_x$. See also Equation (2.9).

The variance–covariance matrix determines the shape and form of the confidence ellipsoid and this motivates us to measure the efficiency of a design based on its variance–covariance matrix $\text{Cov}(\hat{\beta})$. In the literature, various different optimality criteria have been put forward. Among them are the so-called alphabetic optimality criteria (Box, 1982), where each summarizes the variances and covariances of the estimators of parameters in a unique way. Atkinson and Donev (1992, Chapter 10) reviewed the most familiar design criteria that include the D-, A-, G- and the E-optimality criteria.

2.5.1   D-optimality criterion

Uncertainty of a set of parameter estimators can be expressed by the *volume* of a confidence ellipsoid; the smaller the volume, the more accurate the estimators. The determinant or D-optimality criterion minimizes the product of the squared lengths of the axes of the ellipsoid and is proportional to the volume of the confidence ellipsoid. It is defined as the determinant of the variance–covariance matrix $\text{Cov}(\hat{\beta})$, that is:

$$\text{D-criterion} = \text{Det}[\text{Cov}(\hat{\beta})]. \tag{2.19}$$

It is relatively easy to calculate the D-optimality criterion for the variance–covariance matrix in Equation (2.18). Since the determinant for a 2×2 matrix $\begin{bmatrix} a & c \\ c & b \end{bmatrix}$ is equal to $(ab - c^2)$, the determinant of the matrix in Equation (2.18) is

$$\text{D-criterion} = \frac{\sigma_\varepsilon^4}{N^2 \text{SS}_x^2} \left[N \sum x_i^2 - \left(\sum x_i \right)^2 \right] = \frac{\sigma_\varepsilon^4}{N \text{SS}_x}. \tag{2.20}$$

Of course, when this matrix becomes large, one will need a computer program to calculate the determinant.

It can be seen from Equation (2.20) that an increase of both the sample size N and the variation SS_x will reduce the value of the D-optimality criterion. Maximizing the sample size or the SS_x will minimize the D-criterion value. Likewise, minimizing the error variance σ_ε^2 has the same effect on this criterion.

The D-optimality criterion has some useful properties. Firstly, the D-optimal design is invariant under linear transformation of the scale of the independent variable. This means that if we change the design interval, we can deduce the D-optimal design directly from the one constructed over the original design interval. Such property may not hold for the other optimality criteria, and this is perhaps one reason why the D-optimality criterion is applied so often. Secondly, the optimality criterion is proportional to the volume of the confidence ellipsoid for the parameters. This property gives the criterion a natural interpretation in terms of a confidence interval for the parameters in the regression model.

There is some evidence that the D-optimality criterion produces optimal designs that are also highly efficient with respect to other criteria (Goos, 2002,

p. 37). This is perhaps due to the fact that the alphabetic optimality criteria depend on the eigenvalues of the variance–covariance matrix $\text{Cov}(\hat{\beta})$. The D-optimality criterion does have some drawbacks. The value of the D-optimality criterion may not be easy to calculate using a pocket calculator when the number of parameters is larger than two. The second drawback is that minimization of the determinant of a covariance matrix may lead to elongation in the direction of one axis of the confidence ellipsoid. This happens, for example, when the length of the confidence interval of only one of the parameters is short and the others are all long, leading to the situation that only one of the parameters is estimated efficiently while the others are not. Finally, a D-optimal design may be optimal simultaneously for each of the individual parameters in the model, but because of the correlations among the parameter estimators, the design may be very inefficient for estimating certain linear combinations of the parameters.

2.5.2   A-optimality criterion

The average or A-optimality criterion minimizes the sum of the squared lengths of the axes, which indirectly measures the size of the ellipsoid. This is the same as minimizing the trace of the variance–covariance matrix $\text{Cov}(\hat{\beta})$:

$$\text{A-criterion} = \text{Trace}[\text{Cov}(\hat{\beta})]. \tag{2.21}$$

For the 2×2 variance–covariance matrix in Equation (2.18), the A-criterion is just the sum of the two main diagonal elements:

$$\text{A-criterion} = \frac{\sigma_\varepsilon^2}{N\,\text{SS}_x} \left[\sum x_i^2 + N \right]. \tag{2.22}$$

Here, again, an increase in the sample size N or SS_x will result in a decrease of the A-criterion value. Likewise, a decrease in the error variance σ_ε^2 will decrease the criterion value. We iterate that minimizing the A-criterion means that the quantity $\left[\text{var}(\hat{\beta}_0) + \text{var}(\hat{\beta}_1) \right]$ is minimized.

A major drawback of the A-optimality criterion is that it is not invariant under linear transformation of the scale of the independent variables, that is, each scale may lead to another optimal design. In some applications, however, this drawback may not be that important. Another problem of this criterion is that it can be misleading in the sense that the variances of some parameters may have very different magnitudes, and so simply averaging the diagonal elements may not accurately reflect the equal interest in each parameter.

2.5.3   G-optimality criterion

A third criterion is the global or G-optimality criterion, which may be useful when a researcher is interested in predicting the outcome variable Y as efficiently as possible over the design space. The predicted value $\hat{y}_0$ at an arbitrary value x_0 of the independent variable is given by $\hat{y}_0 = \hat{\beta}_0 + \hat{\beta}_1 x_0$, and this predicted value is normally distributed with variance equal to

$$\text{var}(\hat{y}_0) = \sigma_\varepsilon^2 \left\{ \frac{1}{N} + \frac{(x_0 - \bar{x})^2}{\text{SS}_x} \right\}. \tag{2.23}$$

In practice, this variance is usually standardized as $s(x, \xi) = N \, \text{var}(\hat{y}_0)/\sigma_\varepsilon^2$. Clearly, this standardized variance $s(x, \xi)$ depends on x_0, the design points (x values) and the particular design ξ, for which the variance–covariance matrix $\text{Cov}(\hat{\beta})$ is computed. To obtain an accurate prediction of the responses across the design space, we want to have a design that gives us the smallest possible standardized variance of the predicted response across the design space. A G-optimal design is a design that minimizes the maximum standardized variance of the predicted response over the design space Δ. In other words, it selects a design that minimizes

$$\text{G-criterion} = \max_{d \in \Delta} \left[\frac{N \times \text{var}(\hat{y}_0)}{\sigma_\varepsilon^2} \right]. \tag{2.24}$$

Although this criterion is not frequently used in practice and seems only relevant when the prediction of the overall responses is desired, it has an important theoretical property that connects it to the D-optimality criterion. In a ground-breaking paper, Kiefer and Wolfowitz (1960) showed that D- and G-optimal designs are the same when the errors all have constant variance. It is known that the standardized variance for a continuous G-optimal design ξ^* is always less than or equal to the number of parameters p in the model, that is, $s(x, \xi^*) \leq p$ with equality at the design points. This simple inequality can then be used to check whether a design is D-optimal or not. Figure 2.3 displays the standardized variance functions for the simple linear model $y_i = \beta_0 + \beta_1 x_i + \varepsilon_i$ using *Design* 4 and for the quadratic model $y_i = \beta_0 + \beta_1 x_i + \beta_2^2 + \varepsilon_i$ using the

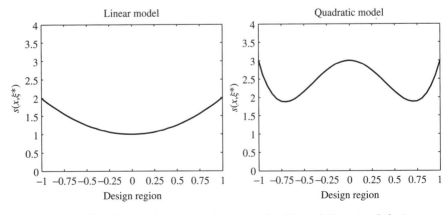

Figure 2.3 Standardized variance functions for D- and G-optimal designs.

design equally supported at -1, 0 and 1. The figure shows that these designs have standard variance functions that satisfy $s(x, \xi^*) \leq p$ with equality at the design points. Here $p = 2$ for the simple linear model and $p = 3$ for the quadratic model. This confirms that the D- and G-optimal design points are equally supported at $[-1, +1]$ for the simple linear model and at $[-1, 0, +1]$ for the quadratic regression model.

2.5.4   E-optimality criterion

We now discuss the E-optimality criterion, where 'E' stands for the extreme axis of a confidence ellipsoid. This criterion minimizes the squared length of the 'largest' axis of the confidence ellipsoid. The rationale for this criterion is that the parameters are estimated least accurately in the direction of the largest axis. By minimizing the squared length of the largest axis of the ellipsoid, the criterion minimizes the 'worst case' scenario, that is, it minimizes the maximal possible variance from any standardized linear combination of the two estimated parameters. Here, standardized means that the sum of squares of the two coefficients in the linear combination should equal 1. Algebraically, it can be shown that minimizing the squared length of the 'largest' axis of the confidence ellipsoid is the same as minimizing the maximum eigenvalue of the 2×2 variance–covariance matrix $\text{Cov}(\hat{\beta})$. The formula is

$$\text{E-criterion} = \text{Root}_{\max}[\text{Cov}(\hat{\beta})], \tag{2.25}$$

where $\text{Root}_{\max}$ is the largest root or eigenvalue of the variance–covariance matrix $\text{Cov}(\hat{\beta})$. This criterion is not often applied in practice. It has the disadvantage that not all information on the parameters in the model is used, because only the maximum root is considered, and that it is sensitive to the scale of the independent variable.

2.5.5   Number of distinct design points

Different optimal designs may require different number of design points. It is clear that in order to estimate p parameters in any model, at least p distinct design points are needed, and for many models and optimality criteria, the optimal number of distinct design points will be p. An interesting result called *Carathéodory's theorem* (Silvey, 1980, Appendix 2) provides us with an upper bound on the number of design points needed for an optimal design. For many design problems with p parameters, this number is $p(p + 1)/2$. Thus, the optimal number of distinct design points is between p and $p(p + 1)/2$ (Pukelsheim, 1993, p.190). Serious loss of efficiency may result in designs with more than $p(p + 1)/2$ distinct design points and less than p distinct design points may not enable us to estimate all the p parameters in the model. Finally, it should be noted that the upper bound does not hold for the Bayesian design criteria (Atkinson, Donev and Tobias, 2007, Chapter 18).

2.6   Relative efficiency

Relative efficiency is a measure that enables us to compare the efficiencies of two designs. For example, if the D-criterion is of interest and the given model has p parameters, the relative efficiency of a design ξ with respect to the D-optimal design ξ^* is

$$
\text{RE}_D = \left\{ \frac{\text{Det}[\text{Cov}(\hat{\beta}^*)]}{\text{Det}[\text{Cov}(\hat{\beta})]} \right\}^{1/p},
\tag{2.26}
$$

where $\text{Cov}(\hat{\beta}^*)$ and $\text{Cov}(\hat{\beta})$ are the covariance matrices of the estimators from the optimal design ξ^* and the design ξ, respectively. The pth root is used in the above ratio to facilitate interpretation. In terms of number of subjects, this expression enables us to determine the number of extra subjects that is needed for design ξ to have the same efficiency as the optimal design ξ^*. For example, if the design ξ has a relative efficiency $\text{RE}_D = 0.597$, then the sample size has to be multiplied by $\text{RE}_D^{-1} = 1.67$, to obtain the same efficiency as the optimal design. This can be rephrased as $(0.597^{-1} - 1)100\%$ more subjects than for the optimal design will be needed, that is, 67% more subjects.

Similar REs for the other optimality criteria can be formulated, but now without the power $(1/p)$. The relative efficiency for the A-criterion is

$$
\text{RE}_A = \left\{ \frac{\text{Trace}[\text{Cov}(\hat{\beta}^*)]}{\text{Trace}[\text{Cov}(\hat{\beta})]} \right\},
\tag{2.27}
$$

and that for the E-criterion is

$$
\text{RE}_E = \left\{ \frac{\text{Root}_{\max}[\text{Cov}(\hat{\beta}^*)]}{\text{Root}_{\max}[\text{Cov}(\hat{\beta})]} \right\}.
\tag{2.28}
$$

A similar expression for the relative efficiency of the G-optimality criteria can be given.

It should be emphasized that the relative efficiencies for these three criteria are generally unequal. They are also not the one presented in Section 2.3, which was computed for estimating only the single parameter β_1. Here, RE_D, RE_A and RE_E are used for the simultaneous estimation of both regression parameters β_0 and β_1, and each contains a different amount of information.

2.7   Matrix formulation of designs for linear regression

In this section, the topics and computations of the previous sections are briefly explained in terms of matrix algebra. Readers not familiar with matrix algebra may skip this section.

The simple linear regression model in Equation (2.7) can be reformulated in matrix notation as

$$y = X\beta + \varepsilon$$

$$
\begin{bmatrix} y_1 \\ y_2 \\ \vdots \\ y_N \end{bmatrix}
=
\begin{bmatrix} 1 & x_1 \\ 1 & x_2 \\ \vdots & \vdots \\ 1 & x_N \end{bmatrix}
\begin{bmatrix} \beta_0 \\ \beta_1 \end{bmatrix}
+
\begin{bmatrix} \varepsilon_1 \\ \varepsilon_2 \\ \vdots \\ \varepsilon_N \end{bmatrix},
\tag{2.29}
$$

where the data vector y is modelled as the product of a design matrix X and a vector of regression parameters $\beta = (\beta_0, \beta_1)'$. The $N \times 1$ column vector y consists of the responses on the dependent variable Y for the N subjects in the study. The $N \times 1$ column vector ε consists of the error terms. The $N \times 2$ design matrix X has two columns: the first consists of 1s, corresponding to the regression parameter β_0 and the second column of X contains the values of the independent variable X. The design matrices X for the four different designs in Table 2.2, discussed in the radiation-dosage study, are shown in Table 2.3.

Table 2.3   Design matrices for the four radiation-dosage designs.

Design 1: $X_1' = \begin{bmatrix} 1 & 1 & 1 & 1 & 1 & 1 & 1 & 1 & 1 & 1 & 1 & 1 & 1 & 1 & 1 & 1 \\ 1 & 1 & 2 & 3 & 3 & 4 & 4 & 4 & 5 & 5 & 5 & 5 & 6 & 7 & 8 & 8 \end{bmatrix}$

Design 2: $X_2' = \begin{bmatrix} 1 & 1 & 1 & 1 & 1 & 1 & 1 & 1 & 1 & 1 & 1 & 1 & 1 & 1 & 1 & 1 \\ 1 & 1 & 2 & 2 & 3 & 3 & 4 & 4 & 5 & 5 & 6 & 6 & 7 & 7 & 8 & 8 \end{bmatrix}$

Design 3: $X_3' = \begin{bmatrix} 1 & 1 & 1 & 1 & 1 & 1 & 1 & 1 & 1 & 1 & 1 & 1 & 1 & 1 & 1 & 1 \\ 1 & 1 & 1 & 2 & 2 & 3 & 3 & 4 & 5 & 6 & 6 & 7 & 7 & 8 & 8 & 8 \end{bmatrix}$

Optimal Design: $X_0' = \begin{bmatrix} 1 & 1 & 1 & 1 & 1 & 1 & 1 & 1 & 1 & 1 & 1 & 1 & 1 & 1 & 1 & 1 \\ 1 & 1 & 1 & 1 & 1 & 1 & 1 & 1 & 8 & 8 & 8 & 8 & 8 & 8 & 8 & 8 \end{bmatrix}$

The least squares estimators $\hat{\beta} = (\hat{\beta}_0, \hat{\beta}_1)'$ are given by

$$\hat{\beta} = (X'X)^{-1}X'y, \tag{2.30}$$

and the variance–covariance matrix of $\hat{\beta}$ in matrix notation is

$$\mathrm{Cov}(\hat{\beta}) = \sigma_\varepsilon^2 (X'X)^{-1}. \tag{2.31}$$

We observe from Equation (2.31) that when the error variance is given, the variance–covariance matrix of the regression parameter estimators can be calculated without actually knowing the responses y. For example, consider Design 1 in Table 2.3 with design matrix given by

$$X_1' = \begin{bmatrix} 1 & 1 & 1 & 1 & 1 & 1 & 1 & 1 & 1 & 1 & 1 & 1 & 1 & 1 & 1 & 1 \\ 1 & 1 & 2 & 3 & 3 & 4 & 4 & 4 & 5 & 5 & 5 & 5 & 6 & 7 & 8 & 8 \end{bmatrix} \quad (2.32)$$

The variance–covariance matrix of the regression parameters can be computed in two steps:

$$X_1' X_1 = \begin{bmatrix} 16 & 71 \\ 71 & 385 \end{bmatrix} \quad (2.33)$$

and

$$\text{Cov}(\hat{\beta}) = \sigma_\varepsilon^2 \begin{bmatrix} 16 & 71 \\ 71 & 385 \end{bmatrix}^{-1} = \sigma_\varepsilon^2 \begin{bmatrix} 0.3441 & -0.0634 \\ -0.0634 & 0.0143 \end{bmatrix}. \quad (2.34)$$

In Table 2.4, we list the variance–covariance matrices and values of the optimality criteria. As can be seen, the optimal design has the smallest volume (D-optimality) and the smallest sum of squared axes (A-optimality) of the confidence ellipse of β_0 and β_1. This means that if we have a quantitative independent variable in a simple regression model, the parameters β_0 and β_1 can simultaneously be estimated as efficiently as possible if the design points were chosen at

Table 2.4   Variance–covariance matrices for the four radiation-dosage designs.

Design 1

$\text{Cov}(\hat{\beta}) = \sigma_\varepsilon^2 (X_1' X_1)^{-1}$

$$= \sigma_\varepsilon^2 \begin{bmatrix} 0.3441 & -0.0634 \\ -0.0634 & 0.0143 \end{bmatrix}$$

D-criterion: $8.9367 \times 10^{-4} \times \sigma_\varepsilon^4$

A-criterion: $0.3584 \times \sigma_\varepsilon^2$

E-criterion: $0.3558 \times \sigma_\varepsilon^2$

Design 2

$\text{Cov}(\hat{\beta}) = \sigma_\varepsilon^2 (X_2' X_2)^{-1}$

$$= \sigma_\varepsilon^2 \begin{bmatrix} 0.3036 & -0.0536 \\ -0.0536 & 0.0119 \end{bmatrix}$$

D-criterion: $7.4405 \times 10^{-4} \times \sigma_\varepsilon^4$

A-criterion: $0.3155 \times \sigma_\varepsilon^2$

E-criterion: $0.3131 \times \sigma_\varepsilon^2$

Design 3

$\text{Cov}(\hat{\beta}) = \sigma_\varepsilon^2 (X_3' X_3)^{-1}$

$$= \sigma_\varepsilon^2 \begin{bmatrix} 0.2500 & -0.0417 \\ -0.0417 & 0.0093 \end{bmatrix}$$

D-criterion: $5.7870 \times 10^{-4} \times \sigma_\varepsilon^4$

A-criterion: $0.2593 \times \sigma_\varepsilon^2$

E-criterion: $0.2570 \times \sigma_\varepsilon^2$

Optimal design

$\text{Cov}(\hat{\beta}) = \sigma_\varepsilon^2 (X_0' X_0)^{-1}$

$$= \sigma_\varepsilon^2 \begin{bmatrix} 0.1658 & -0.0230 \\ -0.0230 & 0.0051 \end{bmatrix}$$

D-criterion: $3.1888 \times 10^{-4} \times \sigma_\varepsilon^4$

A-criterion: $0.1709 \times \sigma_\varepsilon^2$

E-criterion: $0.1690 \times \sigma_\varepsilon^2$

both ends of the x scale and if the observations would be equally divided over these two design points.

For simplicity sake, it was assumed in the breast cancer example that there were no ethical objections to assigning patients to each of the different dosage levels and that each of the four designs would be equally feasible in practice. However, this is not that simple and clinical reasons may prevent a researcher from assigning half of the sample of patients to the lowest and half of the sample to highest radiation-dosage levels. But it should be kept in mind that efficiency of a treatment effect estimator $\hat{\beta}_1$ is improved as more patients are assigned to the lower and higher dosage levels. Moreover, as shown in Section 2.3, it would require almost three times as many patients in *Design* 1 to have the same efficiency for estimating β_1 as the *optimal design* has. A researcher trying to establish a significant treatment effect could therefore choose an optimal design that requires fewer patients having to undergo an often very unpleasant radiation treatment.

In matrix notation, the variance of the predicted response $\hat{y}_0 = \hat{\beta}_0 + \hat{\beta}_1 x_0$ for a linear model is

$$\text{var}(\hat{y}_0) = f'(x_0)\text{Cov}(\hat{\beta})f(x_0), \qquad (2.35)$$

where $\text{Cov}(\hat{\beta})$ is the 2×2 variance–covariance matrix of parameter estimators in Equation (2.31). For the linear model the function, $f'(x_0)$ is given by $f'(x_0) = [1\ x_0]$. This notation indicates that the x values coincide with the design points. When we have a quadratic model, the predicted response is $\hat{y}_0 = \hat{\beta}_0 + \hat{\beta}_1 x_0 + \hat{\beta}_2 x_0^2$ and the function $f'(x_0)$ in Equation (2.35) is $f'(x_0) = [1\ x_0\ x_0^2]$. The $\text{Cov}(\hat{\beta})$ is a 3×3 variance–covariance matrix for the three parameter estimators $\hat{\beta}_0$, $\hat{\beta}_1$ and $\hat{\beta}_2$.

In Figure 2.4, the standardized variance function $s(x, \xi)$ for the four radiation-dosage designs are plotted for all x values. *Design* 1 has the largest value for $s(x, \xi)$ and the function is not completely symmetric. The standardized variance for the first dosage level is $s(x = 1, \xi_1) = 3.7033$, while the standardized variance for the last dosage level is $s(x = 8, \xi_1) = 3.9035$. This is caused in part by the fact that the original design is not symmetrically weighted. The functions for the other three designs are, however, symmetric. The fourth design is clearly the optimal design, having the lowest values for the maximum standardized variance over the range of dosage levels with a maximum for $s(x = 1, \xi_4) = s(x = 8, \xi_4) = 2$. It should be noted that these design points $x = 1$ and $x = 8$ are both G-optimal and D-optimal. Moreover, it can be shown that for this simple linear model the equally weighted design at both ends of the design interval is both A- and E-optimal as well.

Finally, we compare relative efficiencies of these designs under different optimality criteria. Their values for *Designs* 1, 2 and 3 with respect to the optimal *Design* 4 are in Table 2.5. For example, the $\text{RE}_D = (3.1888/8.937)^{1/2} = 0.5973$, where the D-criterion values 3.1888 and 8.937 can be found in Table 2.4. Table 2.5 shows that the criteria have different RE values. This difference is

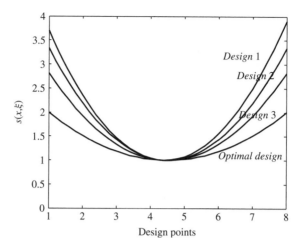

Figure 2.4 Standardized variance functions for the four radiation-dosage designs.

Table 2.5   Relative efficiencies for three designs.

	RE_D	RE_A	RE_E
Design 1	0.5973	0.4768	0.4750
Design 2	0.6547	0.5417	0.5398
Design 3	0.7423	0.6591	0.6576

due to the fact that each optimality criterion measures uncertainty about the parameters in a different way. Recall that the D-optimality criterion is proportional to the volume of the confidence ellipse of the parameters, implemented as the product of the squared lengths of the axes, while the A-criterion is related to the sum of the squared lengths of the axes. The E-criterion is based on the largest axis only. One could generally say that the E-criterion uses less information than the A- and D-criterion. Note that since the D-criterion is invariant under linear transformation of the scale of the independent variables, the RE_D values are also invariant. It should be emphasized that this property does not hold for RE_A and RE_E values. This means that differences in values for these last two REs will be encountered when the scale of the independent variable is changed.

Table 2.5 shows that *Design* 3 has the highest relative efficiency for all three criteria. *Design* 1 surely has the smallest relative efficiency values for all three criteria and if the researcher has a choice, this design should *not* be chosen, because it has less power of finding an effect of radiation dosage than the other designs.

2.8   Summary

There are at least three ways to design a more efficient study for the simple linear regression model. First, efficiency can be improved by measuring the dependent variable more accurately. In this chapter, we show that there is a direct relation between the measurement error variance σ_ε^2 and the variance of the treatment effect estimator $\hat{\beta}_1$. Reduction of the range of the errors by one-half will reduce the var($\hat{\beta}_1$) by a factor of four. A second way to improve efficiency is to increase the total sample size N. But this will of course lead to more costs for collecting the data and running the study. A third way to improve efficiency is to select the levels of the independent variable X in such a way that their var(x) will be as large as possible. Careful selection of the levels of the independent variable can be done in more experimentally controlled studies. The first two ways (more accurate measurement and sample size increase) can be applied in observational studies with less control over the levels of the independent variable.

In order to measure efficiency, a number of optimality criteria have been proposed. In general, the D-optimality criterion has a number of advantages. Because of its invariance under linear transformations of the scale of the independent variable, we recommend the D-optimality criterion, especially when different scales of the independent variable are implemented in different studies.

Finally, we recommend that we should use small number of design points to the extent possible. This number should be between p and $p(p+1)/2$, where p is the number of parameters in the linear model.

3

Designs for multiple linear regression analysis

3.1 Design problem for multiple linear regression

In Chapter 2, we focused on design problems for a simple linear regression model with two parameters. Multiple regression models have more than two parameters in the mean function and oftentimes include more than one independent variable. So, instead of looking at the levels of one independent variable, a multiple regression design problem has to consider selecting levels of more than one independent variable.

Consider the following example taken from educational research. The response variable (Y) is the rate of increase in vocabulary (i.e. vocabulary growth) among pupils in a certain county. The purpose of the research is to study how vocabulary growth is related to the school grade (X_1) the pupil is in and to the income (social) class (X_2) of the pupil's parents. A multiple regression model may be used to study their relationship using X_1 and X_2 as predictors (or independent variables) and vocabulary growth (Y) as the outcome variable.

There are two immediate key questions for the design of this study. The first relates to the choice of number of levels for each of the independent variables. Usually a sample of subjects is selected for each combination of levels of both independent variables. But how many combinations of levels should we use and how do we choose the combination levels so that we have an efficient design? In the vocabulary-growth study, these questions translate to asking how to select the number of different grades and how many different income (social) classes to

An Introduction to Optimal Designs for Social and Biomedical Research   M. P. F. Berger & W. K. Wong
© 2009, John Wiley & Sons, Ltd

include in the study? Would only choosing high and low income classes be best? Or would a selection of only the higher income classes and the higher grades be the most efficient choice?

The second design question, which is closely related to the first question, concerns the number of units assigned to each combination of levels. For example, for the vocabulary study, how many pupils will be needed from each grade level and from each income category? Given a fixed total sample size, would it be more efficient to just select pupils from both the lowest and the highest school grades or would it be more efficient to select an equal number of pupils from each available grade level? These two design questions together constitute the basic design problem for a multiple linear regression model. How many levels of the independent variables are needed and how many units to sample from each combination of levels of the independent variables? The goal is to select a design that meets the objective or objectives of the study. If model parameters estimation is the main goal, the design should provide smallest possible variances for the estimated parameters or sufficient power for tests of hypotheses for the parameters.

3.1.1   The design

We now expand the definition of a design for a simple regression model from Chapter 2 to a multiple regression model. Suppose that we have two independent variables X_1 and X_2, where each has values ranging from -1 to 1 (after suitable recoding). Figure 3.1 shows a two-dimensional space Δ_2. A design point d_j in Figure 3.1 represents a combination of values of the two independent variables and is a point in the two-dimensional space, that is, $d_j \in \Delta_2$.

Figure 3.1 shows six design points for two independent variables X_1 and X_2: $d_1 = (-0.55, 0.7), d_2 = (-1, 0), d_3 = (-0.5, -0.1), d_4 = (0.45, 0.25),$

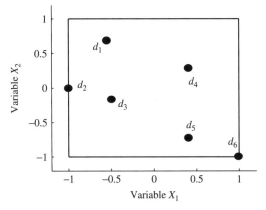

Figure 3.1 A two-dimensional design space Δ_2 scaled to $[-1, 1] \times [-1, 1]$ with six design points $d_1, d_2, \ldots, d_6$.

$d_5 = (0.45, -0.7)$ and $d_6 = (1, -1)$. These design points represent different combinations of values for X_1 and X_2. For example, the first design point $d_1 = (-0.55, 0.7)$ means the value of X_1 is set equal to -0.55 and the value of X_2 is set equal to 0.7. If we have a sample of $N = 10$ observations from these design points, we could have two observations from each of the design points d_1 and d_2, one observation from each of the design points d_3, d_4, and d_4 and three observations from d_6. This design can be schematically represented as an exact design:

$$\xi_{10} = \left\{ \begin{array}{cccccc} d_1 & d_2 & d_3 & d_4 & d_5 & d_6 \\ 2/10 & 2/10 & 1/10 & 1/10 & 1/10 & 3/10 \end{array} \right\}. \tag{3.1}$$

In general, we may denote an exact design ξ_N for a multiple regression model with $(p-1)$ independent variables with n_j observations at the design point d_j, $j = 1, \ldots, m$, by

$$\xi_N = \left\{ \begin{array}{ccccc} d_1 & d_2 & d_3 & \ldots & d_m \\ n_1/N & n_2/N & n_3/N & \ldots & n_m/N \end{array} \right\}, \tag{3.2}$$

where $d_j \in \Delta_{p-1}$ for $j = 1, \ldots, m$ and Δ_{p-1} is a $(p-1)$-dimensional space of combinations of levels of $(p-1)$ independent variables and $\sum_j n_j = N$. This definition is similar to that for a simple linear regression, except that now each design points d_j represents a different combination of levels from the quantitative independent variables. The corresponding notation for a generic approximate or continuous design with m design points is

$$\xi = \left\{ \begin{array}{ccccc} d_1 & d_2 & d_3 & \ldots & d_m \\ w_1 & w_2 & w_3 & \ldots & w_m \end{array} \right\}, \tag{3.3}$$

where $\sum_j w_j = 1$ and $0 \leq w_j \leq 1$, for all j's.

In practice, the values of each of the independent variables are usually discrete and take on a few possible values. This means the total number of possible design points is small and the design problem is simplified. For example, with two independent variables, X_1 with four levels and X_2 with three levels, there are $m = 12$ different combinations of levels and these combinations form the candidate set of design points for the problem. Table 3.1 shows the design just described with $d_1, d_2, \ldots, d_{12}$ as design points. The design question in this case consists of how many observations to assign to these design points.

Table 3.1   Twelve different combinations of a design with two independent variables X_1 and X_2.

	X_{11}	X_{12}	X_{13}	X_{14}
X_{21}	d_1	d_4	d_7	d_{10}
X_{22}	d_2	d_5	d_8	d_{11}
X_{23}	d_3	d_6	d_9	d_{12}

We remind the reader that the ordering $j = 1, 2, \ldots, m$ of these 12 design points d_j is usually arbitrary. This is unlike the case when there is only a single quantitative independent variable where the ordering may be made according to the value of the variable. When there are two or more independent variables, it is less clear how one may order the design points in a meaningful way. In our example, we assume that each combination of the levels of the two independent variables may be independently replicated. In other words, a number of independent observations n_j may be assigned to each combination such that $\sum_j n_j = N$.

3.1.2   The multiple linear regression model

The multiple linear regression model with $(p - 1)$ independent variables X_l, $l = 1, \ldots, (p - 1)$, is an extension of the simple regression model with only one independent variable. Depending on the discipline, the words 'independent variables' may be replaced by 'predictors' and vice versa; so are the words 'response variable' and 'outcome variable'. Here and throughout this book, we do not make a distinction in their terminology.

The multiple linear regression model is usually used as a first approximation of the true relation between a response variable Y and a set of $(p - 1)$ predictor variables X_l. This is because multiple linear models are simple to understand and are quite flexible. In particular, each independent variable X_l may be replaced in the model by a function of itself when desired. Some frequently used functions are the logarithmic function (log X_l), the reciprocal function ($1/X_l$) and the quadratic function $\left(X_l^2\right)$.

A multiple regression model relating the values y_i of a response variable Y to the $(p - 1)$ values x_ls of the predictor variables X_l is given by

$$y_i = \beta_0 + \beta_1 x_{1i} + \beta_2 x_{2i} + \cdots + \beta_{p-1} x_{p-1,i} + \varepsilon_i. \tag{3.4}$$

The total number of parameters in the model is p. β_0 is the intercept and the remaining $(p - 1)$ regression parameters β_l's are the coefficients of the $(p - 1)$ independent variables X_l. The intercept coefficient β_0 is the mean value of the responses y_i when the values of all the independent variables are set equal to 0. Depending on the context of the problem, this interpretation may or may not be meaningful. The regression coefficient β_l represents the mean change in the response variable due to one unit change of the independent variable X_1 when the other independent variables are held constant, $l = 1, \ldots, (p - 1)$. The random errors ε_i's are all assumed to be independent and normally distributed with mean zero and variance σ_ε^2.

3.1.3   Estimation of parameters and efficiency

The least squares method is often used to estimate all the p regression parameters $\beta_l, l = 0, \ldots, (p - 1)$. The least squares method selects those estimators $\hat{\beta}_l$,

$l = 0, \ldots, (p-1)$, among all possible estimators, for which the sum of squared errors is minimized. This is the same as finding estimators $\hat{\beta}_l, l = 0, \ldots, (p-1)$ that minimize the sum of all squared differences between each observed response y_i and its postulated mean $\hat{y}_i$, that is:

$$\sum_i^N \varepsilon_i^2 = \sum_i^N \left[y_i - \hat{\beta}_0 - \hat{\beta}_1 x_{1i} - \hat{\beta}_2 x_{2i} - \cdots - \hat{\beta}_{p-1} x_{p-1,i} \right]^2 = \text{minimum}.$$

(3.5)

These parameter estimators $\hat{\beta}_0, \hat{\beta}_1, \hat{\beta}_2, \ldots, \hat{\beta}_{p-1}$ are called *least squares estimators* and we denote them collectively using the vector $\hat{\beta} = (\hat{\beta}_0, \hat{\beta}_1, \hat{\beta}_2, \ldots, \hat{\beta}_{p-1})'$. The prime notation indicates that $\hat{\beta}$ is a column vector with the individual parameter estimators stacked underneath each other. The minimum value of $\sum_i^N \varepsilon_i^2$ is referred to as the *sum of squared errors* or *residual sum of squares*, SS_e. The estimator of the error variance σ_ε^2 is $MS_e = SS_e/(N-p)$, with p equal to the total number of parameters in the model. Notice that this estimator for σ_ε^2 is a direct generalization of the formula for the simple linear model when $p = 2$ in Chapter 2. Other types of estimates are possible. For example, the method of maximum likelihood (ML) estimation can also be applied to obtain ML estimators. In this case, it can be shown that the ML estimators for $\beta_l, l = 0, \ldots, (p-1)$ are the same as least squares estimators, but the ML estimator for σ_ε^2 is $(N-p)/N$ times that of the least squares estimator given by MS_e.

As is in the case for the simple linear model, these least squares estimators for $\beta_l, l = 0, \ldots, (p-1)$ and σ_ε^2 are all unbiased, that is with mean values equal to the parameter values $-E(\hat{\beta}) = (\beta_0, \beta_1, \beta_2, \ldots, \beta_{p-1})'$ and $E(MS_e) = \sigma_\varepsilon^2$. The variance–covariance matrix for the $p \times 1$ vector $\hat{\beta}$ is the $p \times p$ symmetric matrix symbolized by

$$\text{Cov}(\hat{\beta}) = \begin{bmatrix} \text{var}(\hat{\beta}_0) & \text{cov}(\hat{\beta}_0, \hat{\beta}_1) & \ldots & \text{cov}(\hat{\beta}_0, \hat{\beta}_{p-1}) \\ \text{cov}(\hat{\beta}_0, \hat{\beta}_1) & \text{var}(\hat{\beta}_1) & \ldots & \text{cov}(\hat{\beta}_1, \hat{\beta}_{p-1}) \\ \vdots & \vdots & \vdots & \vdots \\ \text{cov}(\hat{\beta}_0, \hat{\beta}_{p-1}) & \text{cov}(\hat{\beta}_1, \hat{\beta}_{p-1}) & \ldots & \text{var}(\hat{\beta}_{p-1}) \end{bmatrix}.$$

(3.6)

The variances of the estimators are on the main diagonal of $\text{Cov}(\hat{\beta})$ and the off-diagonal elements are the covariances among the estimators. The elements of $\text{Cov}(\hat{\beta})$ contain all information about the uncertainty of the estimators.

We can test the individual effect of each independent variable after adjusting for the presence of other independent variables in the model. The null hypothesis for testing whether X_l is a significant variable in the model is formulated as $H_0 : \beta_l = 0$. The corresponding test statistic is $t_l = \hat{\beta}_l \sqrt{\widehat{\text{var}}(\hat{\beta}_l)}$, where $\widehat{\text{var}}(\hat{\beta}_l)$ is the estimated variance of $\hat{\beta}_l$. Clearly, the power of such a test increases as the variance of the individual parameter estimator decreases.

3.1.3.1 Measures of uncertainty

The actual formulas for the elements of the matrix $\text{Cov}(\hat{\beta})$ are more complicated than those presented for the simple linear regression model. These formulas all have the same components, namely, the error variance σ_ε^2, the sample size N, the variances of the independent variables $\text{var}(x_l)$ and the correlations between the scores of the independent variables. Conclusions similar to those in Chapter 2 can be drawn, namely that the efficiency of the estimators and power of the test of $H_0 : \beta_l = 0$ increase as

- the variance of the errors σ_ε^2 decreases, and/or

- the variances $\text{var}(x_l)$ increase, and/or

- the sample size N increases, and/or

- the correlation between the scores of the independent variables decreases.

The first three components were discussed in Chapter 2, and they perform a similar role in multiple linear regression. The last component comprising pair-wise correlations between the independent variables is added here because we have now more than one independent variable.

The uncertainty of the estimators of the individual parameters $\beta_0, \beta_1, \beta_2, \ldots, \beta_{p-1}$ can also be expressed in terms of individual confidence intervals. The $100(1 - \alpha)\%$ confidence interval corresponding to the null hypothesis $H_0 : \beta_l = 0$ is

$$\hat{\beta}_l - t_{\alpha/2, N-p}\sqrt{\widehat{\text{var}}(\hat{\beta}_l)} \le \beta_l \le \hat{\beta}_l + t_{\alpha/2, N-p}\sqrt{\widehat{\text{var}}(\hat{\beta}_l)}, \qquad (3.7)$$

where $t_{\alpha/2, N-p}$ is the corresponding critical value of a t distribution. The shorter the confidence interval, the more efficient the parameter estimator and the more powerful the test.

Another way to express the uncertainty of the parameter estimators in a multiple regression model is by using the variance $\text{var}(\hat{y}_0)$ of the predicted response $\hat{y}_0$. This variance is a function of the sample size, the variation of the independent variables and the variance of the errors in the multiple regression model. A decrease in $\text{var}(\hat{y}_0)$ can be accomplished by a decrease in SS_e and/or an increase in the variation of the independent variables X_l and an increase in the sample size N.

In the following sections, we revisit the vocabulary-growth study and use it to illustrate how the variances and covariances in the matrix $\text{Cov}(\hat{\beta})$ in Equation (3.6) can guide us to design a more efficient study. We also develop various ways of measuring the worth of a design and use them to compare competing designs.

3.2   Designs for vocabulary-growth study

Let us suppose that we sampled pupils from the 8th, 9th, 10th and 11th school grades in the vocabulary-growth study. This strategy was used in a comparable study reported by Bock (1975, p. 449–468). The pupils from each grade are then divided into three groups depending on the family income levels. For illustrative purposes, we make a further simplifying assumption that both school grade and the family income level are quantitative variables, and the three family income levels are equidistant from one another. Accordingly, we code the three family income levels as 1, 2 or 3 for low, medium and high income levels, respectively. This gives a total of 12 different groups of pupils. If the total number of pupils is N, then the growth in vocabulary of these pupils can be described using a multiple regression model with two predictors: school grade (X_1) and income category (X_2):

$$y_i = \beta_0 + \beta_1 x_{1i} + \beta_2 x_{2i} + \varepsilon_i, \text{ for } i = 1, \ldots, N. \tag{3.8}$$

For simplicity, the model does not include interaction between school grade and family income category. Such a model is called an *additive model*, and for convenience we call the two independent variables additive. We will consider models with interactions later on.

The vector of all parameters of Model (3.8) is $\beta = (\beta_0, \beta_1, \beta_2)'$. Although the intercept β_0 is a parameter in this model, we are often not interested in testing a null hypothesis for the intercept β_0. So, let us assume that we are only interested in the relationship between school grade and income category and their effect on vocabulary growth, that is, in testing the null hypotheses $H_0 : \beta_1 = 0$ and $H_0 : \beta_2 = 0$. The variance–covariance matrix of the three parameter estimators $\hat{\beta} = (\hat{\beta}_0, \hat{\beta}_1, \hat{\beta}_2)'$ can be divided into sub-matrices:

$$\text{Cov}(\hat{\beta}) = \begin{bmatrix} \text{Cov}_{11} & \text{Cov}_{12} \\ \text{Cov}_{21} & \text{Cov}_{22} \end{bmatrix} = \begin{bmatrix} \text{var}(\hat{\beta}_0) & \vdots & \text{cov}(\hat{\beta}_0, \hat{\beta}_1) & \text{cov}(\hat{\beta}_0, \hat{\beta}_2) \\ \text{cov}(\hat{\beta}_0, \hat{\beta}_1) \vdots & \text{var}(\hat{\beta}_1) & \text{cov}(\hat{\beta}_1, \hat{\beta}_2) \\ \text{cov}(\hat{\beta}_0, \hat{\beta}_2) \vdots & \text{cov}(\hat{\beta}_1, \hat{\beta}_2) & \text{var}(\hat{\beta}_2) \end{bmatrix}, \tag{3.9}$$

where the sub-matrix Cov_{11} reduces to a single number representing the variance $\text{var}(\hat{\beta}_0)$. The variances and covariance of $\hat{\beta}_1$ and $\hat{\beta}_2$ are in the lower right sub-matrix of the covariance matrix in Equation (3.9):

$$\text{Cov}_{22} = \begin{bmatrix} \text{var}(\hat{\beta}_1) & \text{cov}(\hat{\beta}_1, \hat{\beta}_2) \\ \text{cov}(\hat{\beta}_1, \hat{\beta}_2) & \text{var}(\hat{\beta}_2) \end{bmatrix}. \tag{3.10}$$

Further algebra shows that

$$\text{Cov}_{22} = \frac{\sigma_\varepsilon^2}{N} \begin{bmatrix} \dfrac{1}{(1 - r_{12}^2)\text{var}(x_1)} & -\dfrac{\text{cov}(x_1 x_2)}{(1 - r_{12}^2)\text{var}(x_1)\text{var}(x_2)} \\ -\dfrac{\text{cov}(x_1 x_2)}{(1 - r_{12}^2)\text{var}(x_1)\text{var}(x_2)} & \dfrac{1}{(1 - r_{12}^2)\text{var}(x_2)} \end{bmatrix}, \tag{3.11}$$

where $\text{var}(x_1)$, $\text{var}(x_2)$ and $\text{cov}(x_1 x_2)$ are the variances of the independent variables X_1 and X_2, respectively, and their covariance. The correlation between the independent variables X_1 and X_2 is $r_{12} = \text{cov}(x_1 x_2)/\sqrt{\text{var}(x_1)\text{var}(x_2)}$.

From the elements of Cov_{22}, we observe that the efficiency of the estimator $\hat{\beta}_l (l = 1, 2)$ increases if one or more of the following conditions hold:

- the variance of the errors σ_ε^2 decreases;

- the variance $\text{var}(x_l)$ increases;

- the sample size N increases;

- the squared correlation r_{12}^2 between the scores of X_1 and X_2 decreases.

Designs for the regression model are based on combinations of the various levels of X_1 and X_2 and on the number of pupils in each of these combinations. Table 3.2 displays 12 different combinations ($m = 12$) of levels of the independent variables for four different designs. These combinations of levels are the design points and for this example, all the sample sizes are equal to the number of design points, that is, $N = m = 12$. So each of the four designs has only one observation (score of a pupil) per design point.

Design 1 is the vocabulary-growth design with one pupil in each of the 12 combinations. The variance of the estimator of the effect of school grade

Table 3.2   Twelve combinations of the levels of X_1 and X_2 for the vocabulary-growth study.

Design points	Design 1		Design 2		Design 3		Design 4	
	X_1	X_2	X_1	X_2	X_1	X_2	X_1	X_2
d_1	8	1	8	1	8	1	8	1
d_2	8	2	8	2	8	1	8	1
d_3	8	3	8	3	8	3	8	1
d_4	9	1	8	1	8	1	8	3
d_5	9	2	8	2	8	2	8	3
d_6	9	3	8	3	8	3	8	3
d_7	10	1	11	1	11	1	11	1
d_8	10	2	11	2	11	2	11	1
d_9	10	3	11	3	11	3	11	1
d_{10}	11	1	11	1	11	1	11	3
d_{11}	11	2	11	2	11	3	11	3
d_{12}	11	3	11	3	11	3	11	3
$\text{var}(x_l)$	1.250	0.667	2.250	0.667	2.250	0.833	2.250	1.000
r_{12}	0.000		0.000		0.183		0.000	
$\text{var}(\hat{\beta}_l)$	0.067	0.125	0.037	0.125	0.038	0.103	0.037	0.083

Note: It is assumed that the error variance $\sigma_\varepsilon^2 = 1$ and that there is a total of $N = 12$ observations, one observation for each combination. The $\text{var}(x_l)$'s are computed by dividing the sum of squares by N.

on vocabulary growth and the variance of the estimator of the effect of income category on vocabulary growth are $\text{var}(\hat{\beta}_1) = 0.067$ and $\text{var}(\hat{\beta}_2) = 0.125$, respectively. Table 3.2 shows that as the variances of the independent variables increase, the variances of the estimators of the regression parameters decrease, and vice versa. In addition, among the four designs, *Design* 4 is the most efficient for estimating either of the parameters because this design gives the smallest values for $\text{var}(\hat{\beta}_1) = 0.037$ and $\text{var}(\hat{\beta}_2) = 0.083$. *Design* 3 does not have an equal number of replications at the design points and is the only design that has a non-zero correlation between the two independent variables. The correlation is $r_{12} = 0.183$ and we know from the above discussion that the larger this correlation, the less efficient will *Design* 3 be for estimating either of the parameters.

Figure 3.2 is a graphical display of the four designs in Table 3.2. The black rectangles in Figure 3.2 represent the two-dimensional design space Δ_2 and the black dots in each rectangle represent the location where the $N = 12$ observations were taken in each of the designs. This figure, along with Table 3.2, provides us an insight on how the distribution of the observations affects the efficiency of the design for estimating the parameters. As discussed above, *Design* 4 is the

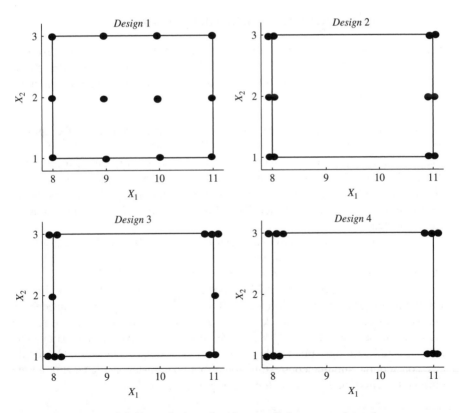

Figure 3.2 Four designs for the vocabulary-growth study.

optimal design among the four for estimating the parameters and the design has all its observations taken at the extreme points in the design space. This suggests that to have high efficiency for estimating the parameters, design points should be spread out as far as possible in the design space and all design points should have equal number of observations.

Design 4 is the most efficient design of these four designs. It can be shown from optimal design theory that Design 4 is also the optimal design among all possible designs for a linear regression model with two independent and additive variables. Design 4 has maximum variances var(x_l)'s for both independent variables and the design is orthogonal in the sense that $r_{12} = 0$ with an equal number of replications at each design point. Sometimes physical or ethical constraints may prevent the use of the optimal design in practice. We remind the reader that when this happens, usually a slight modification of the optimal design or an alternative design may be used instead of Design 4. For instance, Table 3.2 shows that Designs 2 and 3 are good substitutes for the optimal design, Design 4, because of their high efficiencies for estimating the parameters relative to Design 1.

3.3   Relative efficiency and sample size

The efficiencies of Design 1 relative to the optimal design for estimating the parameters are $RE(\hat{\beta}_1) = 0.037/0.067 = 0.555$ and $RE(\hat{\beta}_2) = 0.083/0.125 = 0.667$. This means that, depending on the parameter, about $(0.667^{-1} - 1)100\% = 50\%$ or $(0.555^{-1} - 1)100\% = 80\%$ more subjects are needed for Design 1 to be as efficient as the optimal design, Design 4. In terms of sample sizes required for Design 1 to have the same efficiency as the optimal design, it means that we require at least $N = 18$ subjects for estimating β_1 or at least $N = 22$ subjects for estimating β_2 versus $N = 12$ subjects in the optimal design. In practice, we may require more than the minimal number of subjects given above to have the same efficiency as the optimal design. For this problem, we would have 24 subjects (rather than 22 or 18) so that each combination level has 2 subjects.

The results in Table 3.2 also show that Design 2 or Design 3 is clearly more efficient than Design 1. For instance, the efficiencies of Design 1 relative to Design 3 for estimating the two parameters are $RE(\hat{\beta}_1) = 0.038/0.067 = 0.567$ and $RE(\hat{\beta}_2) = 0.103/0.125 = 0.824$, respectively. This means that depending on which parameter is of interest, about 20% or 76% more subjects are needed by Design 1 to be as efficient as Design 3. Of course, if cost permits, we should increase the sample size for Design 1 by at least 76% so that the parameters are estimated with the same level of efficiency provided by Design 3. One could also consider the simultaneous estimation of both parameters using an appropriate optimality criterion. This is the topic in the next section.

3.4   Simultaneous inference

As mentioned in Chapter 2, when we are interested in more than one parameter
in the statistical model, it is preferable to estimate the parameters simultaneously
so that the covariances among the estimators are also accounted for. Analogous
to the case for the simple linear regression model, we can construct an elliptically
shaped confidence region for estimating the parameters in a multiple regression
model. For a multiple regression model with two independent and additive variables, the confidence ellipsoid for the vector of parameters $\beta = (\beta_0, \beta_1, \beta_2)'$ is
three dimensional. This makes it a bit harder to visualize and interpret the plot
than a two- or one-dimensional plot. Of course, higher dimensional plots for
more than three parameters are even more difficult to interpret and appreciate.
Fortunately, sometimes all parameters are not of interest, and this allows us to
focus on only the parameters of interest. For example, for this additive model,
estimation of the intercept is usually not of interest, in which case, the ellipsoid is
two dimensional for the parameters $\beta_s = (\beta_1, \beta_2)'$. Note that we have a subscript
's' in β_s to emphasize that it is a subset of the full set of model parameters β.

Figure 3.3 is an exemplary plot of a confidence ellipse for $\beta_s = (\beta_1, \beta_2)'$.
From the design perspective, we want to have a design that provides accurate
joint estimates for β_1 and β_2 simultaneously, and this is signified by having as
small an area for the ellipse as possible. For this particular ellipse, we observe that
the ellipse is more elongated along the axis of β_1 than that of β_2. This suggests
that the design has engendered more uncertainty in the estimation of β_1 than that
for β_2. As always, our interest in this book is concerned with how design can
affect statistical inference and how one can come up with an improved design

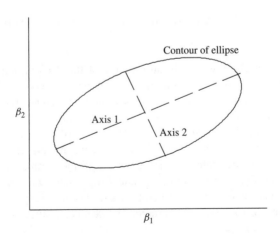

Figure 3.3 Contour of an ellipse for the parameters $\beta_s = (\beta_1, \beta_2)'$.

that more accurately reflects reality. The next section presents design optimality criteria for the multiple linear regression models.

3.5   Optimality criteria for a subset of parameters

In Chapter 2, Section 2.5, we discussed the D-, A- and E-optimality criteria. These criteria are based on the variance–covariance matrix $\text{Cov}(\hat{\beta})$ of all p parameters in the regression model. However, not all parameters are usually of interest. In the vocabulary-growth study, for example, researchers are often only interested in inferences about the parameters β_1 and β_2 because these parameters represent the effects of the school grade and family income level on vocabulary growth. The parameter β_0 is usually of secondary interest and is typically estimated for the purpose of predicting the response. In our case, and many other cases as well, this parameter can also be hard to interpret. This is because the intercept is the mean response when all the independent variables in the model are set to take on zero values and for our model, this means β_0 is the average vocabulary growth of pupils from grade 0 and their family income level is 0. The latter quantities are both undefined and unrealistic.

Accordingly, we should use design optimality criteria that target on the parameters of interest. For our example, the subset of parameters of interest is (β_1, β_2) and we leave out the intercept term. The variance–covariance matrix of the estimators $\hat{\beta}_1$ and $\hat{\beta}_2$ for the multiple regression model (Equation 3.8) is

$$\text{Cov}_{22} = \frac{\sigma_\varepsilon^2}{N} \begin{bmatrix} \dfrac{1}{(1 - r_{12}^2)\text{var}(x_1)} & -\dfrac{\text{cov}(x_1 x_2)}{\left(1 - r_{12}^2\right)\text{var}(x_1)\text{var}(x_2)} \\[2em] -\dfrac{\text{cov}(x_1 x_2)}{\left(1 - r_{12}^2\right)\text{var}(x_1)\text{var}(x_2)} & \dfrac{1}{\left(1 - r_{12}^2\right)\text{var}(x_2)} \end{bmatrix}.$$

(3.12)

This 2×2 matrix is the lower right sub-matrix of the 3×3 variance–covariance matrix of $\hat{\beta} = (\hat{\beta}_0, \hat{\beta}_1, \hat{\beta}_2)'$ in Equation (3.9). A natural modification of the D-, A- and E-optimality criteria can now be based on the variance–covariance matrix Cov_{22}. The D-optimality criterion for estimating a subset of the parameters β_s is referred to as the $D_s - optimality\ criterion$ (Atkinson and Donev, 1992, p. 109). The notation D_s means that it is the determinant of a sub-matrix of the variance–covariance matrix of estimators of all the parameters in the model. Similarly, we call the corresponding A- and E-optimality criteria for estimating a subset of the parameters A_s- and E_s-optimality criteria, respectively.

The D_s-criterion for this 2×2 variance–covariance matrix is

$$D_s\text{-criterion} = \text{var}(\hat{\beta}_1)\text{var}(\hat{\beta}_2) - \text{cov}(\hat{\beta}_1, \hat{\beta}_2)^2 = \frac{\sigma_\varepsilon^4}{N^2(1 - r_{12}^2)\text{var}(x_1)\text{var}(x_2)}.$$

(3.13)

This design criterion for estimating $\beta_s = (\beta_1, \beta_2)'$ is the *determinant* of the covariance matrix in Equation (3.12), and can be shown to be proportional to the volume of the confidence ellipse for β_s. An advantage here is that the estimators of the subset (β_1, β_2) have a 2×2 covariance matrix and this makes computation of the determinant relatively easy. For larger matrices, a computer program is needed to compute and minimize the determinant.

The D-optimality criterion has a number of interesting properties. First, it is known that the D-optimal design for estimating all the parameters β in the linear multiple regression model is also the D_s-optimal design for estimating all non-intercept parameters in the model; a recent proof is available in Goos (2002, p.15). For our example, this implies that the D-optimal design for estimating β is also the D_s-optimal design for estimating β_s, and vice versa. Other optimality criteria do not enjoy this property. Another interesting property of D_s-optimal design is that the D_s-criterion is not always invariant to scale transformation. This is unlike D-optimal designs that are invariant to scale transformations mentioned in Section 2.5.

The A_s-criterion is similar to the A-criterion and it is simply the sum of the diagonal elements of the variance–covariance matrix of the two parameter estimators in Equation (3.12):

$$A_s\text{-criterion} = \text{var}(\hat{\beta}_1) + \text{var}(\hat{\beta}_2) = \frac{\sigma_\varepsilon^2}{N} \left[\frac{1}{(1 - r_{12}^2)\text{Var}(x_1)} + \frac{1}{(1 - r_{12}^2)\text{Var}(x_2)} \right].$$
(3.14)

This criterion is the *trace* of the variance–covariance matrix and is proportional to the sum of the lengths of the major axes of the confidence ellipse. Consequently, minimizing the A_s-criterion is another measure of the smallness of the ellipse, and hence the accuracy of the estimates. The A_s-criterion, like the A-optimality criterion, is a special case of the linear or L-optimality criterion discussed in Atkinson and Donev (1992, p. 113–114). In Chapters 9 and 10, we discuss and apply the L-optimality criterion to design studies to estimate the turning point in a quadratic regression model and also to estimate the percentile in a logistic regression model. The A- or A_s-criterion is not invariant under linear transformations of the scale of the independent variables.

The E-criterion is also based on the length of the major axes of the confidence ellipsoid. It finds a design that minimizes the maximum length of the axes in the ellipsoid. The rationale is that if the largest length of the axis in the ellipsoid is minimized, the volume will also be small, and hence more accurate estimates for the parameters of interest. Mathematically, this criterion minimizes the largest *eigenvalue* or *root* of the relevant variance–covariance matrix; in our case, for estimating $\beta_s = (\beta_1, \beta_2)'$ in the two-predictor additive model, we use the variance–covariance matrix in Equation (3.12) and the criterion is

$$E_s\text{-criterion} = \text{Root}_{\max} \{\text{Cov}_{22}\}.$$
(3.15)

When all parameters are of interest, we simply replace the 2×2 matrix Cov_{22} in the above equation by the full $p \times p$ variance–covariance matrix $\mathrm{Cov}(\hat{\beta})$. The E- or E_s-criterion is not invariant under linear transformations of the scale of the independent variables.

It should be noted that the D_s-criterion is defined as the product of the squared lengths of the axes, which is proportional to the volume of the two-dimensional ellipse. This enables the lengths of the axes to compensate for each other. The product of an extreme long axis and a very short one may be the same as the product of the lengths of two equally long axes. The A_s-criterion, however, is defined as the sum of the squared lengths of the axes. It can also compensate for the differences in axis lengths, but in a different way than the D_s-optimality criterion. The E_s-criterion considers only the longest axis of the ellipse and as such does not use all variance contained by the ellipse.

Table 3.3 shows the values for the D_s-, A_s-, and E_s -criteria for each of the four designs in the vocabulary-growth study. The values for these criteria were computed using the values for $\mathrm{var}(x_l)$, $\mathrm{var}(\hat{\beta}_l)$ and r_{12}, which are copied from Table 3.2. Table 3.3 shows that the fourth design has the smallest values for all the three criteria. This means that of these four designs, *Design* 4 is the most efficient one for estimating the two parameters simultaneously.

Table 3.3   Characteristics of the four designs for estimating only (β_1, β_2) in the vocabulary-growth study with $\sigma_\varepsilon^2 = 1$ and $N = 12$ observations.

	Design 1		Design 2		Design 3		Design 4	
	X_1	X_2	X_1	X_2	X_1	X_2	X_1	X_2
$\mathrm{var}(x_l)$	1.250	0.667	2.250	0.667	2.250	0.833	2.250	1.000
r_{12}	0.000		0.000		0.183		0.000	
$\mathrm{var}(\hat{\beta}_l)$	0.067	0.125	0.037	0.125	0.038	0.103	0.037	0.083
D_s-criterion	0.0083		0.0046		0.0038		0.0031	
A_s-criterion	0.1917		0.1620		0.1418		0.1204	
E_s-criterion	0.1250		0.1250		0.1054		0.0833	

3.6   Relative efficiency

The various efficiencies of a design for estimating a subset of the model parameters β_s can be quantified in the same way as was done in Chapter 2, Section 2.6. The relative efficiency (RE) compares a criterion value of the design with that of the D_s-optimal design and sometimes we need to do it in a standardized way for easy interpretation. If $\hat{\beta}_s^*$ is the estimate for β_s from the optimal design and $\hat{\beta}_s$ is the estimate from the design ξ under comparison, the various formulae for evaluating the different types of REs of a design are

$$D_s\text{-efficiency:} \quad \text{RE}_{D_s} = \left(\frac{\text{Det}[\text{Cov}(\hat{\beta}_s^*)]}{\text{Det}[\text{Cov}(\hat{\beta}_s)]} \right)^{1/p_s},$$

$$A_s\text{-efficiency:} \quad \text{RE}_{A_s} = \frac{\text{Trace}[\text{Cov}(\hat{\beta}_s^*)]}{\text{Trace}[\text{Cov}(\hat{\beta}_s)]} \quad \text{and} \tag{3.16}$$

$$E_s\text{-efficiency:} \quad \text{RE}_{E_s} = \frac{\text{Root}_{\max}[\text{Cov}(\hat{\beta}_s^*)]}{\text{Root}_{\max}[\text{Cov}(\hat{\beta}_s)]}.$$

The variance–covariance matrices in Equation (3.16) are based on the appropriate sub-matrix of the full variance–covariance matrix. In the formula for RE_{D_s}, we use the p_sth *root of the ratio*, where p_s is the number of parameters in the subset of interest. These efficiency measures give us an indication of how many observations a design ξ will need to have the same efficiency as the optimal design ξ^*. For example, Table 3.4 shows REs of each of the three designs compared to *Design* 4 for the vocabulary-growth study designs. The interpretations are similar to those we had before; for example, for the D_s-criterion, *Design* 1 needs about $(0.6111^{-1} - 1)100\% = 63\%$ more observations to have the same efficiency as *Design* 4. We also notice that the three types of REs for the same design are different. This can be explained by the fact that each criterion is a different function of the squared lengths of the axes of the confidence ellipsoid, and so measures the goodness of the design in a different way.

Table 3.4   Relative efficiencies for the vocabulary-growth study designs.

	Design 1	*Design* 2	*Design* 3	*Design* 4
RE_{D_s}	0.6111	0.8209	0.9032	1.0000
RE_{A_s}	0.6280	0.7432	0.8491	1.0000
RE_{E_s}	0.6664	0.6664	0.7903	1.0000

3.7   Designs for polynomial regression model

Polynomial models are a special class of multiple regression models. In a polynomial regression model, there are usually only one or two independent variables, along with higher-order functions of these variables. For example, consider again the linear regression model expressed by $y_i = \beta_0 + \beta_1 x_{1i} + \beta_2 x_{2i} + \varepsilon_i$ (Equation 3.8). If the second independent variable in this model is replaced by $x_{2i} = x_{1i}^2$, this leads to a quadratic regression model:

$$y_i = \beta_0 + \beta_1 x_{1i} + \beta_2 x_{1i}^2 + \varepsilon_i. \tag{3.17}$$

In general, a $(p - 1)$th degree polynomial regression model is represented as

$$y_i = \beta_0 + \beta_1 x_{1i} + \beta_2 x_{1i}^2 + \cdots + \beta_{p-1} x_{1i}^{p-1} + \varepsilon_i. \tag{3.18}$$

Again, it is assumed that the errors in these models are independent and normally distributed, each with zero mean and variance σ_ε^2. The total number of parameters in the model is p, but there is only one independent variable X_1 involved and it takes on values x_1's in this model.

In practice, polynomial models are widely used because they are flexible and usually provide a reasonable approximation to the true relationship among the variables. We recommend working with polynomial models with low order whenever possible. High-order polynomial models may provide a better fit to the data and hence an improved approximation to the true relationship, but the numerous coefficients in such models make them difficult to interpret. In addition, working with high-degree polynomial models may sometimes lead to *ill-conditioning* of the variance–covariance matrix of the parameter estimators. This means that the estimated variance–covariance matrix may be numerically unstable or even impossible to compute.

Sometimes, polynomial models are used after an appropriate transformation has been applied on the independent variable to lessen the degree of nonlinearity. Examples of such transformations are the logarithm and the square transformations. Box and Cox (1984), Carroll and Ruppert (1984) and Neter, Wasserman and Kutner (1983) gave a detailed discussion on the use and properties of various transformations for improving fit in linear regression models.

The D-optimal design points for polynomial regression models with independent errors were first given by Smith (1918) and Guest (1958). Table 3.5 lists the D-, A- and E-optimal designs for polynomial regression models of degree 1, 2, 3, 4 or 5 on the prototype design interval scaled between -1 and $+1$. Optimal designs for higher-order polynomials are given in Pukelsheim (1993, Chapter 9), where the theoretical justifications for these optimal designs are also given. We emphasize here that these optimal designs are optimal among ALL designs on the scaled interval $[-1, 1]$ and not just optimal among a few designs. Of course, if we are only interested in finding the optimal design among a few designs, this can always be done easily by comparing the criterion value of each design.

Table 3.5 shows that the optimal design points for a linear and quadratic polynomial are the same for the three optimality criteria, but the distribution of the weights for each optimal design is different. D-optimal designs are equally weighted, but A- and E-optimal designs are generally not. When the polynomial model has degree $(p - 1) = 3$ or higher, the design points of all three criteria are different and not equally spaced. For all three criteria, however, the optimal number of distinct design points is equal to p, the number of parameters in the polynomial model.

Table 3.6 reports the values of the three optimality criteria $\text{Det}[\text{Cov}(\hat{\beta})]$, $\text{Trace}[\text{Cov}(\hat{\beta})]$ and $\text{Root}_{\max}[\text{Cov}(\hat{\beta})]$ for the D-, A- and E-optimal designs when the degree of the polynomial model is 1, 2, 3, 4 or 5. The table shows that the

Table 3.5   Optimal designs for polynomial regression models.

Degree $(p-1)$	D-optimal designs in design interval $[-1, 1]$					
1	−1 (0.5)					1 (0.5)
2	−1 (0.333)		0 (0.333)			1 (0.333)
3	−1 (0.25)		−0.447 (0.25)	0.447 (0.25)		1 (0.25)
4	−1 (0.20)	−0.655 (0.20)		0 (0.20)	0.655 (0.20)	1 (0.20)
5	−1 (0.167)	−0.765 (0.167)	−0.285 (0.167)	0.285 (0.167)	0.765 (0.167)	1 (0.167)

Degree $(p-1)$	A-optimal designs in design interval $[-1, 1]$					
1	−1 (0.5)					1 (0.5)
2	−1 (0.25)		0 (0.50)			1 (0.25)
3	−1 (0.151)		−0.464 (0.349)	0.464 (0.349)		1 (0.151)
4	−1 (0.105)	−0.677 (0.25)		0 (0.29)	0.677 (0.25)	1 (0.105)
5	−1 (0.080)	−0.789 (0.188)	−0.291 (0.232)	0.291 (0.232)	0.789 (0.188)	1 (0.080)

Degree $(p-1)$	E-optimal designs in design interval $[-1, 1]$					
1	−1 (0.5)					1 (0.5)
2	−1 (0.20)		0 (0.60)			1 (0.20)
3	−1 (0.127)		−0.50 (0.373)	0.50 (0.373)		1 (0.127)
4	−1 (0.093)	−0.707 (0.248)		0 (0.318)	0.707 (0.248)	1 (0.093)
5	−1 (0.074)	−0.809 (0.180)	−0.309 (0.246)	0.309 (0.246)	0.809 (0.180)	1 (0.074)

Note: p = number of polynomial parameters
Weights are within brackets

Table 3.6   Values of the three optimality criteria for the polynomial model of degree $p - 1$.

		Degree ($p - 1$)			
		1	2	3	4	5
D-optimal design	Det[Cov($\hat{\beta}$)]	1.000	6.752	195.281	2.327×10^4	1.127×10^7
	Trace[Cov($\hat{\beta}$)]	2.000	9.000	43.994	224.870	1.182×10^3
	Root$_{max}$[Cov($\hat{\beta}$)]	1.000	6.843	33.563	174.425	927.394
A-optimal design	Det[Cov($\hat{\beta}$)]	1.000	8.000	275.687	3.774×10^4	2.041×10^7
	Trace[Cov($\hat{\beta}$)]	2.000	8.000	37.520	188.695	982.522
	Root$_{max}$[Cov($\hat{\beta}$)]	1.000	5.236	25.856	133.151	702.594
E-optimal design	Det[Cov($\hat{\beta}$)]	1.000	10.417	352.103	4.728×10^4	2.499×10^7
	Trace[Cov($\hat{\beta}$)]	2.000	8.333	38.704	194.533	1.013×10^3
	Root$_{max}$[Cov($\hat{\beta}$)]	1.000	5.000	25.000	129.000	681.005

value of the optimality criterion for the optimal design is smallest. For instance, when we have a quadratic model, the D-criterion value for the D-optimal design is 6.752 and this value is smaller than the corresponding values of 8.000 and 10.417 for the A- and E-optimal designs. We also observe that for the simple linear model (with $p = 2$), the D-, A- and E-optimal designs all have the same value for the D-, A- or E-optimality criteria. The reason for this is that the D-, A- and E-optimal designs are all the same, suggesting that when we have a simple linear model and we want to estimate the parameters, it will be impossible to outperform the design that places equal observations at the extreme ends.

It is interesting to see how the efficiency of an optimal design varies across different optimality criteria. Figure 3.4 displays the D-, A- and E REs, each as a function of the degree of the polynomial. These REs are computed directly using their definitions. As an illustration, let us compute the A-efficiency of the D-optimal design of the second degree polynomial. We observe from Table 3.6 that for this model, the A-criterion value for the A-optimal design is 8.0 and that for the D-optimal design is 9.0. The A-efficiency of the D-optimal design is therefore given by $RE_A = 8.0/9.0 = 0.8889$. Likewise, one can verify that the E-efficiency of an A-optimal design for the cubic polynomial model is $RE_E = 25.000/25.856 = 0.9669$.

From Figure 3.4, we infer that when polynomials have degrees 2 or higher, the efficiencies remain fairly the same as the degree of the polynomials increases. All REs are higher than 0.8, except for the E-efficiencies (RE_E) of D-optimal designs which can reach 0.7. This may be explained by the fact that the E-optimality criterion does not use all information because it is only based on the largest root of the variance–covariance matrix of parameter estimators. The A-optimal designs

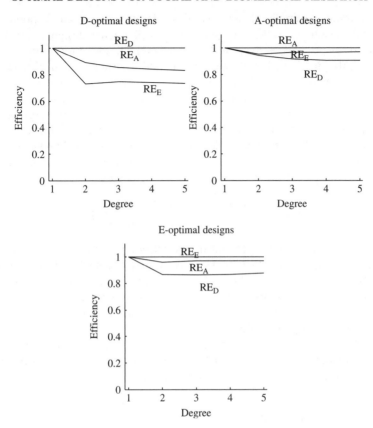

Figure 3.4 Relative efficiency plots for three types of optimality criteria.

are highly efficient in terms of both the D- and E- criteria. Their efficiencies are all higher than 0.9. The E-optimal designs all seem highly efficient in terms of both D- and A-optimality criterion; see Wong (1994) for a more extended comparison of these optimality criteria for polynomial regression models.

3.7.1   Exact D-optimal designs for a quadratic regression model

The optimal designs in Table 3.5 are all approximate designs and are formulated without any reference to the actual sample size N. For instance, the D-optimal design for the quadratic model is equally weighted at the extreme points and at the middle point. If the design interval is $[-1, 1]$, this means that we pick one-third of the total observations N at -1, one-third at 0 and one-third at 1. As indicated in the previous chapters, there are compelling reasons to work with approximate designs in practice. Approximate designs need to be rounded for implementation, but they are relatively easy to find and study. In contrast, exact optimal designs are much more difficult to find and describe.

Here is an example of an analytical description of an N-point exact D-optimal design. Consider the quadratic model on the design interval $[-1, 1]$ defined by $y_i = \beta_0 + \beta_1 x_{1i} + \beta_2 x_{1i}^2 + \varepsilon_i$ and all the error terms ε_i's are independently distributed with zero mean and constant variance. The exact design problem of interest here is how to select N values for the X variable from $[-1, 1]$ to observe the responses y's, in such a way that we get the best possible estimates for the three parameters β_0, β_1 and β_2. This design problem is fully addressed in Gaffke and Krafft (1982) where they used number-theoretic tools and obtained the following description for the N-point D-optimal design for the quadratic polynomial model. Let $N = 3a + b$ and let the N-point exact D-optimum design for the quadratic model be denoted by

$$\xi_N^* = \left\{ \begin{array}{ccc} d_1 = -1 & d_2 = 0 & d_3 = 1 \\ n_1/N & n_2/N & n_3/N \end{array} \right\}. \tag{3.19}$$

If $N = 3a$, the N-point D-optimum design takes equal number of observations at the three design points $[-1, 0, 1]$. If $N = 3a + b$ with $b = 1$ or 2, then the exact D-optimum design is obtained by choosing the design points in such a way that $3 - b$ of the points -1, 0 and 1 occur a times and the remaining b of them $a + 1$ times. Table 3.7 shows N-point exact D-optimal designs for the quadratic model on $[-1, 1]$ for selected values of N.

Recall from Table 3.5 that the approximate D-optimal design for the quadratic regression model simply takes one-third of the observations at each of the design points -1, 0 and 1. The exact optimal designs in Table 3.7 deviate from the

Table 3.7 N-point exact D-optimal designs for quadratic regression model.

N	a	b	$d_1 = -1$	$d_2 = 0$	$d_3 = 1$
3	1	0	1	1	1
4	1	1	1	1	2
			1	2	1
			2	1	1
5	1	2	1	2	2
			2	1	2
			2	2	1
6	2	0	2	2	2
7	2	1	2	2	3
			2	3	2
			3	2	2
8	2	2	3	3	2
			3	2	3
			2	3	3

Note. Entries under the columns of the design points are values for n_1, n_2 and n_3, respectively.

equally weighted approximate optimal design only by a single observation. The difference in efficiency between the exact optimal designs and the approximate optimal design for the quadratic regression model, like all other models, will decrease as the sample size N increases.

3.7.2 Scale dependency of A- and E-optimality criteria

We have mentioned that the D-optimal designs are invariant under linear transformation of the scale of the independent variable. This advantage, however, does not account for the A- and E-optimality criteria. These two criteria are scale dependent.

To illustrate their scale dependency, compare the A-optimal designs shown in Table 3.8 for five polynomial models with degree $(p - 1) = 1, 2, 3, 4$ and 5 on two design intervals $[0,1]$ and $[0,10]$ with those on the interval $[-1, 1]$ for the same polynomials shown in Table 3.5. We observe that not only the design points differ, which is of course to be expected, but that the weight distributions also differ for the three scales. Table 3.5 shows that the weights for the A-optimal designs on the interval $[-1, 1]$ are symmetrically distributed around the centre of the scale, that is, around zero, while in Table 3.8 one can observe that more weight is assigned to the smaller design points. For example, the table shows that almost 91% of the weight in the A-optimal design on $[0, 10]$ is given to the smaller design point in the simple linear model $(p - 1 = 1)$.

The weights for a two-point A-optimal design for the simple linear model (i.e. $p = 2$) on the interval $[a, b]$ are given by

$$w_1^* = \frac{\sqrt{(1 + b^2)}}{\sqrt{(1 + a^2)} + \sqrt{(1 + b^2)}} \quad \text{and} \quad w_2^* = 1 - w_1^*, \qquad (3.20)$$

see Dette (1997, p. 106). For the design interval $[0, 1]$, these optimal weights are $w_1^* = 0.586$ and $w_2^* = 0.414$ and for the design interval $[0, 10]$, these optimal weights are $w_1^* = 0.910$ and $w_2^* = 0.090$ (Table 3.8). Similar formulas for higher-order polynomials do not exist and likewise weights for E-optimal designs are not available analytically. In these cases, one can only compute the optimal weights numerically. Of course, the weights of the D-optimal designs are usually easier to determine; in particular, they are equally weighted when the number of optimal design points is equal to the number of parameters in the model. To this end, a computer program PSI(p) available on the website http://optimal-design.biostat.ucla.edu/optimal/, described in Chapter 11, can be used to generate many types of optimal designs, including A-, D- and E-optimal designs on an arbitrary interval.

3.8 The Poggendorff and Ponzo illusion study

The Poggendorff and Ponzo illusion experiments were reported by Bock (1975) and are described in Chapter 1. The original design for the two experiments

Table 3.8   A-optimal designs for polynomial regression models.

Degree ($p-1$)	Design interval [0, 1]				
1	0 (0.586)				1 (0.414)
2	0 (0.322)		0.5 (0.486)		1 (0.192)
3	0 (0.22)	0.252 (0.375)		0.748 (0.282)	1 (0.123)
4	0 (0.167)	0.147 (0.302)	0.499 (0.242)	0.853 (0.197)	1 (0.091)
5	0 (0.135)	0.096 (0.252)	0.346 (0.213)	0.654 (0.176)	0.904 (0.152)	1 (0.072)

Degree ($p-1$)	Design interval [0, 10]				
1	0 (0.910)				10 (0.090)
2	0 (0.682)		0 (0.264)		10 (0.054)
3	0 (0.5)	2.316 (0.347)		7.412 (0.112)	10 (0.041)
4	0 (0.384)	1.423 (0.374)	4.921 (0.131)	8.509 (0.078)	10 (0.033)
5	0 (0.321)	0.955 (0.367)	3.4 (0.15)	6.509 (0.083)	9.035 (0.061)	10 (0.028)

Note: Designs are computed with PSI(p) program on website: http://optimal-design.biostat.ucla.edu/optimal/
      p = number of polynomial parameters
      Weights are within brackets

consists of nine age groups, each with 16 children. This design can be summarized as follows:

$$\xi_{144} = \left\{ \begin{array}{ccccccccc} 5 & 6 & 7 & 8.5 & 10.5 & 12.5 & 14.5 & 16.5 & 19.5 \\ 16 & 16 & 16 & 16 & 16 & 16 & 16 & 16 & 16 \end{array} \right\}. \qquad (3.21)$$

The Poggendorff illusion and Ponzo illusion can be adequately described by the quadratic $y_i = 2.777 - 0.242x_i + 0.008x_i^2 + \varepsilon_i$ and cubic $y_i = -1.113 + 0.382x_i - 0.031x_i^2 + 0.001x_i^3 + \varepsilon_i$ regression models, respectively. The question of interest here is whether one can find a more efficient design than the original design for estimating the model parameters.

Table 3.5 shows the D-optimal designs for the quadratic and cubic polynomials with the design points scaled between -1 and 1. The corresponding D-optimal designs for the original two illusion experiments are presented in Table 3.9 with design points in brackets and displayed as vertical dotted lines in Figure 3.5. These D-optimal designs with only 3 and 4 design points are most efficient for estimating the parameters in the quadratic and cubic polynomial models, respectively. This suggests that the original design with nine different age groups may not be very efficient. To estimate the efficiency loss of the original design, we first write down its design matrix under the cubic polynomial model. Its transpose is

$$
X_0' = \begin{bmatrix} 1 & 1 & 1 & \cdots & 1 \\ 5 & 6 & 7 & \cdots & 19.5 \\ 5^2 & 6^2 & 7^2 & \cdots & 19.5^2 \\ 5^3 & 6^3 & 7^3 & \cdots & 19.5^3 \end{bmatrix}.
$$

Table 3.9 Equally weighted D-optimal design points for the quadratic and cubic polynomial regression.

Degree $(p-1)$		Scale		
1	$-1(5)$			$1(19.5)$
2	$-1(5)$		$0(12.25)$	$1(19.5)$
3	$-1(5)$	$-0.4472(9.0078)$	$0.4472(15.4922)$	$1(19.5)$

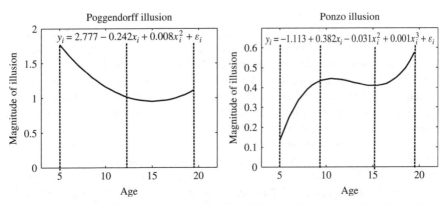

Figure 3.5 Magnitude of the Poggendorff and Ponzo illusions as function of age (data from Bock, 1975, chapter 4).

For comparison purposes, it is convenient to transform the original scale in years ranging from 5 to 19.5 years to a scale between -1 and 1. This can be accomplished by means of the transformation $d'_j = (d_j - \tilde{d})/(d_{max} - \tilde{d})$ with $\tilde{d} = d_{min} + (d_{max} - d_{min})/2$, where d_{max} and d_{min} are the maximum and minimum value of the original scale. For example, the age of the third age group can be transformed into $d_3' = (7 - 12.25)/(19.5 - 12.25) = -0.7241$. The corresponding quadratic and cubic terms are $(-0.7241)^2 = 0.5243$ and $(-0.7241)^3 = -0.3797$. The transposed design matrix of the original design for a cubic polynomial is

$$X_0' = \begin{bmatrix} 1 & 1 & 1 & 1 & 1 & 1 & 1 & 1 & 1 \\ -1 & -0.8621 & -0.7241 & -0.5172 & -0.2414 & 0.0345 & 0.3103 & 0.5862 & 1 \\ 1 & 0.7432 & 0.5243 & 0.2675 & 0.0583 & 0.0012 & 0.0963 & 0.3436 & 1 \\ -1 & -0.6407 & -0.3797 & -0.1383 & -0.0141 & 0.0000 & 0.0299 & 0.2014 & 1 \end{bmatrix}.$$

The corresponding design matrices for the D-optimal designs on $[-1, 1]$ under the quadratic and cubic models are, respectively,

$$X_q^{*'} = \begin{bmatrix} 1 & 1 & 1 \\ -1 & 0 & 1 \\ 1 & 0 & 1 \end{bmatrix} \text{ and } X_c^{*'} = \begin{bmatrix} 1 & 1 & 1 & 1 \\ -1 & -0.4472 & 0.4472 & 1 \\ 1 & 0.2000 & 0.2000 & 1 \\ -1 & -0.0894 & 0.0894 & 1 \end{bmatrix}.$$

These design matrices enable us to calculate the variance–covariance matrix of the estimators $\text{Cov}(\hat{\beta})$ and the relative efficiencies RE_Ds for these two models. Table 3.10 lists the variance–covariance matrices of the parameter estimators from the original design and D-optimal designs under the quadratic and the cubic regression models, along with the REs of the original design. The results in Table 3.10 show that the original design is inefficient when compared to the D-optimal designs. The relative D-efficiency of the original design computed for the quadratic and cubic models was $\text{RE}_\text{D} = 0.7243$ and $\text{RE}_\text{D} = 0.8020$, respectively. This means that for the Poggendorff experiment, the original design needs $(0.7243^{-1} - 1)100\% = 38\%$ more observations to estimate the parameters as well as the D-optimal design's estimates. Likewise, for the Ponzo experiment, the original design needs to have about $(0.8020^{-1} - 1)100\% = 25\%$ more observations to perform as well as the D-optimal design. The practical implication is that resources could have been saved if the D-optimal designs were used in both studies instead of the original design.

In practice, these D-optimal design points may not be feasible. For instance, when the design points from the D-optimal designs for the cubic model are translated to the original scale, we would require children with exact ages of 9.0078 or 15.4922 years and they may not be available. In that case, we make approximations to the optimal design points. If age groups that were used in the original design are only available, D-optimal designs for the Poggendorff and Ponzo experiments can be approximated by rounding off the D-optimal design points to the nearest year group that was used in the original study. This means that the designs we recommend for the studies are

$$\text{Poggendorff experiment: } \xi_{144} = \left\{ \begin{matrix} 5 & 12.5 & 19.5 \\ 48 & 48 & 48 \end{matrix} \right\} \tag{3.22}$$

Table 3.10 Covariance matrices of parameter estimators for the quadratic and cubic models.

Original design for quadratic model

$$\text{Cov}(\hat{\beta}) = \left[\frac{X_o'X_o}{9}\right]^{-1}$$

$$= \begin{bmatrix} 2.4859 & 0.1063 & -3.2773 \\ 0.1063 & 2.4097 & 0.6072 \\ -3.2773 & 0.6072 & 7.5238 \end{bmatrix}$$

D-criterion = 17.7616

$RE_D = 0.7243$

D-optimal design for quadratic model

$$\text{Cov}(\hat{\beta}) = \left[\frac{X_q'X_q}{3}\right]^{-1}$$

$$= \begin{bmatrix} 3.0 & 0 & -3.0 \\ 0 & 1.5 & 0 \\ -3.0 & 0 & 4.5 \end{bmatrix}$$

D-criterion = 6.7500

$RE_D = 1.0$

Original design for cubic model

$$\text{Cov}(\hat{\beta}) = \left[\frac{X_o'X_o}{9}\right]^{-1}$$

$$= \begin{bmatrix} 2.4881 & 0.2908 & -3.2746 & -0.2438 \\ 0.2908 & 17.6336 & 0.8364 & -20.1162 \\ -3.2746 & 0.8364 & 7.5272 & -0.3028 \\ -0.2438 & -20.1162 & -0.3028 & 26.5807 \end{bmatrix}$$

D-criterion = 472.1146

$RE_D = 0.8020$

D-optimal design for cubic model

$$\text{Cov}(\hat{\beta}) = \left[\frac{X_c'X_c}{4}\right]^{-1}$$

$$= \begin{bmatrix} 3.25 & 0 & -3.75 & 0 \\ 0 & 15.75 & 0 & -16.25 \\ -3.75 & 0 & 6.25 & 0 \\ 0 & -16.25 & 0 & 18.75 \end{bmatrix}$$

D-criterion = 195.2807

$RE_D = 1.0$

Note. The computations assume that the error variance $\sigma_\varepsilon^2 = 1$.

and

$$\text{Ponzo experiment:} \quad \xi_{144} = \left\{ \begin{array}{cccc} 5 & 8.5 & 16.5 & 19.5 \\ 36 & 36 & 36 & 36 \end{array} \right\}. \qquad (3.23)$$

These approximations to the D-optimal designs are extremely good with hardly any loss in D-efficiency. Their REs for the two studies are 0.9992 and 0.9627, respectively.

   The above illustration shows that if we have prior knowledge on the relationship among the variables of interest, it is desirable to incorporate such information at the design stage. In this case, the original experimental design for the two studies could have been chosen more efficiently if we knew that there is a quadratic or cubic relation between the magnitude of the Poggendorff and Ponzo illusions and age. The two designs in Equations (3.22) and (3.23) provide improved D-efficiencies over the original design by roughly 30%.

3.9   Uncertainty about best fitting regression models

Let us revisit the example in Chapter 1, where an epidemiologist is interested to describe the body weight (Y) of children with nutritional deficiency by a regression model with predictors height (X_1) and age (X_2) and there are five possible models:

Model 1 : $y_i = \beta_0 + \beta_1 x_{1i} + \beta_2 x_{2i} + \varepsilon_i.$
Model 2 : $y_i = \beta_0 + \beta_1 x_{1i} + \beta_2 x_{2i} + \beta_3 x_{1i} x_{2i} + \varepsilon_i.$
Model 3 : $y_i = \beta_0 + \beta_1 x_{1i} + \beta_2 x_{2i} + \beta_3 x_{1i} x_{2i} + \beta_4 x_{1i}^2 + \varepsilon_i.$
Model 4 : $y_i = \beta_0 + \beta_1 x_{1i} + \beta_2 x_{2i} + \beta_3 x_{1i} x_{2i} + \beta_4 x_{1i}^2 + \beta_5 x_{2i}^2 + \varepsilon_i.$
Model 5 : $y_i = \beta_0 + \beta_1 x_{1i} + \beta_2 x_{2i} + \beta_4 x_{1i}^2 + \beta_5 x_{2i}^2 + \varepsilon_i.$

In the design stage of this study, it is not known which of these five models will provide the best fit to the data. Since each regression model generally has its own optimal design, the researcher is faced with the problem to select an efficient design at the design stage without knowing which model is the best fitting one. In the sequel, we will refer to the best fitting model as the 'true' model. We want to compare the optimal designs for the five models and to ascertain how much efficiency is lost when a wrong optimal design is selected.

A first step is to scale the scores of the independent variables height (X_1) and age (X_2) between -1 and 1. This will facilitate comparisons among designs. A second step is to select the optimality criterion. Since we are interested in finding an optimal design that does not depend on the scales of the independent variables, we will choose the D-optimality criterion.

Figure 3.6 displays the D-optimal designs for the five models. The bold dots in the rectangular design space are the design points with the corresponding weights in brackets. Figure 3.6 also shows that a D-optimal design for one statistical model may or may not be D-optimal for another. For example, models 1 and 2 differ only by an interaction term, but the D-optimal designs for the two models are identical with design points at the corners of the design space. This design is, however, not D-optimal for any of the more complicated models 3, 4 or 5. One explanation for this is that the latter three models contain one or two quadratic terms and we need at least three design points to estimate all the three parameters associated with the variable that has the quadratic term. We also observe that, depending on the model, the D-optimal design points may not have the same weight.

In practice, it is not known which of these five models is the most appropriate model. One simple strategy is to assume that they are all equally likely to be a good fitting model. This assumption may be checked by consulting the experts in the field. Even then, the question remains which of the D-optimal designs in

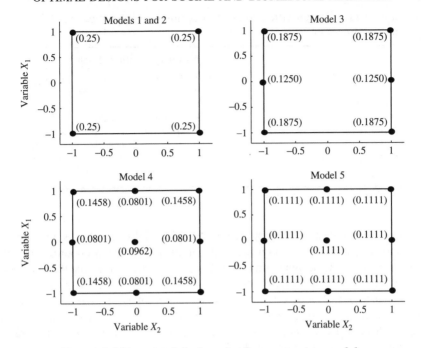

Figure 3.6 D-optimal designs for five regression models.

Figure 3.6 should we use? Clearly, if model 3 is the best fitting model, then the optimal designs for models 1 and 2 should not be chosen, because these designs do not enable estimation of the quadratic term in model 3. Likewise, if model 5 is the best fitting model, then it would be inappropriate to use the D-optimal designs for models 1, 2 and 3 because there are not enough design points to estimate the quadratic terms.

One way to determine which one of the D-optimal designs to use in light of the competing choices is to measure the performance of the D-optimal design relative to one another under different model assumptions. To this end, one assumes one of these models is the true model, and measures the worth of D-optimal design relative to that of the D-optimal for the assumed true model. The RE is given by

$$\mathrm{RE_D} = \left[\frac{\mathrm{Det}\left[\mathrm{Cov}\left(\hat{\beta}^*_{\mathrm{True}}\right)\right]}{\mathrm{Det}\left[\mathrm{Cov}\left(\hat{\beta}^*_{\mathrm{chosen}}\right)\right]} \right]^{1/p}, \tag{3.24}$$

where $\mathrm{Cov}(\hat{\beta}^*_{\mathrm{True}})$ is the variance–covariance matrix of the parameter estimators for the D-optimal design from the assumed true model and $\mathrm{Cov}\left(\hat{\beta}^*_{\mathrm{chosen}}\right)$ is the corresponding variance–covariance matrix from the model under comparison. The value of p in Equation (3.24) is equal to the number of parameters in the model. Table 3.11 presents the relative D-efficiencies for the D-optimal designs when we sequentially assume one of the five regression models is the true model.

Table 3.11   Design efficiencies of D-optimal designs under model mis-specification.

	Chosen D-optimal designs				
	Design 1	Design 2	Design 3	Design 4	Design 5
'True' Models					
Model 1 $(p = 3)$	1.0	1.0	0.9086	0.8206	0.7631
Model 2 $(p = 4)$	1.0	1.0	0.8660	0.7534	0.6667
Model 3 $(p = 5)$	–	–	1.0	0.8975	0.8391
Model 4 $(p = 6)$	–	–	–	1.0	0.9739
Model 5 $(p = 5)$	–	–	–	0.9773	1.0

Note. Dashes indicate a combination that is not feasible.

For the example at hand, suppose that the epidemiologist believes that model 4 or 5 is the true model. From Figure 3.6, this limits the choice of D-optimal designs to designs with nine design points. The REs in Table 3.11 show that the D-optimal design for model 4 has higher efficiency than that for model 5, irrespective of whether model 1, 2, 3 or 4 is the true one. This implies that the epidemiologist should choose the D-optimal design for model 4. In the worst case scenario, this will result in an RE of $RE_D = 0.7534$ when model 2 is the true model. In terms of sample size, this means that the D-optimal design for model 4 will require about $(0.7534^{-1} - 1)100\% = 33\%$ more observations to provide estimates as accurate as those provided from the D-optimal design for model 2. This may seem costly, but one should keep in mind that the D-optimal design for model 2 requires only four distinct design points, whereas the D-optimal design for model 4 has nine distinct design points. On the other hand, if the epidemiologist decides to choose the D-optimal design for model 5 because he/she prefers a balanced design, then it is easy to calculate the maximum efficiency loss, which is $(0.6667^{-1} - 1)100\% = 50\%$.

A second design question to answer here is how we can implement the exact design when we have the budget to sample a fixed number of units. For illustrative purposes, let us suppose that our total sample size N is fixed and equals to 100. Multiplying the weights in Figure 3.6 for the D-optimal design for model 4 by $N = 100$ and rounding off to the nearest integer value will give an exact design. After dropping two observations from the cell (0, 0) to guarantee that the sum of the weights remains equal to 100, the rounded design is

$$\xi_{100} = \left\{ \begin{matrix} (-1, -1) & (0, -1) & (1, -1) & (-1, 0) & (0, 0) & (1, 0) & (-1, 1) & (0, 1) & (1, 1) \\ 15 & 8 & 15 & 8 & 8 & 8 & 15 & 8 & 15 \end{matrix} \right\}.$$
$$(3.25)$$

This rounded design is almost as efficient as the D-optimal design for model 4 with a relative D-efficiency of $RE_D = 0.9990$. In practice, the exact design has nine different samples of nutritionally deficient children selected from different age groups and different heights as shown in Figure 3.7. These age groups and height requirements are found by recoding the design points -1, 0 and 1 back to their original scale with heights ranging from 50 to 60 cm and ages ranging from 6 to 12 years (Section 1.8.3, Chapter 1).

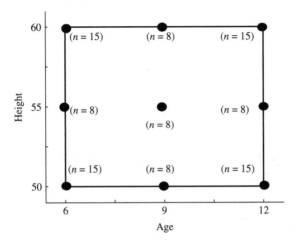

Figure 3.7 Nine sub-samples of nutritionally deficient children.

In summary, in light of model uncertainty for the design problem, it seems most efficient to choose the exact design ξ_{100} (Equation 3.25) and select nine different samples of children for the study. To compensate for possible efficiency loss when model 4 is not the true model, the epidemiologist may also consider increasing the total sample size by about 33% if the budget permits. Further references on designs for these kinds of regression models can be found in Atkinson and Donev (1992) and Goos (2002).

3.10 Matrix notation of designs for multiple regression models

The multiple regression model in Equation (3.4) can be reformulated in matrix notation as

$$y = X\beta + \varepsilon.$$

or in a more explicit form

$$
\begin{bmatrix} y_1 \\ y_2 \\ \vdots \\ y_N \end{bmatrix} = \begin{bmatrix} 1 & x_{11} & x_{21} & \cdots & x_{p-1,1} \\ 1 & x_{12} & x_{22} & \cdots & x_{p-1,2} \\ \vdots & \vdots & \vdots & \vdots & \vdots \\ 1 & x_{1N} & x_{2N} & \cdots & x_{p-1,N} \end{bmatrix} \begin{bmatrix} \beta_0 \\ \beta_1 \\ \vdots \\ \beta_{p-1} \end{bmatrix} + \begin{bmatrix} \varepsilon_1 \\ \varepsilon_2 \\ \vdots \\ \varepsilon_N \end{bmatrix}, \quad (3.26)
$$

where y is the $N \times 1$ data vector, X is the $N \times p$ design matrix, $\beta = (\beta_0, \beta_1, \ldots, \beta_{p-1})'$ is the vector of regression parameters and the $N \times 1$ vector ε contains is the errors which are assumed to be normally distributed, each with mean 0 and variance σ_ε^2. The least squares estimators $\hat{\beta}$ of $\beta = (\beta_0, \beta_1, \ldots, \beta_{p-1})'$ are the estimators for which the sum of squares of errors is minimized; that is, for which $\sum_i^N \hat{\varepsilon}_i^2 = \hat{\varepsilon}'\hat{\varepsilon} = (y - X\hat{\beta})'(y - X\hat{\beta})$ is minimum. These least squares estimators are

$$\hat{\beta} = (X'X)^{-1}X'y, \tag{3.27}$$

and the covariance matrix of the estimators is

$$\mathrm{Cov}(\hat{\beta}) = \sigma_\varepsilon^2 (X'X)^{-1}. \tag{3.28}$$

3.10.1   Design for regression models with two independent variables

The linear regression model for the vocabulary-growth example is $y_i = \beta_0 + \beta_1 x_{1i} + \beta_2 x_{2i} + \varepsilon_i$. The design matrices corresponding to each of the four designs described in Section 3.2 are of the order 12×3 and are presented in transposed form in Table 3.12. The first rows of these transposed design matrices correspond to the intercept β_0, the second rows correspond to the variable X_1 and the third rows correspond to the variable X_2. The values of the elements of the rows are the values of the variables as shown in Table 3.2.

Table 3.12   Design matrices for four vocabulary-growth designs.

Design 1:	$X_1' =$	1	1	1	1	1	1	1	1	1	1	1	1
		8	8	8	9	9	9	10	10	10	11	11	11
		1	2	3	1	2	3	1	2	3	1	2	3
Design 2:	$X_2' =$	1	1	1	1	1	1	1	1	1	1	1	1
		8	8	8	8	8	8	11	11	11	11	11	11
		1	2	3	1	2	3	1	2	3	1	2	3
Design 3:	$X_3' =$	1	1	1	1	1	1	1	1	1	1	1	1
		8	8	8	8	8	8	11	11	11	11	11	11
		1	1	3	1	2	3	1	2	3	1	3	3
Design 4:	$X_4' =$	1	1	1	1	1	1	1	1	1	1	1	1
		8	8	8	8	8	8	11	11	11	11	11	11
		1	1	1	3	3	3	1	1	1	3	3	3

The corresponding variance–covariance matrices and optimality criteria for these designs are displayed in Table 3.13. It can be seen that *Design* 4 has the smallest D-optimality criterion value among the four designs.

In fact, *Design* 4 is D-optimal with the smallest volume of the confidence ellipsoid among all designs on the given design space. The *equivalence theorem*

Table 3.13   Variance–covariance matrices from four designs and values of the D-, A- and E-optimality criteria for the vocabulary-growth study.

Design 1

$$\frac{\text{Cov}(\hat{\beta})}{\sigma_\varepsilon^2} = (X_1'X_1)^{-1} = \begin{bmatrix} 6.6000 & -0.6333 & -0.2500 \\ -0.6333 & 0.0667 & 0.0000 \\ -0.2500 & 0.0000 & 0.1250 \end{bmatrix}$$

D-criterion = 6.9444×10^{-4}
A-criterion = 6.7917
E-criterion = 6.6703

Design 2

$$\frac{\text{Cov}(\hat{\beta})}{\sigma_\varepsilon^2} = (X_2'X_2)^{-1} = \begin{bmatrix} 3.9259 & -0.3519 & -0.2500 \\ -0.3519 & 0.0370 & 0.0000 \\ -0.2500 & 0.0000 & 0.1250 \end{bmatrix}$$

D-criterion = 3.858×10^{-4}
A-criterion = 4.0880
E-criterion = 3.9736

Design 3

$$\frac{\text{Cov}(\hat{\beta})}{\sigma_\varepsilon^2} = (X_3'X_3)^{-1} = \begin{bmatrix} 3.5182 & -0.3410 & -0.0977 \\ -0.3410 & 0.0383 & -0.0115 \\ -0.0977 & -0.0115 & 0.1034 \end{bmatrix}$$

D-criterion = 3.1928×10^{-4}
A-criterion = 3.6600
E-criterion = 3.5540

Design 4

$$\frac{\text{Cov}(\hat{\beta})}{\sigma_\varepsilon^2} = (X_4'X_4)^{-1} = \begin{bmatrix} 3.7593 & -0.3519 & -0.1667 \\ -0.3519 & 0.0370 & 0.0000 \\ -0.1667 & 0.0000 & 0.0833 \end{bmatrix}$$

D-criterion = 2.5720×10^{-4}
A-criterion = 3.8796
E-criterion = 3.7996

Note. The values for D-, A- and E- criteria are computed for $\sigma_\varepsilon^2 = 1$.

from optimal design theory gives us an easy way to check whether a design is D-optimal among all designs. One simply uses the standardized variance function of the predicted response $s(x, \xi) = N \operatorname{var}(\hat{y}_0)/\sigma_\varepsilon^2$ defined in Section 2.5 and uses the fact that the design ξ is D-optimal if and only if it satisfies $s(x, \xi) \leq p$ for all points in the design space with equality at the design points (Kiefer and Wolfowitz, 1960; Kiefer, 1974). Figure 3.8 shows these plots for *Design* 4 and the original *Design* 1 and confirms that *Design* 4 is D-optimal.

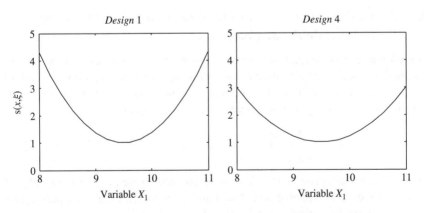

Figure 3.8 Standardized variance functions for Design 1 and Design 4.

It is interesting to note that among the four designs shown in Table 3.13, *Design* 3 has the smallest A-criterion value. This means in terms of the

A-criterion, *Design* 3 is the most efficient for estimating the regression parameters β_0, β_1 and β_2 simultaneously. This is in contrast with the results presented in Table 3.3, where *Design* 4 has the smallest A-criterion value for estimating only the parameters β_1 and β_2. A similar observation can be made for the E-optimality criterion. These observations for the A- and E-optimal designs, however, do not apply to D-optimal designs that remain invariant under linear transformation of the scales of the independent variables and also remain invariant for estimating all parameters or all but the intercept term in linear regression models.

Table 3.14 lists the D-, A- and E REs for the four designs. For the D-criterion, *Design* 4 is used as the reference design, and for the A- and E-optimality criteria *Design* 3 is used as the reference design. The REs show that *Design* 1 is the least efficient design. Depending on the criterion used, *Design* 1 will need between 40 and 80% more observations to become as efficient as the optimal design. Although *Design* 2 has relatively high efficiencies, the most efficient designs are *Designs* 3 and 4. However, if a researcher needs a scale-independent optimal design, *Design* 4 is the best choice.

Table 3.14   Relative D-, A- and E-efficiencies of the four designs.

	Design 1	*Design* 2	*Design* 3	*Design* 4
RE_D	0.7181	0.8736	0.9305	1.0000
RE_A	0.5389	0.8953	1.0000	0.9434
RE_E	0.5328	0.8944	1.0000	0.9354

3.10.2   Design for regression models with two non-additive independent variables

In multiple regression analysis, we have more than one independent variable and the model usually includes interaction terms. It is helpful to recode variables to the prototype range with values between -1 and 1 for easy comparison among designs. This recoding is established by the transformation

$$d'_j = \frac{d_j - \tilde{d}}{d_{\max} - \tilde{d}}, \text{ where } \tilde{d} = d_{\min} + \frac{d_{\max} - d_{\min}}{2}. \qquad (3.29)$$

In the vocabulary-growth study, two variables were considered, namely, the school grade (X_1) variable and the family income variable of pupils (X_2). The school grade variable can be recoded as $d'_j = (d_j - \tilde{d})/(11 - \tilde{d})$, where $\tilde{d} = 8 + (11 - 8)/2 = 9.5$. For the family income variable, the recoding is established by $d'_j = (d_j - \tilde{d})/(3 - \tilde{d})$, where $\tilde{d} = 1 + (3 - 1)/2 = 2.0$.

Suppose that a researcher is now interested to find an efficient design to estimate all the parameters in the model:

$$y_i = \beta_0 + \beta_1 x_{1i} + \beta_2 x_{2i} + \beta_{12} x_{1i} x_{2i} + \varepsilon_i. \tag{3.30}$$

The corresponding design matrices of the four vocabulary-growth designs for the recoded variables are presented in Table 3.15.

Table 3.15   Design matrices for four designs and the regression model with interaction.

Design 1:

$$X_1' = \begin{bmatrix} 1 & 1 & 1 & 1 & 1 & 1 & 1 & 1 & 1 & 1 & 1 & 1 \\ -1 & -1 & -1 & -1/3 & -1/3 & -1/3 & 1/3 & 1/3 & 1/3 & 1 & 1 & 1 \\ -1 & 0 & 1 & -1 & 0 & 1 & -1 & 0 & 1 & -1 & 0 & 1 \\ 1 & 0 & -1 & 1/3 & 0 & -1/3 & -1/3 & 0 & 1/3 & -1 & 0 & 1 \end{bmatrix}$$

Design 2:

$$X_2' = \begin{bmatrix} 1 & 1 & 1 & 1 & 1 & 1 & 1 & 1 & 1 & 1 & 1 & 1 \\ -1 & -1 & -1 & -1 & -1 & -1 & 1 & 1 & 1 & 1 & 1 & 1 \\ -1 & 0 & 1 & -1 & 0 & 1 & -1 & 0 & 1 & -1 & 0 & 1 \\ 1 & 0 & -1 & 1 & 0 & -1 & -1 & 0 & 1 & -1 & 0 & 1 \end{bmatrix}$$

Design 3:

$$X_3' = \begin{bmatrix} 1 & 1 & 1 & 1 & 1 & 1 & 1 & 1 & 1 & 1 & 1 & 1 \\ -1 & -1 & -1 & -1 & -1 & -1 & 1 & 1 & 1 & 1 & 1 & 1 \\ -1 & -1 & 1 & -1 & 0 & 1 & -1 & 0 & 1 & -1 & 1 & 1 \\ 1 & 1 & -1 & 1 & 0 & -1 & -1 & 0 & 1 & -1 & 1 & 1 \end{bmatrix}$$

Design 4 :

$$X_4' = \begin{bmatrix} 1 & 1 & 1 & 1 & 1 & 1 & 1 & 1 & 1 & 1 & 1 & 1 \\ -1 & -1 & -1 & -1 & -1 & -1 & 1 & 1 & 1 & 1 & 1 & 1 \\ -1 & -1 & -1 & 1 & 1 & 1 & -1 & -1 & -1 & 1 & 1 & 1 \\ 1 & 1 & 1 & -1 & -1 & -1 & -1 & -1 & -1 & 1 & 1 & 1 \end{bmatrix}$$

The first rows of the transposed design matrices all contain 1s for the intercept. The second rows of all design matrices contain the scaled codes for variable X_1 (school grades with four levels) and the third rows contain the codes for variable X_2 (family income category with three levels). Finally, the last rows of the transposed design matrices contain the products of the corresponding codes for variables X_1 and X_2 to represent the interaction between these two variables. The model contains four parameters $\beta = (\beta_0, \beta_1, \beta_2, \beta_{12})'$ and so the covariance matrices $\text{Cov}(\hat{\beta}) = \sigma_\varepsilon^2 (X'X)^{-1}$ for the four designs are of the order 4×4. These covariance matrices are all presented in Table 3.16 along with the values of the optimality criteria. Because *Design* 4 has the smallest D-, A- or E-criterion values among all four designs, Design 4 is optimal for all three criteria. Table 3.16 also shows that although Design 3 has an unequal number of observations assigned to the 12 design points, *Design* 3 is more efficient than the original *Design* 1 and should be preferred over *Design* 1.

Table 3.16 Variance–covariance matrices and optimality criteria for four vocabulary-growth study designs and regression model with interaction.

Design 1

$$\frac{\text{Cov}(\hat{\beta})}{\sigma_\varepsilon^2} = (X_1{'}X_1)^{-1} = \begin{bmatrix} 0.0833 & 0 & 0 & 0 \\ 0 & 0.1500 & 0 & 0 \\ 0 & 0 & 0.1250 & 0 \\ 0 & 0 & 0 & 0.2250 \end{bmatrix}$$

D-criterion $= 3.5158 \times 10^{-4}$
A-criterion $= 0.5833$
E-criterion $= 0.2250$

Design 2

$$\frac{\text{Cov}(\hat{\beta})}{\sigma_\varepsilon^2} = (X_2{'}X_2)^{-1} = \begin{bmatrix} 0.0833 & 0 & 0 & 0 \\ 0 & 0.0833 & 0 & 0 \\ 0 & 0 & 0.1250 & 0 \\ 0 & 0 & 0 & 0.1250 \end{bmatrix}$$

D-criterion $= 1.0851 \times 10^{-4}$
A-criterion $= 0.4167$
E-criterion $= 0.1250$

Design 3

$$\frac{\text{Cov}(\hat{\beta})}{\sigma_\varepsilon^2} = (X_3{'}X_3)^{-1} = \begin{bmatrix} 0.0862 & 0 & 0 & -0.0172 \\ 0 & 0.0862 & -0.0172 & 0 \\ 0 & -0.0172 & 0.1034 & 0 \\ -0.0172 & 0 & 0 & 0.1034 \end{bmatrix}$$

D-criterion $= 7.4316 \times 10^{-5}$
A-criterion $= 0.3793$
E-criterion $= 0.1141$

Design 4

$$\frac{\text{Cov}(\hat{\beta})}{\sigma_\varepsilon^2} = (X_4{'}X_4)^{-1} = \begin{bmatrix} 0.0833 & 0 & 0 & 0 \\ 0 & 0.0833 & 0 & 0 \\ 0 & 0 & 0.0833 & 0 \\ 0 & 0 & 0 & 0.0833 \end{bmatrix}$$

D-criterion $= 4.8225 \times 10^{-5}$
A-criterion $= 0.3333$
E-criterion $= 0.0833$

Note. The values for D-, A- and E- criteria are computed for $\sigma_\varepsilon^2 = 1$.

3.11   Summary

In summary, the following conclusions and recommendations can be drawn from this chapter.

The efficiency of a design for estimating parameters in a linear regression model increases as σ_ε^2 decreases. The same is true as var(x) increases and/or as the variance var($\hat{\beta}$) decreases. We do not have much control over the value of σ_ε^2, but we do have over var(x) and var($\hat{\beta}$) by careful design. Optimal design selects design points and allocates the required number of subjects to each level combination of the independent variables to attain the smallest possible value of var($\hat{\beta}$) as measured by the optimality criterion of interest. In addition, we observe that efficiency of a design tends to increase as the correlation between the levels of the independent variables decreases.

The design points of a D-optimal design for a linear regression model with main effects and interactions are placed at the corners of the design space. If quadratic terms are added to the model, then additional design points are placed in the centre of the scales of the corresponding independent variables. In general, the optimal number of distinct design points for a linear regression model does not exceed the number of parameters p in the model. This information is helpful when we search for an optimal design. Simplification in the search for an optimal design may be possible depending on the criterion. For instance, with D-optimality, we can frequently limit our search within the class of equally weighted designs with number of design points equal to the number of parameters in the model.

4

Designs for analysis of variance models

4.1 A typical design problem for an analysis of variance model

In experimental designs, a quantitative (continuous) dependent variable is usually measured from different groups or combinations of groups. Changes in the measurements are considered to be the result of the effects of one or more of the groups and measurement error. In analysis of variance (ANOVA) terminology, these different groupings are often referred to as *factors* and the different categories of the factor are referred to as its *levels*.

There are two parts in the choice of a design for an ANOVA model. The first part concerns the selection of the number of factors and the number of levels for each factor to be used in the design. Here is an example. Worchel and Shackelford (1991) were interested in the group performance of 'groups of students' in stressful and non-stressful environments while working together on a task. 'Groups of students' were assigned to a positive and a negative working environment, where students were informed that the environment would inhibit performance. In this case, the environment variable has only two levels, a positive and a negative level or a non-stressful and stressful environment condition. The 'groups of students' are the (experimental) units of analysis and the group performance of these units while working on a task is the dependent variable. Worchel and Shackelford (1991) extended this design by including a third noisy environment condition and a fourth crowded environment condition. The working environment variable now has four levels. In addition to these four working

An Introduction to Optimal Designs for Social and Biomedical Research   M. P. F. Berger & W. K. Wong
© 2009, John Wiley & Sons, Ltd

environment conditions, Worchel and Shackelford (1991) also included another factor, namely, a group structure factor, where groups were either given structure or left unstructured before starting to work on the task. Both the 'group structure' and 'working environment' factors are nominal variables.

Figure 4.1 shows six designs that may be used for an ANOVA model. The top three designs have one Factor B (one-way design) with two-, three- and four-levels, respectively, and the bottom three designs are two-factorial (two-way) designs with an additional Factor A with only two levels.

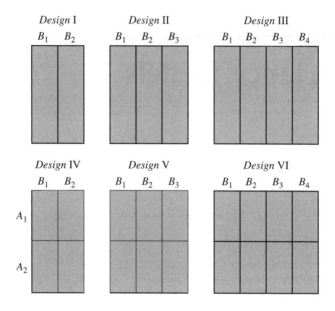

Figure 4.1 Analysis of variance designs.

The second part of the design problem for ANOVA models concerns the number of units assigned to each level of the independent variable or combination of levels of the independent variables. Specifically, how many experimental units are to be included in the study and how to distribute them to the various combination levels? Usually such studies are set up by assigning the same number of units to each condition, but in practice the groups may have unequal sizes because of missing observations and dropouts that cannot be easily controlled by the investigator. For the Groups under Stress study, the experimental units (groups of students) were assigned to a total of eight combinations of the levels of the 'working environment' and 'group structure' factors.

In principle, the ANOVA design problem is similar to the design problem for a regression analysis model in Chapters 2 and 3. The difference here is that the independent variables are not quantitative but qualitative and that there is no natural ordering among the different levels of these factors. The codes for the levels of factors A and B in Figure 4.1 are only used to distinguish the different levels from each other, and they may be reversed or changed without much harm.

4.1.1   The design

In the Groups under Stress study, Factor B has four conditions (positive, negative, noisy and crowded environment) and they are crossed with the two group structure levels (structured and unstructured) of the Factor A. This 2×4 factorial design is shown as *Design* VI in Figure 4.1. Table 4.1 shows that the design has a total of $m = 8$ distinct levels (design points) with n_j observations (within brackets) at the jth combination level, $j = 1, 2, \ldots, 8$.

The exact design is given by

$$\xi_N = \left\{ \begin{matrix} d_1 & d_2 & d_3 & d_4 & d_5 & d_6 & d_7 & d_8 \\ n_1/N & n_2/N & n_3/N & n_4/N & n_5/N & n_6/N & n_7/N & n_8/N \end{matrix} \right\}, \quad (4.1)$$

and the total number of observations is equal to $N = \sum_j n_j$. We emphasize that although variable B (environment conditions) may be considered to have an ordering, we will assume that variable B is a qualitative variable and that the combinations d_1 through d_8 do not have any natural ordering. The symbols d_1 through d_8 are only used to distinguish the eight combinations of levels of Factors A and B.

Table 4.1   Eight design points of a design with two qualitative independent variables.

Variable A (group structure)	Variable B (working environment)			
	Positive	Negative	Noisy	Crowded
Structured	d_1 (n_1)	$d_3(n_3)$	$d_5(n_5)$	$d_7(n_7)$
Unstructured	$d_2(n_2)$	$d_4(n_4)$	$d_6(n_6)$	$d_8(n_8)$

Note. Number of observations are in brackets.

With this design, the social psychologist can at least investigate three effects: (i) the main effect of environment on the performance of the units (groups of students); (ii) the main effect of the group structure on the performance of the units and (iii) the differential effect of environment on the performance of structured and unstructured groups. The latter is the interaction effect between environment and group structure. For such a design, the ANOVA model is a common and appropriate statistical technique to analyse the data.

We next briefly explain how to formulate the ANOVA model as a regression model. The parameters and the variances of their estimators depend on the specific formulation of the model, and we explain the differences and commonalities of the parameters in the ANOVA model and the corresponding regression formulation with different dummy coding schemes.

4.1.2 The analysis of variance model

ANOVA is a statistical technique for comparing the group means. Frequently, we simply compare the group means using the estimated variance of an effect and the estimated error, but sometimes, depending on the problem at hand, more complicated ways may be required. ANOVA models can always be written as a regression model and one might even conclude that an ANOVA model is a special case of a regression model.

In an ANOVA model, the independent variables are usually nominal or qualitative variables, whereas in regression models these independent variables are often quantitative. An important distinction between ANOVA and regression models is that linear regression analysis assumes a linear relation between the dependent and independent variables, whereas ANOVA does not necessarily assume such a linear relation.

The factors in an ANOVA model may be crossed or nested. Factors are crossed if all levels of one factor are combined with all levels of the other factor, and factors are nested if each level of one factor appears with one and only one level of the other factor. Nested factors have a hierarchical relationship. In addition, we may distinguish a random factor from a fixed factor. In the former case, the levels of the factor are assumed to be a random sample from a larger set of levels. The implication is that we can study only a few levels of the factor but we want to make inferences on the population levels for the factor. On the other hand, fixed factors include only a specific number of levels and inferences are desired only for these specific levels; see for example, Kleinbaum *et al.* (1998) and Kirk (1995) for more details.

The observations in each level combination of an ANOVA model can be represented as deviations from the cell mean. Figure 4.1 shows six typical designs for an ANOVA model. Each design, I through III, has C different conditions, where $C = 2$, 3 and 4, respectively. The score y_{ic} of the ith subject in condition c can be modelled as the sum of the mean response from condition c, that is, μ_c, and an error term:

$$y_{ic} = \mu_c + \varepsilon_{ic}. \tag{4.2}$$

This model has one factor and can be expressed in terms of ANOVA effects as

$$y_{ic} = \mu + b_c + \varepsilon_{ic}, \tag{4.3}$$

where μ is the overall mean of all scores and $b_c = (\mu_c - \mu)$ is the effect of condition c defined as the difference between the mean response from condition c and the overall mean response μ. The error ε_{ic} incurred for the ith subject in condition c is assumed to be normally distributed with mean 0 and variance σ_ε^2. Because the model is over-parameterized, it is not possible to estimate the overall mean μ and all the effects b_c separately. For this reason, we follow convention and impose a constraint such as $\sum_c^C b_c = 0$ so that we can estimate all the parameters of interest.

Designs IV through VI in Figure 4.1 are exemplary designs for a two-factor ANOVA model. Here, the score y_{irc} of the ith subject in cell rc is expressed as the sum of the mean μ_{rc} of cell rc and the error term:

$$y_{irc} = \mu_{rc} + \varepsilon_{irc}. \tag{4.4}$$

This model can be rewritten in terms of ANOVA effects as

$$y_{irc} = \mu + (\mu_r - \mu) + (\mu_c - \mu) + (\mu_{rc} - \mu_r - \mu_c + \mu) + \varepsilon_{irc}$$

or $\tag{4.5}$

$$y_{irc} = \mu + a_r + b_c + ab_{rc} + \varepsilon_{irc},$$

where $a_r = (\mu_r - \mu)$ is the effect of the rth level of Factor A and $b_c = (\mu_c - \mu)$ is the effect of the cth level of Factor B. Design VI has $2 \times 4 = 8$ cells and the ANOVA model has eight interaction effects $ab_{rc} = (\mu_{rc} - \mu_r - \mu_c + \mu)$, where μ_r and μ_c are the row and column means of level r of Factor A and level c of Factor B, respectively. As usual, we assume that the error terms ε_{irc} in each cell are normally distributed with mean 0 and common variance σ_ε^2. For estimating main and interaction effects, restrictions must be imposed on the effects. The following restrictions are commonly used:

$$\sum_r^R a_r = 0, \ \sum_c^C b_c = 0, \ \sum_r^R ab_{rc} = 0, \ \text{for all } c, \ \sum_c^C ab_{rc} = 0, \ \text{for all } r. \tag{4.6}$$

More details are available in Chapters 16–19 in Kleinbaum *et al.* (1998) and Kirk (1995).

4.1.3   Formulation of an ANOVA model as a regression model

ANOVA models can be written as a regression model by means of dummy variables. The model parameters and the estimated variances and covariances depend on the specific formulation of a model. This means that the design will also have to take into account the specific formulation of the model. To improve the efficiency of the design for an ANOVA model, we first explain the differences between the sets of parameters that arise from two different but common dummy coding schemes.

A dummy variable is a variable that enables identification of different categories of a nominal or qualitative variable. The values of a dummy variable are usually coded as '0', '1' or '−1'. In a regression model with intercept β_0, a total of $(C-1)$ dummy variables are needed to describe one nominal variable with C distinct categories. Therefore, the regression model with an intercept and a nominal variable with C categories has a total of C parameters.

Consider, for example, Design II in Figure 4.1, where Factor B has three levels, namely a positive, noisy and negative environment. For the single-factor

ANOVA model in Equation (4.3), we have $C = 3$ effects, namely, $b_1 = (\mu_1 - \mu)$, $b_2 = (\mu_2 - \mu)$ and $b_3 = (\mu_3 - \mu)$. A total of $(C - 1) = 2$ dummy variables D_1 and D_2 is then needed. The expected value of the scores for this design can be expressed in terms of an ANOVA model notation and a regression model notation:

$$\text{ANOVA model:}\quad E(y_{ic}) = \mu + b_c$$
$$\text{Regression model:}\quad E(y_i) = \beta_0 + \beta_1 D_1 + \beta_2 D_2. \tag{4.7}$$

In both models, we assume that the expected values of the error terms are $E(\varepsilon_{ic}) = E(\varepsilon_i) = 0$.

The dummy variables D_1 and D_2 may be coded in different ways. The two most commonly used coding schemes are the so-called *dummy coding* and the *effect coding* scheme shown in Table 4.2.

Table 4.2   Two coding schemes for qualitative Factor B in *Design* II of Figure 4.1.

Dummy coding	Effect coding
$D_1 = \begin{cases} 1 & \text{if score } y_i \text{ belongs to level } B_1 \\ 0 & \text{otherwise} \end{cases}$	$D_1 = \begin{cases} 1 & \text{if score } y_i \text{ belongs to level } B_1 \\ -1 & \text{if score } y_i \text{ belongs to the last level } B_3 \\ 0 & \text{otherwise} \end{cases}$
$D_2 = \begin{cases} 1 & \text{if score } y_i \text{ belongs to level } B_2 \\ 0 & \text{otherwise} \end{cases}$	$D_2 = \begin{cases} 1 & \text{if score } y_i \text{ belongs to level } B_2 \\ -1 & \text{if score } y_i \text{ belongs to the last level } B_3 \\ 0 & \text{otherwise} \end{cases}$

The dummy coding scheme assigns dummy values '1' or '0' to each level. A '1' is assigned to the dummy variable when the score y_i belongs to the corresponding level of the independent variable. The last (third) level does not require another dummy variable because it is already accounted for when both dummies take on zero values, that is $D_1 = 0$ and $D_2 = 0$. This last level is often called the *reference group*. This reference group is usually selected in some meaningful or natural way. For example, in dose–response studies where a few treatment groups are compared with the control group, the reference group is usually the control group.

An alternative way for coding the qualitative variable is to use contrasts as in the effect coding scheme for the ANOVA model. In this coding scheme contrasts are formulated between each level and the reference level. Although some computer programs use other coding schemes by default, the dummy coding scheme is the most often applied. In the following sections, we show that effect coding is especially useful for obtaining a one-to-one relation between ANOVA effects and regression parameters.

The two coding schemes reformulate the original ANOVA effects in terms of the regression parameters β_0, β_1 and β_2. Table 4.3 shows these relationships between the two sets of parameters in the ANOVA and in the regression set-up for the model in Equation (4.7) with one qualitative variable having 3 levels. The dummy coding scheme results in an intercept β_0 being equal to the mean of the scores in the reference (third) level. The regression parameters β_1 and β_2 correspond to the dummy variables D_1 and D_2, and they can be expressed as the difference between the effects of level 1 and the reference level 3 and the difference between levels 2 and 3, respectively. Under the effect coding scheme, the intercept term β_0 is equal to the overall mean μ and the regression parameters β_1 and β_2 are the group effects b_1 and b_2, respectively.

Table 4.3   Correspondence between regression parameters and ANOVA effects of *Design* II in Figure 4.1.

Dummy coding	Effect coding
$\beta_0 = \mu + b_3$	$\beta_0 = \mu$
$\beta_1 = (b_1 - b_3)$	$\beta_1 = b_1$
$\beta_2 = (b_2 - b_3)$	$\beta_2 = b_2$

As another illustration of the two coding schemes, consider the 2×4 factorial Design VI in Figure 4.1 for the two-factor ANOVA model with interaction. The expected value of the dependent variable score in cell rc is

$$E(y_{irc}) = \mu_{rc} = \mu + a_r + b_c + ab_{rc}. \tag{4.8}$$

The model includes $R = 2$ effects $a_r = (\mu_r - \mu)$ for Factor A and $C = 4$ effects $b_c = (\mu_c - \mu)$ for Factor B. There are $2 \times 4 = 8$ cell means μ_{rc} with corresponding interaction effects $ab_{rc} = (\mu_{rc} - \mu_r - \mu_c + \mu)$, where μ is the overall mean of all scores.

To reformulate this ANOVA model into a regression model, we need $(R - 1) = 1$ dummy variable $D^{(g)}$ for Factor A (group structures) and $(C - 1) = 3$ dummy variables, $D_1^{(s)}, D_2^{(s)}$ and $D_3^{(s)}$ for Factor B (stressful environment conditions). The expected value of the scores expressed in dummy variables is equivalent to the expected value formulated in terms of ANOVA effects in Equation (4.8):

$$E(y_i) = \beta_0 + \beta_1 D^{(g)} + \beta_2 D_1^{(s)} + \beta_3 D_2^{(s)} + \beta_4 D_3^{(s)}$$
$$+ \beta_5 D_1^{(s)} D^{(g)} + \beta_6 D_2^{(s)} D^{(g)} + \beta_7 D_3^{(s)} D^{(g)}. \tag{4.9}$$

For this regression model, β_0 is the intercept, β_1 is the regression parameter for Factor A, and β_2, β_3 and β_4 are the regression parameters for the three dummy variables of Factor B. The parameters β_5, β_6 and β_7 are the regression parameters for the interactions between the dummy variables, that is $D_1^{(s)} D^{(g)}$, $D_2^{(s)} D^{(g)}$, and $D_3^{(s)} D^{(g)}$. A total of eight regression parameters are thus needed to describe the deviation of each observation from the cell mean score. These dummy variables can be coded either with dummy coding or with effect coding. Both coding schemes are given in Table 4.4.

Table 4.4   Two coding schemes for the 2×4 factorial ANOVA model of *Design* VI in Figure 4.1.

Dummy coding	Effect coding
$D^{(g)} = \begin{cases} 1 & \text{if score } y_i \text{ is in} \\ & \text{level } A_1 \\ 0 & \text{otherwise} \end{cases}$	$D^{(g)} = \begin{cases} 1 & \text{if score } y_i \text{ is in level } A_1 \\ -1 & \text{if score } y_i \text{ is in last} \\ & \text{level } A_2 \end{cases}$
$D_1^{(s)} = \begin{cases} 1 & \text{if score } y_i \text{ is in} \\ & \text{level } B_1 \\ 0 & \text{otherwise} \end{cases}$	$D_1^{(s)} = \begin{cases} 1 & \text{if score } y_i \text{ is in level } B_1 \\ -1 & \text{if score } y_i \text{ is in last} \\ & \text{level } B_4 \\ 0 & \text{otherwise} \end{cases}$
$D_2^{(s)} = \begin{cases} 1 & \text{if score } y_i \text{ is in} \\ & \text{level } B_2 \\ 0 & \text{otherwise} \end{cases}$	$D_2^{(s)} = \begin{cases} 1 & \text{if score } y_i \text{ is in level } B_2 \\ -1 & \text{if score } y_i \text{ is in last} \\ & \text{level } B_4 \\ 0 & \text{otherwise} \end{cases}$
$D_3^{(s)} = \begin{cases} 1 & \text{if score } y_i \text{ is in} \\ & \text{level } B_3 \\ 0 & \text{otherwise} \end{cases}$	$D_3^{(s)} = \begin{cases} 1 & \text{if score } y_i \text{ is in level } B_3 \\ -1 & \text{if score } y_i \text{ is in last} \\ & \text{level } B_4 \\ 0 & \text{otherwise} \end{cases}$

The dummy coding scheme assigns a value '1' to the dummy variable if the score is in the corresponding level of the independent variable, and a '0' to the scores in all other levels of that independent variable. The effect coding scheme assigns a '1' if the score is in the corresponding level of the independent variable and a '−1' if the score is in the last level of the independent variable. Otherwise, a '0' is assigned.

The two coding schemes lead to different interpretations of the regression parameters in Model (4.9) and different relations between the regression parameters and the ANOVA effects. These relations for the two dummy coding schemes are shown in Table 4.5.

Table 4.5 shows the correspondence between the regression parameters and ANOVA effects is more complex with dummy coding than with effect coding. For effect coding, the intercept β_0 is equal to the overall mean of all scores and the other regression parameters have a one-to-one relation with the original ANOVA effects. This particular reformulation is based on the assumption that all eight cells have an equal number of observations. If the cells do not have an

Table 4.5   Correspondence between regression
parameters and ANOVA effects for *Design* VI in
Figure 4.1.

Dummy coding	Effect coding
$\beta_0 = \mu + a_2 + b_4 + ab_{24}$	$\beta_0 = \mu$
$\beta_1 = a_1 - a_2 + ab_{14} - ab_{24}$	$\beta_1 = a_1$
$\beta_2 = b_1 - b_4 + ab_{21} - ab_{24}$	$\beta_2 = b_1$
$\beta_3 = b_2 - b_4 + ab_{22} - ab_{24}$	$\beta_3 = b_2$
$\beta_4 = b_3 - b_4 + ab_{23} - ab_{24}$	$\beta_4 = b_3$
$\beta_5 = ab_{11} - ab_{21} - ab_{14} + ab_{24}$	$\beta_5 = ab_{11}$
$\beta_6 = ab_{12} - ab_{22} - ab_{14} + ab_{24}$	$\beta_6 = ab_{12}$
$\beta_7 = ab_{13} - ab_{23} - ab_{14} + ab_{24}$	$\beta_7 = ab_{13}$

equal number of observations (replications), the above correspondence between regression parameters and ANOVA effects will not hold.

In general, different coding schemes will lead to a different correspondence between ANOVA effects and regression parameters. The dummy coding scheme is the most popular, but the effect coding scheme is more likely to produce a one-to-one relationship between ANOVA effects and regression parameters, especially for models with interactions. Researchers interested in estimating and testing the separate ANOVA effects in a regression analysis may therefore be better off by using effect coding instead of dummy coding.

4.2   Estimation of parameters and efficiency

In principle, the reformulation of an ANOVA model into a regression model should yield similar conclusions. As has been shown in the previous section, the parameters in the regression model depend upon the particular coding system that is used for the dummy variables. This means that inferences on individual regression parameters depend on these coding systems. However, the overall explained variance from the regression model and the ANOVA models is the same. The effects in an ANOVA model can be estimated by the method of least squares described in the earlier chapters. Consider the factorial ANOVA model with two factors A and B:

$$y_{irc} = \mu + a_r + b_c + ab_{rc} + \varepsilon_{irc}. \tag{4.10}$$

To estimate the parameters in this model, we minimize the sum of squared deviations of scores from their estimated cell means, that is we want the least squares estimators $\hat{\mu}, \hat{a}_r, \hat{b}_c$, and $a\hat{b}_{rc}$ to satisfy

$$\sum_i \sum_r \sum_c \varepsilon_{irc}^2 = \sum_i \sum_r \sum_c (y_{irc} - \hat{\mu} - \hat{a}_r - \hat{b}_c - a\hat{b}_{rc})^2$$

$$= \text{minimum}, \tag{4.11}$$

where the minimization in Equation (4.11) is taken over all values of the parameters μ, a_r, b_c, and ab_{rc}. According to the Gauss–Markov theorem, these estimators enjoy good statistical properties. Two important properties of least squares estimators are that they are unbiased and their variances are not larger than the variances of any other linear unbiased estimators for the same model.

4.2.1   Measures of uncertainty

We now consider Design VI for the two-way ANOVA model in terms of ANOVA effects and in terms of dummy variables for the corresponding regression model. The expected value of the scores from the rc cell is $E(y_{irc}) = \mu_{rc} = \mu + a_r + b_c + ab_{rc}$ and there are n_{rc} observations in the rc cell to estimate the cell mean. Under the assumption that all error terms have equal variance σ_ε^2, the variance–covariance matrix of the cell mean estimators $\hat{\mu} = (\hat{\mu}_{11}, \hat{\mu}_{21}, \hat{\mu}_{12}, \ldots, \hat{\mu}_{24})'$ can be schematically displayed as a diagonal matrix:

$$\text{Cov}(\hat{\mu}) = \begin{bmatrix} \frac{\sigma_\varepsilon^2}{n_{11}} & 0 & 0 & \cdots & 0 & 0 \\ 0 & \frac{\sigma_\varepsilon^2}{n_{21}} & 0 & \cdots & 0 & 0 \\ 0 & 0 & \frac{\sigma_\varepsilon^2}{n_{12}} & \cdots & 0 & 0 \\ \vdots & \vdots & \vdots & \vdots & \vdots & \vdots \\ 0 & 0 & 0 & \cdots & \frac{\sigma_\varepsilon^2}{n_{14}} & 0 \\ 0 & 0 & 0 & \cdots & 0 & \frac{\sigma_\varepsilon^2}{n_{24}} \end{bmatrix}. \tag{4.12}$$

The matrix $\text{Cov}(\hat{\mu})$ is a diagonal matrix because each estimator for a cell mean is independent of the other cell mean estimators. Obviously, the cell mean estimators depend only on:

- the common error variance σ_ε^2 and
- the number of independent observations in each cell n_{rc}, where $\sum_r \sum_c n_{rc} = N$.

This implies that the smaller the error variance is, the more efficient the parameter estimators will be. Again, we remind readers that here and throughout the book, efficient estimators mean estimators with small variances. A larger sample size N and more observations in each cell will also decrease the variances of the estimators. If one is only interested in estimating a certain cell mean, the number of observations in that cell should be chosen as large as possible. When the total sample size N is fixed and all cell means need to be estimated simultaneously as efficiently as possible, it is best to have an equal number of observations per cell.

Similar conclusions can be drawn from the variance–covariance matrix of the estimators in the corresponding regression model with effect coding. The $p = 8$ regression parameters are $\beta = (\beta_0, \beta_1, \beta_2, \beta_3, \beta_4, \beta_5, \beta_6, \beta_7)'$, where β_1

corresponds to the first ANOVA effect a_1 for Factor A and $(\beta_2, \beta_3, \beta_4)'$ correspond to the three first ANOVA effects for Factor B. The parameters $(\beta_5, \beta_6, \beta_7)'$ are the interactions between the dummy variables. When all cells have an equal number of observations, the variance–covariance matrix of the estimators of these parameters is a blocked diagonal matrix:

$\mathrm{Cov}(\hat{\beta})$

$$
=\begin{bmatrix}
\mathrm{var}(\hat{\beta}_0) & 0 & 0 & 0 & 0 & 0 & 0 & 0 \\
0 & \mathrm{var}(\hat{\beta}_1) & 0 & 0 & 0 & 0 & 0 & 0 \\
0 & 0 & \mathrm{var}(\hat{\beta}_2) & \mathrm{cov}(\hat{\beta}_2, \hat{\beta}_3) & \mathrm{cov}(\hat{\beta}_2, \hat{\beta}_4) & 0 & 0 & 0 \\
0 & 0 & \mathrm{cov}(\hat{\beta}_2, \hat{\beta}_3) & \mathrm{var}(\hat{\beta}_3) & \mathrm{cov}(\hat{\beta}_3, \hat{\beta}_4) & 0 & 0 & 0 \\
0 & 0 & \mathrm{cov}(\hat{\beta}_2, \hat{\beta}_4) & \mathrm{cov}(\hat{\beta}_3, \hat{\beta}_4) & \mathrm{var}(\hat{\beta}_4) & 0 & 0 & 0 \\
0 & 0 & 0 & 0 & 0 & \mathrm{var}(\hat{\beta}_5) & \mathrm{cov}(\hat{\beta}_5, \hat{\beta}_6) & \mathrm{cov}(\hat{\beta}_5, \hat{\beta}_7) \\
0 & 0 & 0 & 0 & 0 & \mathrm{cov}(\hat{\beta}_5, \hat{\beta}_6) & \mathrm{var}(\hat{\beta}_6) & \mathrm{cov}(\hat{\beta}_6, \hat{\beta}_7) \\
0 & 0 & 0 & 0 & 0 & \mathrm{cov}(\hat{\beta}_5, \hat{\beta}_7) & \mathrm{cov}(\hat{\beta}_6, \hat{\beta}_7) & \mathrm{var}(\hat{\beta}_7)
\end{bmatrix}.
$$

$$(4.13)$$

In summary, different coding schemes for the ANOVA model will lead to re-parameterization of the cell means. The structure and elements of the variance–covariance matrix $\mathrm{Cov}(\hat{\beta})$ may be quite different from that in Equation (4.13) if other coding schemes are used. The effect coding of the ANOVA effects will lead to the re-parameterization $\beta = (\beta_0, \beta_1, \beta_2, \beta_3, \beta_4, \beta_5, \beta_6, \beta_7)'$, where the regression parameters have a one-to-one relation with the ANOVA effect parameters, as shown in Table 4.5.

4.3   Simultaneous inference and optimality criteria

To find an optimal design, we first have to select a suitable optimality criterion function of the variance–covariance matrix $\mathrm{Cov}(\hat{\beta})$ of the estimators. The choice should reflect the objective or objectives of the study as close as possible. For models with quantitative independent variables, the usual D-, A- and E-optimality criteria come to mind.

One of the most appealing properties of the D-optimality criterion is that it is invariant under linear scale transformation of the independent variable. This makes the D-optimality criterion especially attractive because a D-optimal design remains the same for different scaling of the independent variables. This property does not hold in general for the A- and E-optimality criteria. They are scale dependent and A- or E-optimal designs can vary considerably for different scaling of the independent variables and coding of the dummy variables.

The advantage of the A-optimality criterion is its relative ease in use. For example, when a researcher is only interested in estimating the main effects $b_1 = \beta_2$, $b_2 = \beta_3$ and $b_3 = \beta_4$ of a suitably parameterized model the optimal design can be found from the sub-matrix of the total variance–covariance matrix in Equation (4.13):

$$\text{Cov}(\hat{\beta}_s) = \begin{bmatrix} \text{var}(\hat{\beta}_2) & \text{cov}(\hat{\beta}_2, \hat{\beta}_3) & \text{cov}(\hat{\beta}_2, \hat{\beta}_4) \\ \text{cov}(\hat{\beta}_2, \hat{\beta}_3) & \text{var}(\hat{\beta}_3) & \text{cov}(\hat{\beta}_3, \hat{\beta}_4) \\ \text{cov}(\hat{\beta}_2, \hat{\beta}_4) & \text{cov}(\hat{\beta}_3, \hat{\beta}_4) & \text{var}(\hat{\beta}_4) \end{bmatrix}, \qquad (4.14)$$

where $\hat{\beta}_s = (\hat{\beta}_2, \hat{\beta}_3, \hat{\beta}_4)'$ is a subset of $\hat{\beta}$. In this case, we add up the diagonal elements and minimize the sum by choice of a design. The D-optimality criterion for such a 3×3 covariance matrix is more complicated to compute. We recall that in Chapter 3, the sum of the diagonal elements of the sub-matrix in Equation (4.14) is referred to as the A_s-optimality criterion, and a D-optimal design for a subset of the parameters in a model is referred to as D_s optimal. In this case where a subset of parameters (main effects only) is considered, both the A_s- and D_s-optimal designs are not scale independent.

4.4   Designs for groups under stress study

The design for the Groups under Stress study performed by Worchel and Shackelford (1991) is a 2×4 factorial design, where Factor A (group structures) has two levels and Factor B (stressful environment conditions) has four levels, that is, a positive, a negative, a noisy and a crowded environment. The ANOVA model with interaction between group structure and environment is $y_{irc} = \mu + a_r + b_c + ab_{rc} + \varepsilon_{irc}$. The corresponding regression model has eight parameters. The design problem is how to determine the most efficient design for estimating the eight parameters. This question can be answered in the following steps.

First, the two factors are qualitative variables and there are eight different combinations of levels. We want each cell to have at least one observation and the design problem is to select an optimal number of observations for each of the eight cells. A generic design with n_j observations in each cell is shown in Table 4.1. Table 4.6 shows the corresponding weights $w_j = n_j/N$, $j = 1, \ldots, 8$ for all the cells with $N = \sum_j^8 n_j$.

The next step is to select an optimality criterion to use. Suppose we choose the D-optimality criterion and want the design to have as many design points as the number of cells in the ANOVA model. In this case, it is straightforward

Table 4.6   Weights of eight combinations of a 2×4 factorial design with qualitative independent variables.

Factor A (group structure)	Factor B (environment)			
	Positive	Negative	Noisy	Crowded
Structured	w_1	w_3	w_5	w_7
Unstructured	w_2	w_4	w_6	w_8

to show that the minimization of the D-optimality criterion depends only on the product of the number of observations in the cells. The diagonal form of Cov($\hat{\mu}$) in Equation (4.12) shows that minimizing Det[Cov($\hat{\mu}$)] is equivalent to

$$\text{minimize} \prod_j^8 n_j^{-1}. \tag{4.15}$$

A formal justification for the equivalence can be found in Silvey (1980, Lemma 5.1.3, p. 42). A minimum in Equation (4.15) is reached when all cells have an equal number of observations. Thus, the volume of the confidence ellipsoid for the eight parameters is minimized when all cells in the 2×4 design have an equal number of observations.

To show how much efficiency is lost when cells do not have an equal number of observations, consider the four 2×4 designs in Table 4.7. The first design has an equal number of observations in the cells and is D optimal. The rest have unequal sample sizes, which may be a priori planned or may be caused by loss of subjects during the course of the study.

Table 4.7 Weights for 2×4 factorial designs with corresponding relative efficiencies.

		Factor B				RE$_D$	RE$_{D_s}$
		1	2	3	4		
Factor A						(8 par.)	(5 par.)
Design 1	1	0.125	0.125	0.125	0.125	1.000	1.000
	2	0.125	0.125	0.125	0.125		
Design 2	1	0.175	0.175	0.175	0.175	0.916	0.966
	2	0.075	0.075	0.075	0.075		
Design 3	1	0.200	0.150	0.100	0.050	0.885	0.907
	2	0.200	0.150	0.100	0.050		
Design 4	1	0.250	0.200	0.150	0.100	0.816	0.902
	2	0.125	0.100	0.050	0.025		

4.4.1 A priori planned unequal sample sizes

A researcher may find it necessary to randomly assign relatively more subjects to some combinations of levels in a study, and fewer to other combinations of levels. The reason may be that there are not enough subjects available for all conditions or that the treatment in one condition is potentially harmful to the subjects and the researcher does not want to expose too many subjects to this treatment.

Design 2 in Table 4.7 reflects such a case with 70% of the total sample size assigned to the first level of Factor *A* and 30% assigned to the second level. *Design* 3 is also an example of such a case, where 40, 30, 20 and 10% of the total sample size is assigned to each of the four levels of Factor *B*, respectively. Such a priori designed or planned unequal sample sizes are not expected to be a great threat to the internal validity of the study. See Cook and Campbell (1979) for more details on threats to internal validity.

4.4.2   Not planned unequal sample sizes

Unequal sample sizes in a study may arise due to subject attrition. The reasons for such a dropout are diverse. These reasons are often related to the subjects themselves, such as illness, non-response and lack of motivation. Cook and Campbell (1979) referred to this threat to the internal validity of a study as 'mortality'. Non-response is a form of self-selection and often casts doubts upon the representativeness of the samples. Unequal sample sizes may also be caused by the conditions of the study itself, such as errors in coding the scores and by the complexity of the treatment or task involved, causing subjects to stop their participation. This artificial selection or loss of subjects has often an irregular pattern and is considered to be not only a potential threat to the internal validity but also a threat to external validity. *Design* 4 in Table 4.7 displays such an irregular pattern of sample sizes. In this case, the researcher should take the validity threats into account when drawing conclusions.

The loss in D efficiency for designs with unequally sized cells is summarized by their relative efficiencies RE_D for estimating the eight parameters. These efficiencies are also presented in Table 4.7. In *Design* 2, the number of observations differs only with respect to the levels of Factor *A*; that is, for the Groups under Stress study, 70% of the data come from the structured group condition and 30% from the unstructured group condition. The loss of efficiency is moderate. *Design* 2 will need $(0.9165^{-1} - 1)\% = 9.1\%$ more observations than the D-optimal *Design* 1 to achieve the same efficiency. This efficiency loss is due to the many more units in the structured group condition than in the unstructured group condition.

A similar effect is found in *Design* 3, where the proportions of the total number of observations for Factor *B* are 0.4, 0.3, 0.2 and 0.1, respectively. This imbalance causes the relative efficiency (RE) to drop to $RE_D = 0.8853$. The greatest efficiency loss occurs in *Design* 4, where the number of observations in each cell varies from 2.5 to 25% of the total sample size. The $RE_D = 0.8160$ and about 22% more observations are needed for *Design* 4 to be as efficient as the D-optimal design.

As an illustration, Table 4.7 shows the loss of efficiency for an ANOVA model with only main effects for Factors A and B. The ANOVA model with only main effects has five parameters to be estimated, namely, μ, a_1, b_1, b_2 and b_3. *Design* 1 is D_s optimal, because it is formulated on the subset of five parameters and Table 4.7 shows that the corresponding REs are a little higher than those for the model with eight parameters.

In conclusion, when a researcher plans a factorial design with qualitative factors, the design should have an equal number of observations in each cell. When this is not possible because subjects are expected to drop out, either planned or not planned, then one should aim for a design that is as balanced as possible. RE is an effective way to compare designs and arrive at an informed decision. For instance, among the designs in Table 4.7, *Design* 4 is the least efficient for estimating all eight parameters in the model and will require 22% more observations to guard against efficiency loss. The other three designs require fewer observations to guard against the efficiency loss.

4.5   Specific hypotheses and contrasts

The optimal designs and RE computations in the previous section were concerned with estimating the whole set of parameters and ascertaining whether there is a significant relationship between the response and the independent variables. For testing purposes, an overall F test may be used, but the test does not answer specific research questions. Guided by theory or by previous research, a researcher may be more interested in testing the so-called a priori hypotheses about the parameters. In that case, the optimal design with equal weights may not be optimal.

Suppose that a researcher is a priori interested in one or more specific differences between the performances of the units (groups of students) in the four different working environments. The single-factor ANOVA model is $y_{ic} = \mu + b_c + \varepsilon_{ic}$. The null hypothesis that the means (or ANOVA effects b_c's) for the environment conditions are equal in the population can be tested with an overall ANOVA F test. More specifically, a priori questions about the environments can be formulated by means of a contrast, which is defined as a weighted sum of means of all conditions, that is $\psi = \left(\sum_c \gamma_c \mu_c\right)$, where the μ_c's are the population means and the γ_c's are the user-selected contrast coefficients such that $\sum_c \gamma_c = 0$. The hypothesis that this contrast is zero can be tested by the statistic

$$F_\psi = \frac{\left(\sum_c \gamma_c \bar{y}_c\right)^2}{\mathrm{MS_e}\left(\sum_c \dfrac{\gamma_c^2}{n_c}\right)}, \tag{4.16}$$

which under the null hypothesis has a F distribution with numerator 1 degree of freedom and denominator error degrees of freedom equal to df_e. In the test statistic, $\bar{y}_c$ is the estimated mean score of condition c with sample size n_c and the mean squared error MS_e is the ANOVA estimator for the error variance σ_ε^2. The sum of squares for the contrast ψ is the ratio $SS_\psi = \left(\sum_c \gamma_c \bar{y}_c\right)^2 / \left(\sum_c \gamma_c^2 / n_c\right)$.

For the Groups under Stress study, a researcher may, a priori, be interested to compare the performances in a positive environment and all three other (negative, noisy and crowded) environments. The interest then is to test the null hypothesis that there is no difference in performances among these groups. The contrast ψ_1 in Table 4.8 may be used to test this hypothesis. A second hypothesis may be to compare the groups in a negative environment with groups in a noisy and crowded environment. The contrast ψ_2 in the same table serves this purpose. Finally, a comparison between groups in a noisy and crowded environment may be formulated using contrast ψ_3 in Table 4.8.

Table 4.8   Contrasts and optimal weights for comparisons between the four environment conditions.

| Contrast | Levels | | | |
| | B_1 | B_2 | B_3 | B_4 |
	Positive environment	Negative environment	Noisy environment	Crowded environment
ψ_1	$\gamma_1 = 3$ $w_1^* = 0.5$	$\gamma_2 = -1$ $w_2^* = 0.1667$	$\gamma_3 = -1$ $w_3^* = 0.1667$	$\gamma_4 = -1$ $w_4^* = 0.1667$
ψ_2	$\gamma_1 = 0$ $w_1^* = 0$	$\gamma_2 = 2$ $w_2^* = 0.5$	$\gamma_3 = -1$ $w_3^* = 0.25$	$\gamma_4 = -1$ $w_4^* = 0.25$
ψ_3	$\gamma_1 = 0$ $w_1^* = 0$	$\gamma_2 = 0$ $w_2^* = 0$	$\gamma_3 = 1$ $w_3^* = 0.5$	$\gamma_4 = -1$ $w_4^* = 0.5$

The design problem for estimating or testing a contrast is to select the optimal sample size for each working environment in such a way that the F_ψ statistic for that contrast becomes as large as possible. This can be done by maximizing the SS_ψ or by minimizing the term $\sum_c (\gamma_c^2 / n_c)$. Intuitively, we expect that when we compare two means of group performances in two working environments, an equal number of units from each condition is optimal. So, for contrast ψ_1, which compares the first mean with the other three means, the first environment condition should have the same sample size as the three other environment conditions together have, that is $n_1 = 3n_2 = 3n_3 = 3n_4$, where $\sum_c n_c = N$. The optimal weights defined as $w_c^* = n_c/N$ for contrast ψ_1 then become $w_1^* = 0.5$, $w_2^* = 0.1667$, $w_3^* = 0.1667$ and $w_4^* = 0.1667$, respectively. A similar line of reasoning for contrast ψ_2 leads to $n_2 = 2n_3 = 2n_4$ with $n_1 = 0$, and for contrast ψ_3 to $n_3 = n_4$ with $n_1 = n_2 = 0$. Table 4.8 lists the optimal weights for these contrasts.

Table 4.8 shows that the optimal sample size (weight) can be quite different for estimating or testing each a priori contrast. For example, for contrast ψ_3, no units (groups of students) need to be assigned to the positive and negative working environment condition, while for contrast ψ_1, units need to be assigned to all four working environment conditions, with three times as many units assigned to the positive environment condition than to the other three environment conditions.

Considering these contrasts, the researcher is faced with the problem of selecting the most efficient number of units (sample sizes). If the researcher is only interested in contrast ψ_3, there is no need to assign units to the positive and negative environment conditions. However, if the researcher is interested in all three contrasts in the same study, then it is not clear which design is preferable. Although a design with an equal number of units assigned to each of the four conditions is D optimal for the simultaneous estimation of all parameters in the model, it is generally not optimal if these three contrasts are of interest.

4.5.1 Loss of efficiency and power

Since each contrast has its own design with optimal weights, finding a design that is highly efficient for all three contrasts is the best thing a researcher can do when all three contrasts are equally important; that is, a design that yields the same (small) loss of efficiency for the three contrasts. A measure to express efficiency loss is the RE. For a given contrast ψ, the RE of a design is defined as

$$
\mathrm{RE}_\psi = \frac{\left(\sum_c \dfrac{\gamma_c^2}{w_c^*} \right)}{\left(\sum_c \dfrac{\gamma_c^2}{w_c} \right)},
\tag{4.17}
$$

where w_c and w_c^* are the weights for that contrast and the optimal weights, respectively. Clearly, the value of RE_ψ in Equation (4.17) lies between 0 and 1.

4.5.1.1 Set of all possible designs

It is instructive to consider the set of all possible designs for the four conditions. This is characterized by the set of weights $[w_1\ w_2\ w_3\ w_4]$ such that $\sum_c w_c = 1$ and $w_c \geq 0$ for all c. Among all these designs, there are three designs that are optimal for estimating the three contrasts ψ_1, ψ_2 and ψ_3 in Table 4.8. Table 4.9 lists eight designs out of the set of all possible designs, including the three optimal designs for estimating the three contrasts.

The REs of these designs for estimating each of the three contrasts are also given. For example, the RE_{ψ_1} value for the design with the first set of weights (Set 1) is 1 because those weights are optimal for estimating the first contrast. The RE of this design for estimating the second contrast ψ_2 is $\mathrm{RE}_{\psi_2} = 0.4445$, which is not very high. Dashes in the cells of Table 4.9 mean that the RE is not

Table 4.9   Designs and their relative efficiencies for the three contrasts.

Set	Design weights				Relative efficiency		
	w_1	w_2	w_3	w_4	RE_{ψ_1}	RE_{ψ_2}	RE_{ψ_3}
1	0.5000	0.1667	0.1667	0.1667	1.0000	0.4445	0.3333
2	0.0000	0.5000	0.2500	0.2500	–	1.0000	0.5000
3	0.0000	0.0000	0.5000	0.5000	–	–	1.0000
4	0.2500	0.2500	0.2500	0.2500	0.7500	0.6667	0.5000
5	0.2000	0.2000	0.3000	0.3000	0.6353	0.6000	0.6000
6	0.1667	0.2222	0.3055	0.3055	0.5535	0.6518	0.6110
7	0.2500	0.3333	0.2084	0.2084	0.7407	0.7407	0.4167
8	0.3000	0.3000	0.2000	0.2000	0.8308	0.6857	0.4000

computed for that design. For example, the RE of contrast ψ_1 corresponding to the second set of weights (Set 2) is not computed because of the zero weight in the first group.

The first three sets of weights are the optimal weights for the three contrasts, respectively. This implies that their RE values are one. The fourth set of weights represents the design where all samples have an equal number of observations, that is, $n_1 = n_2 = n_3 = n_4$. This design is optimal for simultaneous estimation of all parameters. The REs of this design compared to the designs with optimal weights for contrasts ψ_1, ψ_2 and ψ_3 are $RE_{\psi_1} = 0.75$, $RE_{\psi_2} = 0.6667$ and $RE_{\psi_3} = 0.50$, respectively. We observe that the largest efficiency loss is found for contrast ψ_3 because half of the total sample is not used for estimating this contrast.

The remaining sets of weights in Table 4.9 represent four specific designs, namely a maximin design for the three contrasts (Set 5), a design with averaged weights (Set 6), a maximin design for only the first two contrasts (Set 7) and a special weighted design (Set 8). In the following subsections, each of these designs will be explained in more detail.

4.5.1.2   Maximin procedure

A procedure to find a design that is highly efficient for all these three contrasts is the maximin procedure. The name 'maximin' indicates that the procedure selects the best of all worst cases; that is, it selects the maximum out of all the smallest REs of the three contrasts per set of weights. As such, the procedure searches for a design with a sufficiently high level of efficiency for all three contrasts. This can be done in three steps:

1. Compute for the three contrasts the relative efficiencies RE_{ψ_1}, RE_{ψ_2} and RE_{ψ_3} over all possible designs with weights $[w_1 \ w_2 \ w_3 \ w_4]$, such that $\sum_c w_c = 1$ and $w_c \geq 0$, for all c.

2. Select the smallest of these three REs for each design.

3. Select from the whole set of smallest REs the largest RE value. This largest RE value is referred to as the *maximin value* (*MMV*) and the corresponding design is the *maximin design* for these three contrasts.

The MMV is defined as

$$\text{MMV} = \max_{\text{all designs}} \{\min(\text{RE}_{\psi_1}, \text{RE}_{\psi_2}, \text{RE}_{\psi_3})\}, \quad (4.18)$$

For the three contrasts in Table 4.8, the MMV is RE = 0.60 and the maximin design is the one with weights in Set 5. This maximin design was found numerically by computing the REs for the set of all possible designs with different weights. This MMV = 0.6 is higher than $\text{RE}_{\psi_3} = 0.50$ for the weights in Set 4, but may still be unsatisfactorily low in practice.

4.5.1.3 Averaging optimal weights

An alternative (heuristic) procedure is to average the optimal weights for the three contrasts. Although this procedure is easier to apply than the maximin procedure, it is in general not a satisfactory procedure. This can be seen by considering the sixth set of weights in Table 4.9, which are the averaged weights of the three optimal designs for the three contrasts, that is the first three weight sets in Table 4.9. The average optimal weight for the first condition is quite low because of the two zero weights for the second and third contrasts and this may be partly responsible for the smallest relative efficiency $\text{RE}_{\psi_1} = 0.5535$, which is also smaller than the MMV = 0.6.

The only way to improve efficiency is to decrease the number of contrasts to be considered in the maximin procedure. Since the third contrast actually does not use the observations of the first two conditions, it is reasonable to just restrict interest to the first two contrasts. The seventh set of weights in Table 4.9 is found by the maximin procedure for estimating the first two contrasts. The MMV for both contrasts is MMV = 0.7407. It should be emphasized that for only two contrasts, the maximin design is equal to averaging the optimal weights for the two contrasts. Finally, the eighth set of weights is added to show that compared to Set 7, putting a little more weight on the first contrast increases its RE value to $\text{RE}_{\psi_1} = 0.8308$ at the expense of the second contrast.

4.5.1.4 Loss of power

Loss of efficiency can also be explained in terms of power. Figure 4.2 displays the power for testing the prior contrast hypotheses $H_0 : \psi_1 = 0$ and $H_0 : \psi_2 = 0$ as a function of the non-centrality parameter of the F distribution. As an example to show how the power differs for the different sets, the power functions of some of the sets of weights from Table 4.9 are shown in Figure 4.2.

It can be seen that Sets 1 and 2 have the most power for testing contrasts ψ_1 and ψ_2, respectively. However, Set 1 has the least power for testing contrast ψ_2. This can of course be explained by the fact that 50% of the total sample in Set 1 is

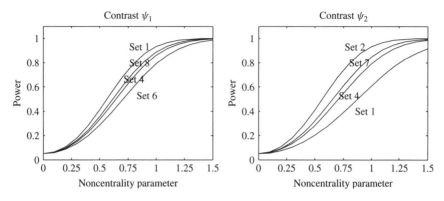

Figure 4.2 Power Functions for the designs in Table 4.9.

not used for testing contrast ψ_2. Set 1 would need $(0.4445^{-1} - 1)100\% = 125\%$ more observations in the study to produce the same power as Set 2 for testing contrast ψ_2.

The functions also show that the power for testing the hypothesis $H_0 : \psi_2 = 0$ is overall a little lower than the power for $H_0 : \psi_1 = 0$. Testing the hypothesis $H_0 : \psi_2 = 0$ does not use observations (information) from the first (positive environment) condition. Sets 7 and 8 have relatively high power for contrasts ψ_1 and ψ_2, respectively, and this is in agreement with the REs of these sets of weights. Set 8 needs about $(0.8308^{-1} - 1)100\% = 20\%$ and $(0.6857^{-1} - 1)100\% = 46\%$, respectively, more observations in the total sample to have optimal power for testing the two contrasts. This may be acceptable for testing contrast ψ_1, but probably not for ψ_2.

In conclusion, efficiency for estimating and testing a priori contrasts can be improved by carefully selecting the number of observations for each condition. As the number of a priori contrasts increases, it will become more difficult to obtain a highly efficient design for these contrasts. Eventually, as more contrasts are considered in a single study, the researcher may be better off by applying a design with equal weights that is optimal for the simultaneous estimation of all group parameters.

4.6   Designs for the composite faces study

The study with composite faces (Langlois and Roggeman, 1990) described in Chapter 1 distinguishes five different composite levels. These composite levels are ordered and we will assume that composite levels are scaled from 1 to 5. We could also have considered the actual number of faces that were used in the study by Langlois and Roggeman (1990) as coding for the composite levels, namely 2-, 4-, 8-, 16- and 32 faces, but then the analysis would have been more complicated because these codings are unequally spaced. We note that a simple $\log_2$ transformation can be applied to equalize the two different scales.

The exact design for the composite faces study can be displayed as

$$\xi_N = \begin{Bmatrix} d_1 = 1 & d_2 = 2 & d_3 = 3 & d_4 = 4 & d_5 = 5 \\ n_1/N & n_2/N & n_3/N & n_4/N & n_5N \end{Bmatrix}, \qquad (4.19)$$

where the design points d_j's are the five composite levels and the n_j's are the sizes of the samples of raters such that $\sum_j n_j = N$. The corresponding one-way ANOVA model is $y_{ij} = \mu_j + \varepsilon_{ij}$.

The design problem is now to choose the number of raters for each composite level so that the estimator of the linear or nonlinear effect of the composite level on the attractiveness score is estimated as efficiently as possible. This problem can be dealt with by fitting polynomials to the data. A polynomial of degree $(p - 1)$ has the form $\beta_0 + \beta_1 x + \beta_2 x^2 + \beta_3 x^3 + \cdots + \beta_{p-1} x^{p-1}$ and requires at least p distinct levels of X to estimate the parameters $\beta_0, \beta_1, \beta_2, \beta_3, \ldots$ and β_{p-1}. For example, to fit a second ($p - 1 = 2$) degree polynomial, we need to have at least $p = 3$ levels of the independent variable X.

Sometimes researchers use orthogonal polynomials to describe the relation between the response and the quantitative variable X. The total amount of variation between the levels of the independent variable X can be divided into separate pieces of variation connected to the linear and curvilinear relations. The between composite face levels sum of squares SS_B, which contains all the linear and curvilinear relations between Y and X, is equal to the sum of squares of the separate linear and curvilinear effects, that is $SS_B = SS_{lin} + SS_{quad} + SS_{cubic} + \cdots$.

Orthogonal polynomial coefficients have been extensively tabulated for different numbers of levels of the independent variable. Fisher and Yates (1963) is the original source, but tabulations can also be found in Bock (1975) and Kirk (1995), among others. An example of orthogonal polynomial coefficients for five levels is given in Table 4.10.

An orthogonal polynomial contrast among means is $\psi = \left(\sum_j \gamma_j \mu_j \right)$, where the μ_j's are the attractiveness group means for the different composite levels and the γ_j's are the corresponding contrast coefficients such that $\sum_j \gamma_j = 0$. For the levels of the composite faces (design points), the following estimated contrasts

Table 4.10   Orthogonal polynomial coefficients γ for five levels.

Polynomials	Levels of composite faces				
	1	2	3	4	5
Linear	-2	-1	0	1	2
Quadratic	2	-1	-2	-1	2
Cubic	-1	2	0	-2	1
Quartic	1	-4	6	-4	1

are of interest:

$$
\begin{aligned}
\hat{\psi}_{\text{lin}} &= (-2)\bar{y}_1 + (-1)\bar{y}_2 + (0)\bar{y}_3 + (1)\bar{y}_4 + (2)\bar{y}_5, \\
\hat{\psi}_{\text{quad}} &= (2)\bar{y}_1 + (-1)\bar{y}_2 + (-2)\bar{y}_3 + (-1)\bar{y}_4 + (2)\bar{y}_5, \\
\hat{\psi}_{\text{cubic}} &= (-1)\bar{y}_1 + (2)\bar{y}_2 + (0)\bar{y}_3 + (-2)\bar{y}_4 + (1)\bar{y}_5, \\
\hat{\psi}_{\text{Quartic}} &= (1)\bar{y}_1 + (-4)\bar{y}_2 + (6)\bar{y}_3 + (-4)\bar{y}_4 + (1)\bar{y}_5,
\end{aligned}
\tag{4.20}
$$

where $\bar{y}_j$ is the estimated group mean of the attractiveness scores for composite level j based on n_j observations. Similar to the standard test of a contrast in ANOVA, the hypothesis that a polynomial contrast is zero can be tested with the F statistic in Equation (4.16). A simple rule can be used to check whether two contrasts are orthogonal. Two contrasts ψ_1 and ψ_2 are orthogonal if the weighted sum of products of their contrast coefficients is zero, that is, if $\sum_j \gamma_{1j}\gamma_{2j}/n_j = 0$, where n_j is the number of observations in group j. For example, if we assume that the groups all have an equal sample size, then the polynomial contrasts $\hat{\psi}_{\text{lin}}$ and $\hat{\psi}_{\text{quad}}$ in Equation (4.20) are orthogonal because $(-2)(2) + (-1)(-1) + (0)(-2) + (1)(-1) + (2)(2) = 0$.

As indicated in Section 4.5, the variance of an estimated contrast $\hat{\psi}$ is minimized when the term $\left(\sum_j \gamma_j^2/n_j\right)$ is minimized. This term is minimized if the sizes of the weights $w_j = n_j/N$ are in the same proportion as the corresponding γ_j values. For example, the linear contrast coefficients for the five composite levels are $\gamma_1 = -2$, $\gamma_2 = -1$, $\gamma_3 = 0$, $\gamma_4 = 1$ and $\gamma_5 = 2$, respectively (Table 4.10). The corresponding optimal weights are in the same proportions: 0.3333, 0.1667, 0, 0.1667, and 0.3333. The weight 0.3333 is to 0.1667 and 0, as the coefficient 2 is to 1 and to 0. These optimal weights are presented in Table 4.11.

Table 4.11   Optimal weights for the polynomial coefficients of five levels.

Polynomials	1	2	3	4	5
Linear	0.3333	0.1667	0	0.1667	0.3333
Quadratic	0.25	0.125	0.25	0.125	0.25
Cubic	0.1667	0.3333	0	0.3333	0.1667
Quartic	0.0625	0.25	0.3750	0.25	0.0625

The optimal weights in Table 4.11 show that there are different optimal weights for different polynomial contrasts. For instance, if the main interest is in the linear contrast, then 67% of the raters should be evenly assigned to the first and last composite level and 33% to the second and fourth levels. This set of weights is different from the optimal allocation of raters for the quadratic polynomial contrast. If two or more contrasts are equally important, the maximin procedure described in the previous section can be used. It should, however, be kept in mind that if the samples of raters for each of the composite levels are not equally sized, the polynomial contrasts will not be orthogonal anymore and by focusing on specific contrasts the sizes of optimal samples will become unequal, thus loosing efficiency for the simultaneous estimation of all the parameters involved in the model.

Another problem that is not incorporated here is that, as shown in Figure 1.4 (Chapter 1), the ratings for the second composite level (four composite faces) have a much higher standard deviation than the ratings for the other levels. In order to increase efficiency, relatively more raters should be assigned to the second composite level to compensate for this high standard deviation.

4.7   Balanced designs versus unbalanced designs

The results in the previous sections suggest that the overall efficiency of parameter estimators is improved when we have an equal number of observations (subjects and replications) per group or combination of levels. This is the same as saying that a balanced design has an equal number of observations in each cell. However, studies with unequal sample sizes are frequently encountered in both experimental and non-experimental research. In the literature, these designs are often referred to as *unbalanced*, that is cells have unequal sample sizes. In this case, it is generally not easy to determine how much of the total sum of squares is attributed by the individual main effects and interactions separately, and the analysis of the data from such unbalanced studies is more complicated. In the ANOVA literature, different procedures to analyse such data have been discussed; see for example, Winer, Brown and Michels (1991) and Kirk (1995). The simultaneous estimation of the parameters will become less efficient when the samples are unequally sized.

4.8   Matrix notation for Groups under Stress study

In this section, we use matrix algebra to explain how the optimality criteria can be applied to the Groups under Stress example. The regression formulation of the ANOVA model in matrix notation is

$$
\begin{bmatrix} y_1 \\ y_2 \\ \vdots \\ y_N \end{bmatrix} = \begin{bmatrix} 1 & x_{11} & x_{21} & \cdots & x_{p-1,1} \\ 1 & x_{12} & x_{22} & \cdots & x_{p-1,2} \\ \vdots & \vdots & \vdots & \vdots & \vdots \\ 1 & x_{1N} & x_{2N} & \cdots & x_{p-1,N} \end{bmatrix} \begin{bmatrix} \beta_0 \\ \beta_1 \\ \vdots \\ \beta_{p-1} \end{bmatrix} + \begin{bmatrix} \varepsilon_1 \\ \varepsilon_2 \\ \vdots \\ \varepsilon_N \end{bmatrix}, \qquad (4.21)
$$

where y is the $N \times 1$ data vector, X is the $N \times p$ design matrix for $p \times 1$ regression parameters in the column vector $\beta = (\beta_0, \beta_1, \ldots, \beta_{p-1})'$ and the $N \times 1$ vector ε consists of errors, which are assumed to be normally distributed with mean 0 and variance σ_ε^2. Table 4.12 shows the design matrix with effect coding

Table 4.12   Design matrix with effect coding for the 2×4 Groups under Stress study.

	Factor A		Factor B			Interaction effects		
	$\beta_0 = \mu$	$\beta_1 = a_1$	$\beta_2 = b_1$	$\beta_3 = b_2$	$\beta_4 = b_3$	$\beta_5 = ab_{11}$	$\beta_6 = ab_{12}$	$\beta_7 = ab_{13}$
n_{11} rows	1	1	1	0	0	1	0	0
n_{21} rows	1	−1	1	0	0	−1	0	0
n_{12} rows	1	1	0	1	0	0	1	0
n_{22} rows	1	−1	0	1	0	0	−1	0
n_{13} rows	1	1	0	0	1	0	0	1
n_{23} rows	1	−1	0	0	1	0	0	−1
n_{14} rows	1	1	−1	−1	−1	−1	−1	−1
n_{24} rows	1	−1	−1	−1	−1	1	1	1

for the 2×4 design for the Groups under Stress study. The number of rows of the design matrix is $N = \sum_r \sum_c n_{rc}$ and the number of columns is $p = 8$.

The covariance matrix of the parameter estimators is

$$\text{Cov}(\hat{\beta}) = \sigma_\varepsilon^2 (X'X)^{-1}. \qquad (4.22)$$

With an equal number of observations in the cells and effect coding, $\text{Cov}(\hat{\beta})$ is a blocked diagonal:

$$\text{Cov}(\hat{\beta}) = \frac{\sigma_\varepsilon^2}{N} \begin{bmatrix} 1 & 0 & 0 & 0 & 0 & 0 & 0 & 0 \\ 0 & 1 & 0 & 0 & 0 & 0 & 0 & 0 \\ 0 & 0 & 3 & -1 & -1 & 0 & 0 & 0 \\ 0 & 0 & -1 & 3 & -1 & 0 & 0 & 0 \\ 0 & 0 & -1 & -1 & 3 & 0 & 0 & 0 \\ 0 & 0 & 0 & 0 & 0 & 3 & -1 & -1 \\ 0 & 0 & 0 & 0 & 0 & -1 & 3 & -1 \\ 0 & 0 & 0 & 0 & 0 & -1 & -1 & 3 \end{bmatrix}. \qquad (4.23)$$

We caution the reader that when the eight cells do not have an equal number of observations, the matrix $\text{Cov}(\hat{\beta})$ will generally not have this blocked diagonal structure.

Table 4.13 shows the efficiencies for the D-, A- and E-optimality criteria of *Designs* 2, 3 and 4 shown in Table 4.7 for the Groups under Stress study relative to the efficiency of *Design* 1, because it has the smallest value for each criterion. We remind the reader that the A- and E criteria depend on the scaling of the variables and other coding schemes may produce different optimal designs for these two criteria.

Table 4.13   Relative efficiencies of designs of the Groups under Stress study.

	RE_D	RE_A	RE_E
Design 1	1.0000	1.0000	1.0000
Design 2	0.9165	0.8400	0.6000
Design 3	0.8853	0.9412	0.9078
Design 4	0.8160	0.7583	0.4746

4.9   Summary

We discussed design problems for the ANOVA model where the independent variables are qualitative. In general, when we want to estimate the main and interaction effects from different groups, we recommend using a balanced design for ANOVA models. However, when it is impossible to implement a balanced design, a properly chosen unbalanced design may be just as effective. In our case, the unbalanced design shows a maximum efficiency loss of about 20%.

A second conclusion is that optimality depends on the criterion that is used. This is not a surprise because the same finding holds true for regression models also. The estimated parameters in the model and their variances and covariances depend on the specific coding of the dummy variables, and the criteria for finding optimal designs may be scale dependent. This is especially true when inferences are based only on a subset of the parameters in the model.

A balanced design is optimal for estimating model parameters, but it may not be efficient for making inference on a specific contrast among the groups of interest. Similarly, the optimal design for estimating a specific contrast may not be optimal for estimating another contrast. In either of the above cases, we have multiple objectives. We showed in this chapter that a simple maximin procedure may be used to find a design that is highly efficient for the two objectives. In Chapters 9 and 10, multiple-objective optimal designs are discussed in greater detail.

5

Designs for logistic regression models

5.1 Design problem for logistic regression

Dichotomous response variables are often encountered in social and biomedical research. For example, many studies seek to investigate the effect of a treatment for a disease and the outcome is whether the patient is cured or not. Other studies may want to estimate the effect of an intervention programme on whether subjects change behaviour or the effect of a new instructional method on the performance of the students and the outcome is whether the students pass the examination or not. It is convenient to denote the binary outcome as simply a 'success' or a 'failure'. If the treatment cures the patient of the disease, the outcome is a 'success'; otherwise, it is a 'failure'. 'Success' is usually coded with the value '1' and 'Failure' is usually coded with a value '0' but other values are possible. The research questions usually concern on the relation between such dichotomous outcomes and one or more predictors, and in our case, we focus on a few design issues.

As in linear regression models, there are two main design questions for the logistic regression models. The first question concerns the selection of the levels of the independent variables. If the independent variable is quantitative, the ordering of the levels must be taken into account. As an example, consider an educational psychologist who wants to study the effect of 'hours of practice' on the 'mastery of a difficult task'. The dependent variable is binary and may take on the value '1' if the student has mastered the difficult task and '0' otherwise. In the study, each student is allowed only up to six practice hours before trying to master a difficult task and so the level is the number of practice hours received.

An Introduction to Optimal Designs for Social and Biomedical Research   M. P. F. Berger & W. K. Wong
© 2009, John Wiley & Sons, Ltd

The design questions comprise what is the optimal number of levels to use in the study, what are these levels and how many students to assign to each of these levels. For example, should the design include an equal number of students who have practised for 0, 2, 4 and 6 hours or an equal number of students who have practised for only 0, 3 and 5 hours? Which is a better design for estimating the effect of 'hours of practice'? Is there an optimal design for estimating the effect of 'hours of practice' on the mastery of the task?

When the independent variable is qualitative, all the levels of the independent variable may be included in the design. We may use dummy variables to indicate the various levels, as was done in Chapter 4 for the analysis of variance model. For example, if students are assigned to one of two different instructional methods (directive and nondirective), these two instructional methods form the two levels of a qualitative independent variable.

The second question concerns the allocation of the subjects (patients and students) to the different level combinations of the independent variables. For instance, in designing a dose–response study, we need to know how many patients are going to be allocated to the different dose levels, and in our example on the instructional methods, we need to know how many students are assigned to each of the two methods. In an experimental study, we allocate the units (subjects, students or patients) to different combinations of levels (conditions) at random. However, when we have quasi-experimental or observational studies, we may not be able to assign units to various conditions at random.

5.2   The design

An exact design for the logistic regression model is characterized by the fixed total sample size N, the levels of the independent variables and the number of units assigned to each combination of levels. As in the case for linear models, we may denote a two-point exact design by

$$\xi_N = \left\{ \begin{array}{cc} d_1 & d_2 \\ n_1/N & n_2/N \end{array} \right\}. \tag{5.1}$$

There is only one qualitative independent variable and the design points d_1 and d_2 may represent the instructional methods. The sample sizes for the two treatments are n_1 and n_2 with $\sum_j n_j = N$. Because the independent variable is qualitative, the codes for the design points d_1 and d_2 are arbitrary and we can change or reverse them with no harm. Clearly, the notation in Equation (5.1) can be easily generalized to a design including more than two instructional methods.

When the exact design has a quantitative independent variable and there are m design points, we can likewise denote such a design by

$$\xi_N = \left\{ \begin{array}{ccccc} d_1 & d_2 & d_3 & \dots & d_m \\ n_1/N & n_2/N & n_3/N & \dots & n_m/N \end{array} \right\}. \tag{5.2}$$

Exact designs, such as ξ_N in Equation (5.2) for the logistic model, assign $n_1, n_2, n_3, \ldots, n_m$ units to the m design points such that $\sum_j n_j = N$. These exact designs can be replaced by their continuous counterparts with weights w_j's ($0 \leq w_j \leq 1$) instead of fractions n_j/N at the design points, and the sum of the w_j's is 1.

5.3 The logistic regression model

The logistic regression model is commonly used to describe the relation between a dichotomous dependent variable Y and one or more independent variables $X_1, X_2, \ldots, X_{p-1}$. The responses of Y are usually coded as $y = 1$ for occurrence (success) and $y = 0$ for no occurrence (failure) and modelled using the logistic model given by

$$\text{Prob}(y = 1) = p(z) = \frac{\exp(z)}{1 + \exp(z)}, \qquad (5.3)$$

where $z = \beta_0 + \sum_l^{p-1} \beta_l x_l$, x_l is the value of the independent variable $X_l, l = 1, 2, \ldots, (p-1)$ and the exponential function is $\exp(z) = e^z \approx 2.7184^z$. The function in Equation (5.3) represents the probability of an occurrence when the z takes on a specified value determined from the linear combination of the independent variables. The probability of no occurrence is $q(z) = [1 - p(z)]$. All observations are assumed to be independent in the logistic model.

Figure 5.1 displays the probability $p(z)$ as a function of z. The logistic model has an S shape and flattens off as z approaches plus or minus infinity. For $z = 0$, we have probability $p(z) = q(z) = 0.5$. Each of the binary outcomes y has a

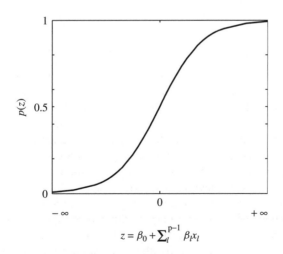

Figure 5.1 The logistic function $p(z)$.

binomial distribution with mean $p(z)$ and variance is $\text{var}(y) = p(z)q(z)$. This means that, unlike linear regression models, the mean and variance of the binary outcome y now depend on the values of the independent variables through z.

The logistic regression model can be linearized by using a so-called logit transformation. The odds are

$$\text{odds} = \frac{p(z)}{[1 - p(z)]} = \exp(z) = \exp\left(\beta_0 + \sum_{l}^{p-1} \beta_l x_l\right),$$

and the logarithm of the odds that are called logits are

$$\text{logit}(p) = \ln\left[\frac{p(z)}{[1 - p(z)]}\right] = z = \beta_0 + \sum_{l}^{p-1} \beta_l x_l, \qquad (5.4)$$

where 'ln' symbolizes the natural logarithm with $e \approx 2.7184$. The logit is often symbolized as $\text{logit}(p)$ and is a linear function of the independent variables. Hosmer and Lemeshow (1989) and Kleinbaum and Klein (2002) give further details.

In the following sections, design issues for logistic regression models with one or more quantitative or qualitative independent variables are illustrated using simple examples.

5.3.1   Design for a single dichotomous independent variable

Consider the study to investigate the effect of two instructional methods (directive and nondirective) on the ability of students to solve a complex problem. Both the dependent and the independent variables are dichotomous. The logistic regression model with two parameters is

$$p(z) = \frac{\exp(z)}{1 + \exp(z)} = \frac{\exp(\beta_0 + \beta_1 x)}{1 + \exp(\beta_0 + \beta_1 x)}, \qquad (5.5)$$

where $z = \beta_0 + \beta_1 x$. This equation represents the probability that a student will solve the problem correctly given the fact that the student has received one of the instructional methods. The two instructional methods are arbitrarily coded as $x = 1$ and $x = -1$. The dependent variable, problem solving, is coded as $y = 1$ for success and as $y = 0$ for failure. We use the zero–one coding for the dependent variable because it is commonly used in logistic regression analysis, and for consistency we retain the same coding for the independent variable with $x = 1$ and $x = -1$ as we have for the other chapters.

Table 5.1 shows the model and the design with the four conditions given. n_1 and n_2 are the number of observations for the two conditions of the independent variable $x = 1$ and $x = -1$, respectively. Following convention, we use the odds ratio (OR) as a suitable measure for the association between two dichotomous variables.

Table 5.1   The design for the logistic model with a dichotomous independent variable.

		Independent variable (instructional methods)	
		$x = 1$ (Method 1)	$x = -1$ (Method 2)
Dependent variable (problem solving)	$y = 1$ (success)	$p_1 = \frac{e^{\beta_0 + \beta_1}}{1 + e^{\beta_0 + \beta_1}}$	$p_2 = \frac{e^{\beta_0 - \beta_1}}{1 + e^{\beta_0 - \beta_1}}$
	$y = 0$ (failure)	$q_1 = \frac{1}{1 + e^{\beta_0 + \beta_1}}$	$q_2 = \frac{1}{1 + e^{\beta_0 - \beta_1}}$
	Total N	n_1	n_2

The odds of a student from the first method group ($x = 1$) solving the problem are

$$\text{odds}(y = 1; x = 1) = \frac{p_1}{(1 - p_1)} = \exp(\beta_0 + \beta_1), \tag{5.6}$$

and the odds for the second method group ($x = -1$) are

$$\text{odds}(y = 1; x = -1) = \frac{p_2}{(1 - p_2)} = \exp(\beta_0 - \beta_1). \tag{5.7}$$

The ratio of these two odds is the OR and for this example this value is given by OR $= \exp(\beta_0 + \beta_1 - \beta_0 + \beta_1) = \exp(2\beta_1)$. Although there are two regression parameters, the association between the independent and dependent variable is dependent only on the parameter β_1. A positive value of β_1 indicates a positive association between the two variables and a negative value of β_1 indicates a negative association. When $\beta_1 = 0$, there is no association between the number of practice hours and the ability of solving the difficult task.

We now construct an interval estimate for the parameter β_1 using a confidence interval and use it to capture the uncertainty of the estimate. A $100(1 - \alpha)\%$ confidence interval for β_1 is given by

$$\hat{\beta}_1 \pm z_{1-\alpha/2}\sqrt{\widehat{\text{var}}(\hat{\beta}_1)}, \tag{5.8}$$

where $\hat{\beta}_1$ is the estimator of β_1 and $\sqrt{\widehat{\text{var}}(\hat{\beta}_1)}$ is its estimated standard error. The critical value is $z_{1-\alpha/2}$, which is the $100(1 - \alpha/2)$th percentile of the standard normal distribution. The confidence interval for the OR can be directly found from the end points of the confidence interval for β_1. Since OR $= \exp(2\beta_1)$, the endpoints of the $100(1 - \alpha)\%$ confidence interval for OR have limits given by exponentiating both limits of the interval given in Equation (5.8). The width of the interval is $2 z_{1-\alpha/2}\sqrt{\widehat{\text{var}}(\hat{\beta}_1)}$ and we want this width to be as short as possible for accurate inference for the regression parameter β_1 or equivalently, the OR.

One way to ensure that we have a short confidence interval is through careful design of the study. An optimal design for estimating β_1 gives us the smallest possible variance of $\hat{\beta}_1$ and hence the shortest possible confidence interval for OR or β_1.

Suppose the coding for the independent variable in logistic regression model in Equation (5.5) is $x_1 = 1$ and $x_2 = -1$ and the sample sizes at these levels are n_1 and n_2, respectively, with $n_1 + n_2 = N$. If the probabilities of a response at these levels are p_1 and p_2, respectively, it can be shown that the variance–covariance matrix of the estimators of the two regression parameters β_0 and β_1 is

$$\mathrm{Cov}(\hat{\beta}) = \begin{bmatrix} \mathrm{var}(\hat{\beta}_0) & \mathrm{cov}(\hat{\beta}_0, \hat{\beta}_1) \\ \mathrm{cov}(\hat{\beta}_0, \hat{\beta}_1) & \mathrm{var}(\hat{\beta}_1) \end{bmatrix}$$

$$= D \begin{bmatrix} n_1 p_1 q_1 + n_2 p_2 q_2 & n_1 p_1 q_1 - n_2 p_2 q_2 \\ n_1 p_1 q_1 - n_2 p_2 q_2 & n_1 p_1 q_1 + n_2 p_2 q_2 \end{bmatrix}, \tag{5.9}$$

where $D = (4 n_1 n_2 p_1 q_1 p_2 q_2)^{-1}$ is the determinant of the matrix $\mathrm{Cov}(\hat{\beta})$.

A D-optimal design minimizes the expression $D = (4n_1 n_2 p_1 q_1 p_2 q_2)^{-1}$ and because we have independent samples, the determinant D is minimized when $n_1 = n_2$, regardless of the values of p_1 and p_2. This means that for the logistic model with one qualitative variable, equal sample sizes for both groups are D-optimal for estimating the two regression parameters.

In practice, a researcher may not be interested in the efficient estimation of the intercept β_0, but only in the efficient estimation of the parameter β_1, or alternatively, in the OR. Then, instead of minimizing the determinant of $\mathrm{Cov}(\hat{\beta})$ in Equation (5.9), we want to minimize $\mathrm{var}(\hat{\beta}_1)$. From Equation (5.9), we have

$$\mathrm{var}(\hat{\beta}_1) = \frac{n_1 p_1 q_1 + n_2 p_2 q_2}{4 n_1 p_1 q_1 n_2 p_2 q_2} = \frac{1}{4}\left(\frac{1}{n_1 p_1 q_1} + \frac{1}{n_2 p_2 q_2}\right). \tag{5.10}$$

When the probabilities are equal, that is $p_1 = p_2$ or when $p_1 = (1 - p_2)$, the minimum of $\mathrm{var}(\hat{\beta}_1)$ is reached when we have equal sample sizes $n_1 = n_2$. If the probabilities are not equal, $p_1 \neq p_2$ or if $p_1 \neq (1 - p_2)$, Dette (2004) provided the optimal weights for minimizing $\mathrm{var}(\hat{\beta}_1)$ and they are functions of the probabilities p_1 and p_2:

$$w_1^* = \frac{\sqrt{p_2(1 - p_2)}}{\sqrt{p_1(1 - p_1)} + \sqrt{p_2(1 - p_2)}} \quad \text{and} \quad w_2^* = (1 - w_1^*). \tag{5.11}$$

This means the optimal design for estimating β_1 requires $N w_1^*$ units at x_1 and $N w_2^*$ units at x_2 subject to $N w_1^* + N w_2^* = N$. Table 5.2 shows the optimal weights computed for a few different combinations of probabilities p_1 and p_2. The minimal values of $\mathrm{var}(\hat{\beta}_1^*)$ from the optimal design for estimating β_1 and the value of $\mathrm{var}(\hat{\beta}_1)$ from the equally weighted design are also shown in Table 5.2.

Table 5.2 also shows that the variances $\mathrm{var}(\hat{\beta}_1)$ of designs with equal weights are not that much larger than the minimum values of $\mathrm{var}(\hat{\beta}_1^*)$ suggesting that equally weighted designs are quite efficient.

Table 5.2   Optimal weights for efficient estimation of β_1 in the logistic model.

p_1	p_2	w_1^*	w_2^*	$\mathrm{var}(\hat{\beta}_1^*)$	$w_1 = w_2$	$\mathrm{var}(\hat{\beta}_1)$
0.3	0.5	0.5218	0.4782	4.3727	0.5	4.3810
0.1	0.6	0.6202	0.3798	7.2215	0.5	7.6389
0.1	0.8	0.5714	0.4286	8.5069	0.5	8.6806
0.3	0.1	0.3956	0.6044	7.6052	0.5	7.9365

5.3.1.1   Local optimality and maximin designs

Table 5.2 shows that the weights depend on the unknown parameters β_0 and β_1 through the probabilities p_1 and p_2. These parameters are of course not known before the actual data have been obtained. This phenomenon is usually referred to as *local optimality*; that is, the optimal design is only optimal for specific values of p_1 and p_2, and not for other values. Since in most studies the real values of p_1 and p_2 are not known before the actual data have been gathered, it may be problematic to apply these optimal designs in practice. The results in Table 5.2 show that a design with equal weights is quite efficient for the tabulated values of p_1 and p_2 because the two variances for the estimated β_1 from the equally weighted design and the optimal design are about the same. This suggests that equally weighted designs may be a good alternative to using a locally optimal design for estimating β_1 when no reliable estimates of β_1 are available.

It is sometimes possible to find a non-optimal design for estimating β_1 that is even more efficient than the equally weighted designs. One such strategy is to employ a maximin design for the logistic model. A *maximin design* is a design that is based on a relative efficiency (RE) measure computed for a whole range of values of p_1 and p_2. Suppose that p_1 and p_2 are in a rectangular sub-region P of unknown probabilities such that $P \subset [0\ 1] \times [0\ 1]$. Then, for a specific design with weights w_1 and w_2, the REs for all combinations of values of p_1 and p_2 can be computed:

$$\mathrm{RE} = \frac{\mathrm{var}(\hat{\beta}_1^*)}{\mathrm{var}(\hat{\beta}_1)}, \tag{5.12}$$

where $\mathrm{var}(\hat{\beta}_1^*)$ and $\mathrm{var}(\hat{\beta}_1)$ are the variances of $\hat{\beta}_1$ from the optimally weighted design and the specific design with weights w_1 and w_2, respectively. The smallest RE for all combinations of p_1 and p_2 values is selected, that is the worst case in terms of efficiency. These smallest REs are computed for all possible designs with weights w_1 and w_2 ($w_1 + w_2 = 1, 0 < w_j < 1$). From the set of minimal REs for all possible designs, the design with the largest minimum RE is selected and referred to as the *maximin design*. This design performs best in terms of the 'worst case' scenario (minimal RE). The RE of the maximin design is called *maximin value (MMV)* and is

$$\mathrm{MMV} = \max_{\text{all designs}} \left\{ \min_{P} (\mathrm{RE}) \right\}. \tag{5.13}$$

The search for such a maximin design may be tedious, but fortunately Dette (2004) provides expressions for the weights of the maximin design for all combinations of probabilities in P. The maximin design has weights:

$$w_1 = \frac{\left(2 + \sqrt{p_{\min}} + \sqrt{p_{\max}}\right)}{2\left(\sqrt{p_{\min}} + \sqrt{p_{\max}}\right) + 2\left(1 + \sqrt{p_{\min} \times p_{\max}}\right)} \text{ and } w_2 = 1 - w_1, \quad (5.14)$$

where

$$p_{\min} = \min_P \left\{\frac{p_1(1 - p_1)}{p_2(1 - p_2)}\right\} \text{ and } p_{\max} = \max_P \left\{\frac{p_1(1 - p_1)}{p_2(1 - p_2)}\right\}.$$

The performance of a maximin design depends on the range of probability values for p_1 and p_2. In order to apply a maximin design, a researcher must a priori specify the range of p_1 and p_2 values. To illustrate how a maximin design behaves for different ranges of the probabilities, consider again the example of the two instructional methods and the probability of solving a problem by students who have received one of the two instructional methods. Suppose that the first method is a new method and that we want to investigate whether the probability of successfully solving the complex problem p_1 with method 1 is greater than the success probability p_2 of method 2. In order to apply the maximin procedure, we will first need to specify a range for p_1 and p_2 values.

In Table 5.3, four sets of ranges of p_1 and p_2 values are presented. The first set reflects a situation where there is not much prior knowledge available for the p_1 and p_2 values. Almost the full range of possible values for p_1 and

Table 5.3   Maximin designs for efficient estimation of β_1 under different assumptions on the response probabilities.

Range of probabilities	$OR_{\min}$	MMV	Weights	Probabilities	RE_{equal}
Set 1					
$p_1 = [0.05\text{–}0.95]$	0.028	0.8663	$w_1 = 0.5000$	$p_2 = 0.5$	0.8663
$p_2 = [0.05\text{–}0.95]$			$w_2 = 0.5000$	$p_2 = 0.05$	
Set 2					
$p_1 = [0.2\text{–}0.8]$	0.0625	0.9878	$w_1 = 0.5000$	$p_1 = 0.5$	0.9878
$p_2 = [0.2\text{–}0.8]$			$w_2 = 0.5000$	$p_2 = 0.8$	
Set 3					
$p_1 = [0.3\text{–}0.9]$	0.2857	0.9012	$w_1 = 0.4643$	$p_1 = 0.9$	0.9412
$p_2 = [0.05\text{–}0.6]$			$w_2 = 0.5357$	$p_2 = 0.5$	
Set 4					
$p_1 = [0.4\text{–}0.9]$	0.4444	0.9577	$w_1 = 0.5208$	$p_1 = 0.9$	0.9412
$p_2 = [0.15\text{–}0.6]$			$w_2 = 0.4792$	$p_2 = 0.5$	

p_2 is considered, namely [0.05, 0.95]. Because of the monotonicity of the odds and the inverse odds, this in turn implies that the smallest possible value of the OR for the study is 0.028. Expressed in another way, this means that all values of the OR > 0.028 are assumed to be possible outcomes for this study. The corresponding maximin design has equal weights and a minimum RE of MMV = 0.8663 (Table 5.3, Set 1).

The second set is a little more restrictive in range, but also reflects a situation where little prior information about the possible values of p_1 and p_2 is available. The corresponding MMV is equal to 0.9878, and the weights are also equal. Actually, the weights of the maximin design can be shown to be equal if the ranges of p_1 and p_2 values are equal. As can be seen in Table 5.3, this is the case for Sets 1 and 2.

The third set reflects the situation where we assume that the OR is > 0.2857, but now the range of p_1 values is chosen to be generally much higher than that for p_2 and with some overlap. In practice, this means that we expect that students who have the first instructional method will have a higher probability of correctly solving the problem than the students who have received the second instructional method. Since the ranges of p_1 and p_2 values are different, the maximin design also has different weights. These weights were computed using Equation (5.14) and are given in Table 5.3. The maximin design has MMV = 0.9012 when $p_1 = 0.9$ and $p_2 = 0.5$. If we compute the RE of an equally weighted design for these probabilities, we have $RE_{equal} = 0.9412$, which is a little higher than the MMV = 0.9012. A similar comparison can be made for Set 4, but now we have MMV = 0.9577 which is a little higher than $RE_{equal} = 0.9412$.

The results in Tables 5.2 and 5.3 show that the REs of an equally weighted design are always very high. In cases where almost no a priori knowledge about the actual p_1 and p_2 values is available, a maximin design can be computed, but this design is also equally weighted when the ranges of p_1 and p_2 values are equal. If more specific information about the ranges of p_1 and p_2 values is available and these ranges are unequal, then the maximin design may be a little more efficient than an equally weighted design.

Dette (2004) also provided optimal weights for relative risk and attributable risk and showed that in cases where costs are different for sampling the two groups, the locally optimal designs can be rather inefficient when the parameters are miss-specified. More details on attributable and relative risks can be found in Walter (1976, 1977).

In conclusion, equally weighted designs for the logistic model with one dichotomous independent variable are D-optimal for the whole variance–covariance matrix $Cov(\hat{\beta})$, and they are highly efficient when the efficient estimation of only the effect parameter β_1 is taken into account. It should be mentioned that although the elements of $Cov(\hat{\beta})$ change for another coding of the independent variable, the equally weighted design will remain D optimal. Therefore, for a logistic model with one dichotomous independent variable, we recommend using an equally weighted design.

5.3.2   Design for multiple qualitative independent variables

Students from the social and medical sciences often find it difficult to solve problems in their statistical courses. Statistics education is generally focused on methods to improve the learning process of these students. Among the factors that are inferred to improve the problem solving abilities of students are directive instructions, motivation and the inclusion of practical exercises in a course. Take as example, an experiment where students are assigned to combinations of the following conditions: directive/nondirective instruction, enhanced/not enhanced motivation, inclusion/no inclusion of practical exercises. It is inferred that the probability of solving a statistical problem will increase when the instruction is more directive, when motivation of the students is enhanced by providing additional information about the importance of finding the right solution to that problem and when practical exercises are included in the problem solving process. The dependent variable is binary, indicating whether a student has correctly solved the statistical problem or not. A logistic regression model is applied to describe the relation between the effects of instruction (X_1), motivation (X_2) and practical exercises (X_3) on problem solving. All variables are dichotomous variables and the logit of the probability of solving the problem is linearly related to

$$z = \beta_0 + \beta_1 x_1 + \beta_2 x_2 + \beta_3 x_3, \tag{5.15}$$

where β_0 and $\beta_1, \beta_2, \beta_3$, are the intercept and the three effect parameters, respectively. The design of this study is displayed in Table 5.4, where the d_j's are the eight design points and the n_j's within brackets are the number of students assigned to each condition or design point, such that $\sum_j n_j = N$. For consistency, we choose the coding of the independent variables to be 1 and -1, respectively, but other coding schemes are also possible.

Table 5.4   Design for the problem solving study with three independent variables.

		Directive instruction	
	Motivation	$x_1 = 1$ (yes)	$x_1 = -1$ (no)
$x_3 = 1$ (yes)	$x_2 = 1$ (yes)	d_1 (n_1/N)	d_2 (n_2/N)
	$x_2 = -1$ (no)	d_3 (n_3/N)	d_4 (n_4/N)
Working exercises			
$x_3 = -1$ (no)	$x_2 = 1$ (yes)	d_5 (n_5/N)	d_6 (n_6/N)
	$x_2 = -1$ (no)	d_7 (n_7/N)	d_8 (n_8/N)

The odds of successfully solving the problem for the first two design points d_1 and d_2 in Table 5.4 are

$$\text{odds}(y = 1; x_1 = 1, x_2 = 1, x_3 = 1) = \exp(\beta_0 + \beta_1 + \beta_2 + \beta_3)$$

and   (5.16)

$$\text{odds}(y = 1; x_1 = -1, x_2 = 1, x_3 = 1) = \exp(\beta_0 - \beta_1 + \beta_2 + \beta_3).$$

Here, the OR $= \exp(2\beta_1)$, and depends only on the parameter β_1 because it describes the association between directive instruction and problem solving. The uncertainty of the parameter estimator $\hat{\beta}_1$ can again be expressed in terms of a $(1 - \alpha)100\%$ confidence interval, namely: $\hat{\beta}_1 \pm z_{1-\alpha/2}\sqrt{\widehat{\text{var}}(\hat{\beta}_1)}$. The odds for the other design points in Table 5.4 can be obtained in a similar way.

Equation (5.15) has four parameters that describe the effects on the problem solving ability of students. The variance–covariance matrix of these four parameter estimators is

$$\text{Cov}(\hat{\beta}) = \begin{bmatrix} \text{var}(\hat{\beta}_0) & \text{cov}(\hat{\beta}_0, \hat{\beta}_1) & \text{cov}(\hat{\beta}_0, \hat{\beta}_2) & \text{cov}(\hat{\beta}_0, \hat{\beta}_3) \\ \text{cov}(\hat{\beta}_0, \hat{\beta}_1) & \text{var}(\hat{\beta}_1) & \text{cov}(\hat{\beta}_1, \hat{\beta}_2) & \text{cov}(\hat{\beta}_1, \hat{\beta}_3) \\ \text{cov}(\hat{\beta}_0, \hat{\beta}_2) & \text{cov}(\hat{\beta}_1, \hat{\beta}_2) & \text{var}(\hat{\beta}_2) & \text{cov}(\hat{\beta}_2, \hat{\beta}_3) \\ \text{cov}(\hat{\beta}_0, \hat{\beta}_3) & \text{cov}(\hat{\beta}_1, \hat{\beta}_3) & \text{cov}(\hat{\beta}_2, \hat{\beta}_3) & \text{var}(\hat{\beta}_3) \end{bmatrix}. \quad (5.17)$$

These variances and covariances of the parameter estimators depend on the actual values of the parameters and on the coding of the independent variables. So, in order to minimize the variances of the parameters, one has to have an accurate estimate of the parameters.

Suppose that in the problem solving study, we believe a value of at least OR $= 2.5$ for any one of the three factors is sufficient evidence that students' ability to correctly solve the problems has improved. From Equation (5.16), this happens when all estimated parameter values of $\hat{\beta}_1, \hat{\beta}_2$ or $\hat{\beta}_3$ are at least $\ln(2.5)/2 = 0.4581$. Assuming that $\hat{\beta}_0 = 1$, the variances of the responses $\text{var}(y_i) = p(z_i)[1 - p(z_i)]$ can be estimated by substituting these parameter values in Equation (5.15), using the codings of the x values corresponding to the eight different design points. These estimated variances are presented in Table 5.5

Table 5.5   Response variances, D-optimal weights and samples for the problem solving study.

	d_1	d_2	d_3	d_4	d_5	d_6	d_7	d_8	RE_D
$\widehat{\text{var}}_j(y_i)$	0.0775	0.1529	0.1529	0.2326	0.1529	0.2326	0.2326	0.2412	
w_j^*	0.0000	0.1216	0.1216	0.1533	0.1216	0.1533	0.1533	0.1751	1.0000
w_j^{cb}	0.1250	0.1250	0.1250	0.1250	0.1250	0.1250	0.1250	0.1250	0.9565
w_j^{pb}	0.0000	0.1429	0.1429	0.1429	0.1429	0.1429	0.1429	0.1429	0.9960

together with the D-optimal weights w_j^* from minimizing the determinant of the variance–covariance matrix of the parameter estimators $\text{Cov}(\hat{\beta})$ in Equation (5.17). It is interesting to see that the D-optimal weights are monotonically related to $\widehat{\text{var}}(y_i)$, and that the smallest $\widehat{\text{var}}(y_i)$ from the first design point receives zero weight. Such a relation was also noted by Tekle, Tan and Berger (2008a), who proposed a restricted optimal design based on this relation.

Table 5.5 lists the weights of the D-optimal design, the weights of a *completely balanced design* and a third design called a *partially balanced design*. The latter design has all cells with equal weights, except the cell with the smallest value for the $\widehat{\text{var}}(y_i)$. This partially balanced design must not be confused with the partially balanced incomplete block design mentioned in Goos (2002, p.34). Table 5.5 displays the weights w_j^{cb} of a completely balanced design (with equal weights for all the design points) and the weights w_j^{pb} of the partially balanced design that has zero weight for the first design point and equal weights for all the remaining design points. Their REs are 0.9565 and 0.9960, respectively, which are actually very high. This means that there is not much efficiency loss when a completely or partially balanced design is used instead of the D-optimal design.

But, even such small efficiency loss can lead to a contradictory decision about a null hypothesis; that is, the D-optimal design may lead to rejection of a null hypothesis while the completely balanced and partially balanced designs may not. This can be illustrated by returning to the confidence interval for a single parameter given by $\hat{\beta}_1 \pm z_{1-\alpha/2}\sqrt{\widehat{\text{var}}(\hat{\beta}_1)}$. Suppose that a sample of $N = 90$ students is available for this study. The optimal number of students for the design points can be obtained by rounding the products $n_j^* = 90 \times w_j^*$ off to the nearest integers in such a way that their sum remains equal to $N = 90$. The optimal sample sizes n_j^* and the sample sizes n_j^{cb} for completely balanced design and n_j^{pb} for a partially balanced design are given in Table 5.6. The 95% confidence intervals for the parameters based on the D-optimal, completely balanced and partially balanced designs can be obtained from

D-optimal design:

$$0.4581 \pm 1.96/\sqrt{\textstyle\sum_j n_j^*\widehat{var}_j(y)} = 0.4581 \pm 0.4565,.$$

Completely balanced design:

$$0.4581 \pm 1.96/\sqrt{\textstyle\sum_j n_j^{\text{cb}}\widehat{\text{var}}_j(y)} = 0.4581 \pm 0.4818 \text{ and}$$

Partially balanced design:

$$0.4581 \pm 1.96/\sqrt{\textstyle\sum_j n_j^{\text{pb}}\widehat{\text{var}}_j(y)} = 0.4581 \pm 0.4618, \text{ respectively.}$$

These 95% intervals imply that with $\hat{\beta}_1 = 0.4581$ and a total sample size of $N = 90$, the optimal design will probably reject the null hypothesis $H_0 : \beta_1 = 0$, whereas the completely balanced design and the partially balanced design would probably fail to reject this hypothesis. However, it must be emphasized that for this example, the differences are quite small at the significance level of $\alpha = 0.05$.

Table 5.6   D-optimal and (partially) balanced sample sizes for the problem solving study.

	d_1	d_2	d_3	d_4	d_5	d_6	d_7	d_8	N
$\widehat{var}_j(y)$	0.0775	0.1529	0.1529	0.2326	0.1529	0.2326	0.2326	0.2412	
n_j^*	0	11	11	14	11	14	14	15	90
n_j^{cb}	12	11	11	11	11	11	11	12	90
n_j^{pb}	0	13	13	13	12	13	13	13	90

Note. The term *completely balanced design* does not really apply here for this exact design due to the conditions that the samples sizes n_j^{cb} and n_j^{pb} were rounded and adjusted to ensure that $\sum_j n_j^{cb} = \sum_j n_j^{pb} = 90$.

We end this section with a remark that for a logistic model, that with number of parameters equal to the number of design points, the D-optimal design is always a balanced design (Silvey, 1980, p.42). For example, if the logistic model has three main effects and all interaction terms, the linear predictor is $z = \beta_0 + \beta_1 x_1 + \beta_2 x_2 + \beta_3 x_3 + \beta_4 x_1 x_2 + \beta_5 x_1 x_3 + \beta_6 x_2 x_3 + \beta_7 x_1 x_2 x_3$, and the D-optimal design for this model is completely balanced.

5.3.3   Design for a single quantitative independent variable

Consider the study on the effect of the number of hours practice on the ability to master a difficult task. The probability that a student will 'master the task' ($y = 1$) is a logistic function of the number of 'hours practice' (X):

$$p(z) = \frac{\exp(z)}{1 + \exp(z)} = \frac{\exp(\beta_0 + \beta_1 x)}{1 + \exp(\beta_0 + \beta_1 x)}. \tag{5.18}$$

Suppose that the design of the study consists of groups of students, each allowed to have a different amount of practice time and that the independent variable 'hours of practice' (X) is a continuous variable. In this case, the design points have a one-on-one relation with the values of X, that is $d_1 = x_1, d_2 = x_2, \ldots, d_m = x_m$. In other words, the design region $d_{min} \le d_j \le d_{max}$ and the region of x values $x_{min} \le x \le x_{max}$ coincide. The design points are bounded between 0 and 6 hours and the binary dependent variable is coded as $y = 1$ if mastery of the task is achieved, and $y = 0$ otherwise. A design for this study is characterized by the design points d_j's and the weights w_j's at these points. For example, a design equally spread out at 0, 1, 2, 3, 4, 5 and 6 hours is written as

$$\xi = \left\{ \begin{array}{ccccccc} d_1 = 0 & d_2 = 1 & d_3 = 2 & d_4 = 3 & d_5 = 4 & d_6 = 5 & d_7 = 6 \\ w_1 & w_2 & w_3 & w_4 & w_5 & w_6 & w_7 \end{array} \right\}, \tag{5.19}$$

where $0 \le w_j \le 1$ and $\sum_j w_j = 1$. With a pre-determined sample of N students, this design roughly allocates Nw_j students to d_j, $j = 0, 1, 2, \ldots, 6$, subject to $\sum_j Nw_j = N$.

The odds of 'mastering the task' after 6 hours of practice are

$$\text{odds}(y = 1; x_7 = 6) = \exp(\beta_0 + 6\beta_1), \tag{5.20}$$

while the odds of 'mastering the task' after 3 hours practice are

$$\text{odds}(y = 1; x_4 = 3) = \exp(\beta_0 + 3\beta_1). \tag{5.21}$$

The corresponding OR is equal to $\exp(\beta_0 + 6\beta_1 - \beta_0 - 3\beta_1) = \exp(3\beta_1)$, which again depends only on the parameter β_1. When the parameter $\beta_1 > 0$, the odds increase as the hours of practice increase, while for $\beta_1 < 0$, the odds decrease with increasing hours of practice. No association exists when OR $= 1$ and $\beta_1 = 0$. Efficient estimation of the OR is equivalent to efficient estimation of the parameter β_1.

The variance–covariance matrix of the estimators of the parameters β_0 and β_1 is

$$\text{Cov}(\hat{\beta}) = \begin{bmatrix} \text{var}(\hat{\beta}_0) & \text{cov}(\hat{\beta}_0, \hat{\beta}_1) \\ \text{cov}(\hat{\beta}_0, \hat{\beta}_1) & \text{var}(\hat{\beta}_1) \end{bmatrix}$$

$$= D \begin{bmatrix} \sum_j w_j x_j^2 p(x_j)q(x_j) & -\sum_j w_j x_j p(x_j)q(x_j) \\ -\sum_j w_j x_j p(x_j)q(x_j) & \sum_j w_j p(x_j)q(x_j) \end{bmatrix}, \tag{5.22}$$

where

$$D = \left\{ \left[\sum_j w_j p(x_j)q(x_j) \right] \left[\sum_j w_j x_j^2 p(x_j)q(x_j) \right] - \left[\sum_j w_j x_j p(x_j)q(x_j) \right]^2 \right\}^{-1}$$

is the determinant of $\text{Cov}(\hat{\beta})$.

Abdelbasit and Plankett (1983), Minkin (1987), Khan and Yazdi (1988) and Mathew and Sinha (2001) are just a few among many others who have reviewed D-optimal designs for estimating the parameters β_0 and β_1 in the logistic model with one continuous independent variable. The D-optimal design is characterized as follows:

$$\xi_N{}^* = \left\{ \begin{array}{cc} d_1^* & d_2^* \\ w_1^* & w_2^* \end{array} \right\}. \tag{5.23}$$

If N is even, then the optimal weights $w_1^* = w_2^* = 0.5N$. If N is odd $w_1^* = 0.5(N - 1)$ and $w_2^* = 0.5(N + 1)$. The optimal design points d_1^* and d_2^* satisfy $d_1^* = (-1.5434 - \beta_0)/\beta_1$ and $d_2^* = (1.5434 - \beta_0)/\beta_1$, respectively.

Figure 5.2 shows a plot of the determinant of $\text{Cov}(\hat{\beta})$ as a function of the probabilities $p(x_1)$ and $p(x_2)$ of two equally weighted design points under the

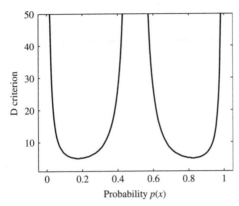

Figure 5.2 The D-optimality criterion as a function of the response probability function p(x).

condition that $p(x_1) = 1 - p(x_2)$. Among these two design points, the minimum value for the determinant of $Cov(\hat{\beta})$ is attained for the two D-optimal design points with $p(d_1^*) = 0.176$ and $p(d_2^*) = 0.824$. The figure also shows that the determinant function roughly remains small for probabilities in the intervals $0.1 < p(x) < 0.3$ and $0.7 < p(x) < 0.9$. Outside these intervals, the determinant function increases rapidly and approaches infinity as $p(x) \to 0$, $p(x) \to 0.5$ or $p(x) \to 1$. This indicates that a relatively small mis-specification of the probability will lead to large variances of the parameter estimators. Figure 5.2 shows that for probabilities close to 0, 0.5 and 1, one cannot expect to obtain efficient parameter estimators, even if an extremely large sample is used.

Given the optimal probabilities $p(d_1^*) = 0.176$ and $p(d_2^*) = 0.824$, the two D-optimal design points d_1^* and d_2^* can be obtained. For example, if the parameters of the logistic model that describes the relation between 'hours of practice' and the probability of 'mastering a task' have values $\beta_0 = -4$ and $\beta_1 = 1.3333$, the optimal design points can be computed from $d_1^* = (-1.5434 + 4.0)/1.3333 = 1.8425$ and $d_2^* = (1.5434 + 4.0)/1.3333 = 4.1577$, respectively.

Figure 5.3 shows the logistic function with these parameter values and the optimal design points d_1^* and d_2^* associated with the optimal probabilities 0.176 and 0.824.

When the total sample size N is fixed for this problem and the initial values for the logistic model parameters are $\beta_0 = -4$ and $\beta_1 = 1.3333$, the D-optimal design for estimating the two parameters is to divide students into two equally sized samples, one having a little less than 2 hours and the other having a bit more than 4 hours of practice. Neither inclusion of students in this study having less than 1 hour or more than 5 hours of practice nor inclusion of students having about 3 hours of practice will improve the efficiency of the parameter estimators very much. If one is only interested in efficient estimation of the single parameters β_0 or β_1, different designs will be required.

Figure 5.4 shows the variances of the two estimated parameters given in the main diagonal of the variance–covariance matrix $Cov(\hat{\beta})$ in Equation (5.22).

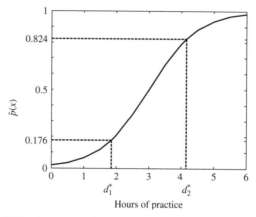

Figure 5.3 Probability function of mastering the task for the logistic model with $\beta_0 = -4$ *and* $\beta_1 = 1.3333$.

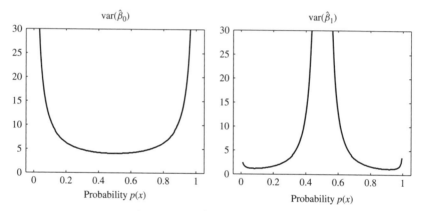

Figure 5.4 Variances $var(\hat{\beta}_0)$ *and* $var(\hat{\beta}_1)$ *as functions of the response probability function* $p(x)$.

These plots show that the location parameter β_0 is most efficiently estimated when $p(x) = 0.5$. This is in contrast to the plot for the variance of the slope parameter estimator, which becomes extremely large as $p(x)$ reaches 0.5. The smallest variance for the estimator of β_1 is attained when $p(x_1) = 0.09$ and $p(x_2) = 0.91$.

In summary, as the probability $p(x) \to 0$ or $p(x) \to 1$, both parameters are estimated with very large variances. A contradiction arises when $p(x) \to 0.5$. Here, the location parameter β_0 is estimated with minimum variance while the slope β_1 cannot be reliably estimated. These plots in Figure 5.4, however, do confirm the conclusion that was drawn from the D-optimality criterion plot in Figure 5.2, namely that the variances of both estimators $\hat{\beta}_0$ and $\hat{\beta}_1$ remain relatively small in the probability intervals $0.1 < p(x) < 0.3$ and $0.7 < p(x) < 0.9$, respectively.

5.3.3.1   Bounded and unbounded design regions

In the cases considered up to now, the design regions have been assumed to be bounded, that is $d_{\min} \leq d_j \leq d_{\max}$. Likewise, the values of the independent variable have been assumed to lie between an upper- and a lower bound $x_{\min} \leq x \leq x_{\max}$. In the 'hours of practice' example, the design region is bounded between 0 and 6 hours. However, the D-optimal design characterized in Equation (5.24) actually holds when the design region is unbounded and the design points lie between $-\infty \leq d_j \leq \infty$. The D-optimal design points d_1^* and d_2^* associated with the probabilities $p(d_1^*) = 0.176$ and $p(d_2^*) = 0.824$ are shown in Figure 5.5 for an unbounded design region.

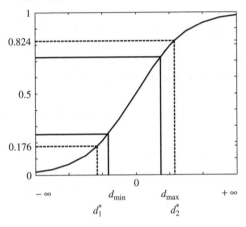

Figure 5.5 Response probability function $p(d_1^*) < p(d_{\min}) < p(d_{\max}) < p(d_2^*)$.

In practice, researchers may want to study a design region that is more restricted. In pharmaceutical research, for example, such a restriction of the design region is often necessary to avoid high doses to prevent unwanted effects. Such a restricted region may also be needed to avoid uniform responses from very low or very high doses. In these cases, not much information about the shape of the response curve is going to be available. We now consider two interesting conditions for the design region and we call them *Case I* and *Case II*.

Case I: Design region $\{d_{\min} \leq d_1^* < d_2^* \leq d_{\max}\}$
When the lower bound $d_{\min} \to -\infty$ and the upper bound $d_{\max} \to +\infty$, the region is said to be unbounded. Silvey (1980, p. 59) and Sebastiani and Settimi (1997), among others, described the D-optimal design points for this case. They can be derived from $d^* = (\pm 1.5434 - \beta_0)/\beta_1$, and are associated with the probabilities $p(d_1^*) = 0.176$ and $p(d_2^*) = 0.824$, respectively.

Case II: Design region $\{d_1^* < d_{\min} < d_{\max} < d_2^*\}$
For this bounded region, it has been shown (Silvey, 1980; Sebastiani and Settimi, 1997) that the D-optimal design points will become equal to $d_{\min}$ and $d_{\max}$. In

Figure 5.5, this case is plotted. It can be easily seen that this case resembles that of the D-optimal design case for the linear regression model. The response curve for the interval $[d_{min}, d_{max}]$ approximately becomes linear. If we would expand the width of the design region $[d_{min}, d_{max}]$, and thus increase the variance of X, we would obtain more efficient parameter estimators. As the width of the design region increases and becomes larger than the interval $[d_1^*, d_2^*]$, the optimal design points will not be at the boundaries of the design region anymore and *Case I* will then apply.

It is interesting to note that the magnitude of the slope parameter β_1 determines where the optimal design points are. A moderate slope results in a more linear shape of the response function and leads to optimal design points closer to the boundaries of the design region, whereas a steeper slope requires that the optimal design points be more in the interior of the design region. For more insights, we refer the reader to Silvey (1980) and Sebastiani and Settimi (1997), who also considered cases where the design region is only bounded on one side.

5.3.4   Design for two independent quantitative variables

When a bank considers whether to provide a loan or not to a business firm, it is important that the bank carefully estimates the risk of bankruptcy. A logistic regression analysis that describes the relation between the probability of bankruptcy of 66 firms and their retained earnings is presented by Chatterjee and Price (1991, p. 147–148). The dependent variable Y is coded as $y = 0$ if a firm went bankrupt after two years and $y = 1$ if a firm remained solvent in that period. The probability of remaining solvent is

$$p(z) = \frac{\exp(z)}{1 + \exp(z)}, \quad \text{where } z = \beta_0 + \beta_1 x_1 + \beta_2 x_2. \tag{5.24}$$

The variable X_1 represents the retained earnings divided by the total assets of the firm and X_2 represents the earnings before interest and taxes divided by the total assets of the firm. The estimated parameters are $\hat{\beta}_0 = -0.550, \hat{\beta}_1 = 0.157$ and $\hat{\beta}_2 = 0.194$. The OR for X_1 given X_2 is

$$\text{OR} = \frac{\text{odds}(y = 1; x_1 = 2, x_2)}{\text{odds}(y = 1; x_1 = 1, x_2)} = \exp(\beta_1), \tag{5.25}$$

and this ratio can be interpreted as the change of the odds for the firms becoming bankrupt when their retained earnings (X_1) increase 1 point on the X_1 scale. The estimated OR for the association between retained earnings (X_1) and remaining solvent is 1.170. The predicted probabilities of the 66 firms for remaining solvent are shown in Figure 5.6 as a function of the estimated linear predictor $\hat{z} = -0.550 + 0.157x_1 + 0.194x_2$. It can be seen from this figure that the probabilities of remaining solvent become extremely small when $\hat{z} < -6$ and these probabilities are close to 1 when $\hat{z} \rangle 6$.

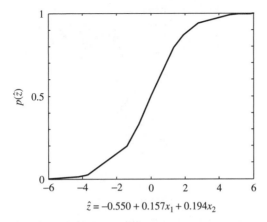

$$\hat{z} = -0.550 + 0.157x_1 + 0.194x_2$$

Figure 5.6 Predicted probabilities of remaining solvent for the bankruptcy example.

The design question is now how can the probability of remaining solvent be estimated as efficiently as possible? Or, which combination of values $[x_1, x_2]$ will provide the most information for the simultaneous estimation of the parameters β_0, β_1 and β_2? Of course, strictly speaking, this is not an experimental study and we cannot manipulate firms beforehand to produce certain combinations of values $[x_1, x_2]$ that provide the most information for modelling bankruptcy. However, banks want to estimate the model parameters as efficiently as possible using an efficient design. In particular, including firms in the sample that do not provide much information is a waste of resources. A proper choice of the design space is therefore in order.

The first step is to find proper bounds for the design region of the two independent variables. Since very little information is available when the probability of remaining solvent is very low or very high, it makes sense to restrict the probabilities of remaining solvent to $0.1 < p(x) < 0.9$. This means that we have $-2.2 < z < +2.2$.

Figure 5.7 is a two-dimensional plot of the estimated probability function for variables X_1 and X_2 with values ranging from -40 to $+40$. Clearly, the probability of remaining solvent decreases when the firm has negative retained earnings and negative earnings before taxes and interest; in particular, this probability is less than 0.1 when both the predictor variables X_1 and X_2 have large negative values. In contrast, this probability exceeds 0.9 when both variables have very large positive values. These combinations of x values are thus not very informative. In terms of the linear predictor, the same conclusions apply when $z < -2.2$ and $z > 2.2$ (Figure 5.6).

Figure 5.8 shows a two-dimensional design space Δ_2 for the bankruptcy example. The upper and lower bounds of the restricted design space are drawn for $z = 2.2$ and $z = -2.2$. The region between these two parallel lines is the region where the probability of remaining solvent lies between 0.1 and 0.9. To reflect

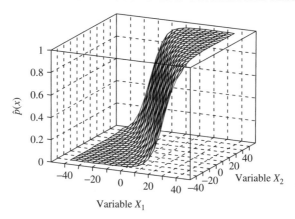

Figure 5.7 Estimated probability of remaining solvent as a function of X_1 and X_2 for the bankruptcy data.

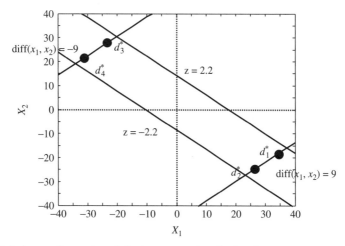

Figure 5.8 A two-dimensional design space for the bankruptcy example with four optimal design points.

reality, we impose a second constraint that banks frequently use. Ordinarily, banks prefer firms not to have too much difference in retained earnings (X_1) and earnings before taxes and interest (X_2), and we assume that the maximum difference between x_1 and x_2 values is about 50. A convenient difference function between x_1 and x_2 is diff $(x_1, x_2) = 0.157x_1 - 0.194x_2 + 0.058$. It can be shown that if we set diff $(x_1, x_2) = \pm 9$, the difference between the two variables is approximately $(x_1 - x_2) \approx \pm 50$.

Sitter and Torsney (1995) used a canonical form to express the design problem for two independent variables and numerically found that there are four distinct optimal design points, namely:

$$\xi^* = \left\{ \begin{array}{cccc} d_1^* & d_2^* & d_3^* & d_4^* \\ 0.25 & 0.25 & 0.25 & 0.25 \end{array} \right\}. \tag{5.26}$$

These D-optimal design points all have an equal weight of 0.25 and are graphically displayed in Figure 5.8. They are the four intersection points of the two lines diff$(x_1, x_2) = \pm 9$ with the two lines for $z = \pm 1.22$, respectively. See Sitter and Torsney (1995), for more details. So, the optimal design point d_1^* is the intersection of diff$(x_1, x_2) = +9$ and $z = 1.22$, while the optimal design point d_4^* is the intersection between the lines diff$(x_1, x_2) = -9$ and $z = -1.22$. For the bankruptcy example, the D-optimal design points are $d_1^* = [34.11; -18.48]$, $d_2^* = [26.34; -24.77]$, $d_3^* = [-23.21; 27.91]$ and $d_4^* = [-30.98; 21.62]$. The response probabilities at design points d_1^* and d_3^* are 0.7721 and the response probabilities at design points d_2^* and d_4^* are 0.2279. This D-optimal design depends on the user-selected restrictions imposed on the design space and different optimal designs will emerge with different restrictions on the design space.

5.3.4.1   Number of support points

The procedure of Sitter and Torsney (1995) resulted in a D-optimal design with four equally weighted design (support) points for the logistic model with only three parameters, namely β_0, β_1 and β_2. It is known that for a model with p parameters, we can find a non-singular D-optimal design among designs with at least p points and at most $p(p + 1)/2$ points. This is a consequence of Charathéodory's Theorem described in Silvey (1980, p.72). In our case, we have a model with three parameters and this means that we can restrict our search for the D-optimal design among designs with at least three and at most six support points. Illustrative examples and details are available in Sebastiani and Settimi (1998). We end this section with a note that D-optimal designs can have unequal weights, and whether this happens or not depends on the restrictions on the boundaries of the design space and the values of the model parameters.

5.4   Approaches to deal with local optimality

The variances of the parameter estimators in the logistic model are functions of the values of the parameters that we are trying to estimate. The practical implication is that if we want to design a study to estimate the model parameters as efficiently as possible, one must actually know the parameter values beforehand. Here, and elsewhere, we remind the reader that efficient estimators are estimators with minimal variance. This problem is comparable to the problem faced by the legendary Baron von Münchhausen when he got stuck in a swamp. How can you design a study to optimally estimate the parameters of your model without knowing the actual values of the model parameters? Baron von Münchhausen solved his problem by pulling himself up from the swamp by his own hair. Surely we cannot do this here, but fortunately, there are several approaches to circumvent the parameter dependency problem. Here are some strategies for overcoming the dependency problem:

- If a researcher has initial values or 'best guesses' for the model parameters, the optimal design can be constructed based on these initial values. Such optimal designs are referred to as *locally optimal designs*.

- If a researcher is able to update estimates of the model parameters in successive stages, the resulting design is called a *sequential optimal design*. Starting with initial values in the first stage, the locally optimal designs are modified using estimated parameters from data obtained in that stage. At each stage, the current optimal design is updated to incorporate the new information from the latest data and improved designs are constructed from the latest estimates for the model parameters. This process continues until convergence is reached. Such sequential two- and multi-stage design procedures have been suggested by Wu (1985) and Sitter and Wu (1999), among others.

- A Bayesian approach assumes a prior distribution for the parameters of interest and uses it to construct the optimal design by averaging over the prior information. Chaloner (1984) and Chaloner and Verdinelli (1995) reviewed the Bayesian approach and design issues for linear and nonlinear models and constructed several types of Bayesian optimal designs. Additional illustrative examples of the Bayesian approach are given in Chapters 9 and 10.

- Another approach to overcome the local optimality problem is the maximin approach. This approach can also be inversely formatted as a minimax approach. Both methods generally require a specification of a certain range of the unknown parameters. A more detailed explanation of the maximin method is given in Section 5.3.1.1 and Chapters 9 and 10.

5.5   Designs for calibration of item parameters in item response theory models

In educational testing, large-scale administrations of tests are conducted every year. For example, the scholastic assessment tests (SAT) in the United States are administered to a couple of million students annually. Such large-scale administrations require a lot of test items to be calibrated. Calibration in this context means that the item parameters, such as item discrimination and item difficulty, have to be estimated. Computerized adaptive testing (CAT) has become an alternative to such large-scale testing programmes, but CAT administration also requires a large item bank of calibrated items. As such, an enormous amount of resources is required to calibrate items on such a large scale. Because CAT is not only used in education but is also used in the military and companies such as

Microsoft and Oracle, it is important to have efficient sampling designs to calibrate the items and estimate the item parameters accurately at minimal cost. An overview of these issues in computer-based testing is given in Wainer (2000) and Van der Linden and Glas (2000). See also Van der Linden (2005) who described optimal test design models with constraints for efficient estimation of students' abilities.

Among all item response theory (IRT) models used in practice, the two-parameter logistic model is probably the most often applied. Its most common form is

$$p_i(\theta_j) = \frac{\exp[a_i(\theta_j - b_i)]}{1 + \exp[a_i(\theta_j - b_i)]},\tag{5.27}$$

where θ_j is a random ability parameter for student j and a_i and b_i are the item discrimination and difficulty parameter, respectively. In order to calibrate the item parameters a_i and b_i, we first calculate Fisher's information given by

$$M = \begin{bmatrix} M_{11} & M_{12} \\ M_{12} & M_{22} \end{bmatrix}$$

$$= \begin{bmatrix} \sum_j w_j(\theta_j - b_i)^2 p(\theta_j)q(\theta_j) & -\sum_j w_j a_i(\theta_j - b_i)p(\theta_j)q(\theta_j) \\ -\sum_j w_j a_i(\theta_j - b_i)p(\theta_j)q(\theta_j) & \sum_j w_j a_i^2 p(\theta_j)q(\theta_j) \end{bmatrix}.$$

$$\tag{5.28}$$

Here, M_{11} and M_{22} are Fisher information for the item parameters a_i and b_i, respectively, and M_{12} is the joint information. The probability $q(\theta_j) = [1 - p(\theta_j)]$ and the weights w_j's are connected to each distinct ability level θ_j, such that $\sum_j w_j = 1$. The inverse of the information matrix M^{-1} is proportional to the asymptotic variance–covariance matrix of the parameter estimators of a_i and b_i. This variance–covariance matrix is comparable to the form given in Equation (5.22) for the estimators of the regression parameters in Model (5.18). In Chapter 9, the inverse relation between Fisher information and the asymptotic variance–covariance matrix of the parameter estimators is further elaborated.

It was shown in Section 5.3.3 that the minimum of $\text{Det}(M^{-1})$ or the maximum for $\text{Det}(M)$ is attained at two distinct equally weighted values of abilities $\theta^* = b_i \pm 1.5434/a_i$. The corresponding probabilities of correctly answering the item are $p(\theta^*) = 0.824$ and $q(\theta^*) = 0.176$, respectively. This means that the D-optimal sampling design for estimating both the item parameters a_i and b_i is to have an equal number of students with probabilities of 0.824 and 0.176 of answering the item correctly. In practice, the abilities of students are unknown and only their estimates are available. In some cases, a teacher may be able to select only two groups of students, one for which the particular item would be too difficult and one for which the item would be too easy. Compared to a single sample of students, such a mixture of two samples with two modes is more likely to improve the efficiency of the parameter estimators.

Figure 5.9 shows a normally distributed sample of abilities and three different bimodal samples. For all four samples, the location parameter $b_i = \bar{\theta}$, which we assume is 0. The standard deviations s_θ of these samples are also more or less equal. We calculate RE_Ds, the efficiencies of these samples relative to the D-optimal design with two equally sized samples of students with abilities $\theta^* = b_i \pm 1.5434/a_i$. This D-optimal design has mean $\bar{\theta}$ and standard deviation $s_\theta = 1.5434/a_i$.

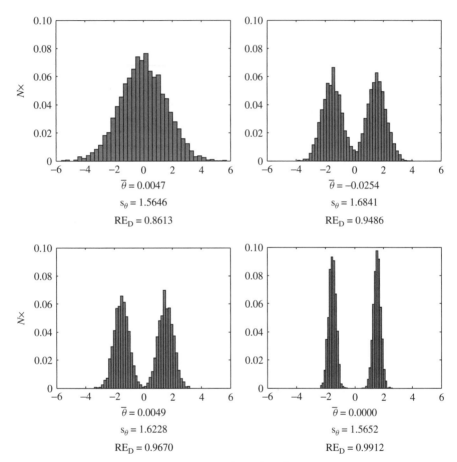

Figure 5.9 Unimodal and bimodal samples of students with same sample size N and approximately the same mean $\bar{\theta}$ and standard deviation s_θ.

The plots show that the (unimodal) normally distributed sample has the smallest $RE_D = 0.8613$, and that this normal sample would have to be increased by about $(0.86^{-1} - 1)\% = 16\%$ to come up with the same efficiency as the D-optimal sampling design. However, this efficiency loss is relatively small in part because D-optimality criterion includes information on both parameters a_i and b_i. The D-optimal design has a contradicting effect on each of the separate item parameter estimators. A normally distributed sample is usually not very efficient for the slope parameter a_i, while it is very efficient for the location parameter if $b_i = \theta$.

In practice, calibration of item parameters is infeasible for single items. Usually, a whole test is administered to a sample of students and the scores are used to calibrate all item parameters together. The optimal form of the sampling design depends on the form of the distributions of the set of location and the slope parameters for all items in the whole achievement test. Interestingly, it has been found that a more or less uniformly distributed sample of abilities generally contains more information about a set of the items parameters in a test than a normally distributed sample of abilities (Stocking, 1990; Berger, 1994). See also Lima Passos and Berger (2004) and Lima Passos, Berger and Tan (2007, 2008) who studied item calibration designs for the nominal response model for CAT administration.

5.6   Matrix formulation of designs for logistic regression

In this section, we use matrix notation to describe the statistical set-up for the logistic regression model with one or more independent variables. We use the 'hours of practice experiment' and the 'problem solving study' examples to demonstrate the matrix calculation involved and display their design matrices and the variance–covariance matrices of the estimated parameters. The reader may skip this section without too much interruption.

Suppose we decide to take a fixed number of N observations in the study and we have a logistic model with $p - 1$ independent variables $X_1, \ldots, X_{p-1}$. For the ith case, let y_i be the binary response and let the linear predictor be $z_i = \beta_0 + \sum_{l=1}^{p-1} \beta_l x_{li}$, where x_{li} is the ith value of the independent variable $X_l, l = 1, 2, \ldots, p - 1$. In matrix form, we write

$$z = X\beta$$

$$
\begin{bmatrix} z_1 \\ z_2 \\ \vdots \\ z_N \end{bmatrix}
=
\begin{bmatrix}
1 & x_{11} & \cdots & x_{p-1,1} \\
1 & x_{12} & \cdots & x_{p-1,2} \\
\vdots & \vdots & \vdots & \vdots \\
1 & x_{1N} & \cdots & x_{p-1,N}
\end{bmatrix}
\begin{bmatrix} \beta_0 \\ \beta_1 \\ \vdots \\ \beta_{p-1} \end{bmatrix}
\tag{5.29}
$$

where z is the $N \times 1$ vector of linear predictors, X is the $N \times p$ design matrix and $\beta = (\beta_0, \beta_1, \ldots, \beta_{p-1})'$ is the column vector of logistic regression parameters.

If L is the likelihood function, the log likelihood function of N dichotomous responses $y_1, y_2, \ldots, y_N$ can be written as

$$\ln(L) = \sum_{i=1}^{N} \{ y_i \ln p(z_i) + (1 - y_i) \ln[1 - p(z_i)] \}, \tag{5.30}$$

where $p(z_1), p(z_2), \ldots, p(z_N)$ are the probabilities of observing the responses $y_1, y_2, \ldots, y_N$ as successes.

Some of the rows of the design matrix X may be the same. This happens when there are replicates; that is, we take observations at the same condition or at the same combination levels of the independent variables. Without loss of generality, we assume that there are m conditions or m distinct combination levels of the independent variables and there are n_j observations (replications) at each of these conditions. This implies that $n_j = N w_j$ where w_j's are the weights at the m conditions, such that all $w_j > 0$, $\sum_j^m w_j = 1$ and $\sum_j^m n_j = N$. Assuming replicates are grouped in conditions, the log likelihood function for the N dichotomous responses $y_1, y_2, \ldots, y_N$ in Equation (5.30) can be formulated as

$$\ln(L) = N \sum_{j=1}^{m} \left\{ w_j y_j \ln p(z_j) + w_j (1 - y_j) \ln[1 - p(z_j)] \right\}. \tag{5.31}$$

The maximum likelihood estimators for $\beta = (\beta_0, \beta_1, \ldots, \beta_{p-1})'$ are usually obtained by maximizing $\ln(L)$ with a Gauss–Newton type of algorithm (Lange, 1999, p. 135). The estimators $\hat{\beta}$ have an asymptotic variance–covariance matrix given by

$$\mathrm{Cov}(\hat{\beta}) = (X'\hat{W}X)^{-1}, \tag{5.32}$$

where the design matrix X now is of order $m \times p$ and $\hat{W}$ is an $m \times m$ diagonal matrix given by

$$\hat{W} = \mathrm{Diag}\left\{ w_1 \hat{p}(z_1)[1 - \hat{p}(z_1)], \ldots, w_m \hat{p}(z_m)[1 - \hat{p}(z_m)] \right\}. \tag{5.33}$$

Note that the $p(z_1), p(z_2), \ldots, p(z_m)$ are the response probabilities for the m distinct conditions with corresponding weights $w_1, w_2, \ldots, w_m$ indicating the replications at each of these conditions. The asymptotic variance–covariance matrix $\mathrm{Cov}(\hat{\beta})$ is inversely related to the information matrix:

$$M = X'\hat{W}X, \tag{5.34}$$

and so minimizing the determinant $\mathrm{Det}[\mathrm{Cov}(\hat{\beta})]$ is equivalent to maximizing $\mathrm{Det}[M]$.

5.6.1 Hours of practice experiment

In Figure 5.3, the estimated probability of mastering a difficult task is plotted as a function of hours of practice for the logistic regression parameters $\beta_0 = -4$ and $\beta_1 = 1.3333$. Students are divided into seven groups each having a different number of hours practice, that is $d_j = 0, 1, 2, 3, 4, 5$ and 6 hours. The design matrix with corresponding diagonal matrix of response variances $\hat{W}$ is

$$X = \begin{bmatrix} 1 & 0 \\ 1 & 1 \\ 1 & 2 \\ 1 & 3 \\ 1 & 4 \\ 1 & 5 \\ 1 & 6 \end{bmatrix} \text{ and}$$

$$\hat{W} = \begin{bmatrix} 0.0177w_1 & 0 & 0 & 0 & 0 & 0 & 0 \\ 0 & 0.0607w_2 & 0 & 0 & 0 & 0 & 0 \\ 0 & 0 & 0.1651w_3 & 0 & 0 & 0 & 0 \\ 0 & 0 & 0 & 0.2500w_4 & 0 & 0 & 0 \\ 0 & 0 & 0 & 0 & 0.1651w_5 & 0 & 0 \\ 0 & 0 & 0 & 0 & 0 & 0.0607w_6 & 0 \\ 0 & 0 & 0 & 0 & 0 & 0 & 0.0177w_7 \end{bmatrix},$$

$$(5.35)$$

where the w_j's are weights for each of the practice groups, such that $\sum_j w_j = 1$. If these weights are all equal to w, the variance–covariance matrix of parameter estimators is

$$\text{Cov}(\hat{\beta}) = (X'\hat{W}X)^{-1} = \frac{1}{w}\begin{bmatrix} 9.2906 & -2.6446 \\ -2.6446 & 0.8815 \end{bmatrix}, \qquad (5.36)$$

and the D-optimality criterion is $\text{Det}[\text{Cov}(\hat{\beta})] = 1.1959/w^2$. Note that an equal number of students is assigned to each of the practice groups and that all weights are equal to $w = 1/7$.

To see how much efficiency can be gained by using the D-optimal design for this study with D-optimal design points $d_1^* = 1.8425$ and $d_2^* = 4.1577$ and corresponding weights $w^* = w_1 = w_2 = 0.5$, we must set up the corresponding design matrix with corresponding diagonal matrix with estimated response variances:

$$X = \begin{bmatrix} 1 & 1.8425 \\ 1 & 4.1577 \end{bmatrix}, \hat{W} = \begin{bmatrix} 0.1451w^* & 0 \\ 0 & 0.1451w^* \end{bmatrix}. \qquad (5.37)$$

The variance–covariance matrix of the parameters estimators now becomes

$$\text{Cov}(\hat{\beta}^*) = (X'\hat{W}X)^{-1} = \frac{1}{w^*}\begin{bmatrix} 26.5920 & -7.7153 \\ -7.7153 & 2.5718 \end{bmatrix} \qquad (5.38)$$

with determinant $\text{Det}[\text{Cov}(\hat{\beta}^*)] = 8.8635/w^{*2}$.

We can now evaluate the efficiency of the design that evenly divides students into seven practice groups with 0, 1, 2, 3, 4, 5 and 6 hours of practice. We do so by computing its efficiency relative to the D-optimal design with an equal number of students with $d_1^* = 1.8425$ and $d_2^* = 4.1577$ hours of practice. This RE is

$$\mathrm{RE_D} = \left\{ \frac{\mathrm{Det}[\mathrm{Cov}(\hat{\beta}^*)]}{\mathrm{Det}[\mathrm{Cov}(\hat{\beta})]} \right\}^{1/2} = \left\{ \frac{8.8635 \times 4}{1.1959 \times 49} \right\}^{1/2} = 0.7778. \qquad (5.39)$$

This means that compared to the D-optimal design, the original design with seven distinct practice groups would require $(\mathrm{RE_D^{-1}} - 1)100\% = 28\%$ more students to have the same efficiency as the D-optimal design has.

5.6.2 Problem solving study

In the problem solving study, students could be assigned to any one of the $N = 2^3 = 8$ conditions. Table 5.4 shows the design for this study, which has eight design points d_j with corresponding weights n_j/N, $j = 1, \ldots, 8$. The corresponding design matrix X now has dimension 8×4 and is shown below. Each of the independent variables, instruction type, motivation level and practical exercise level is coded as 1 and -1, as discussed in Section 5.3.2.

The first column of X consists of 1's only for the intercept and the second, third and fourth columns correspond to levels of the three variables, respectively:

$$X = \begin{bmatrix} 1 & 1 & 1 & 1 \\ 1 & -1 & 1 & 1 \\ 1 & 1 & -1 & 1 \\ 1 & -1 & -1 & 1 \\ 1 & 1 & 1 & -1 \\ 1 & -1 & 1 & -1 \\ 1 & 1 & -1 & -1 \\ 1 & -1 & -1 & -1 \end{bmatrix}. \qquad (5.40)$$

Table 5.7 shows the matrices $\hat{W}$ for the D-optimal design, the completely balanced and the partially balanced design. For completeness, the weights for the three designs are also provided, along with their corresponding determinants $\mathrm{Det}[\mathrm{Cov}(\hat{\beta})]$. The table also includes the D efficiencies of the other designs relative to the D-optimal design ($\mathrm{RE_D}$s). As expected, the D-optimal design obviously has the smallest determinant criterion value. It is interesting to note that the $\mathrm{RE_D}$s of the completely balanced and the partially balanced designs are very high. The completely balanced design with $N = 2^3 = 8$ cells is D optimal if the logistic model contains $p = 8$ parameters. Table 5.7 also shows that the completely balanced design is also an efficient alternative for a model with only $p = 4$ parameters. Moreover, an even more efficient design is the partially balanced design, where the cell corresponding to the smallest response variance has no observations and all the other cells have equal numbers of observations.

Table 5.7   The matrices $\hat{W}$ and weights w for three designs of the problem solving study.

D-optimal design

$\hat{W} = \text{Diag} [0.0000\ 0.0186\ 0.0186\ 0.0358\ 0.0186\ 0.0358\ 0.0358\ 0.042]$
$w^* = [0.0000\ 0.1216\ 0.1216\ 0.1533\ 0.1216\ 0.1533\ 0.1533\ 0.1751]'$

$\text{Det}[\text{Cov}(\hat{\beta})] = 797.6874,\quad \text{RE}_D = 1$

Completely balanced design:

$\hat{W} = \text{Diag} [0.0097\ 0.0191\ 0.0191\ 0.0291\ 0.0191\ 0.0291\ 0.0291\ 0.0301]$
$w^{cb} = [0.1250\ 0.1250\ 0.1250\ 0.1250\ 0.1250\ 0.1250\ 0.1250\ 0.1250]'$

$\text{Det}[\text{Cov}(\hat{\beta})] = 952.8497,\quad \text{RE}_D = 0.9565$

Partially balanced design:

$\hat{W} = \text{Diag} [0.0000\ 0.0219\ 0.0219\ 0.0332\ 0.0219\ 0.0332\ 0.0332\ 0.0345]$
$w^{pb} = [0.0000\ 0.1429\ 0.1429\ 0.1429\ 0.1429\ 0.1429\ 0.1429\ 0.1429]'$

$\text{Det}[\text{Cov}(\hat{\beta})] = 810.5884,\quad \text{RE}_D = 0.9960$

5.7   Summary

In this chapter, we have described design problems for logistic regression models. For simplicity, we focused on D-optimal designs estimating all or specific parameters in the model; other types of optimal designs for the logistic model are discussed in Chapter 10. Optimal designs for logistic models depend on the specific values of the regression parameters. This means that the regression parameters have to be known in advance, before an optimal design can be implemented. For a logistic model with a single dichotomous independent variable, the weights of the optimal design are a function of the proportions in each category. In particular, the locally D-optimal design has equal weights (i.e. equal number of observations) assigned to the conditions of the independent variable only if the proportions are equal. When the logistic model has several qualitative independent variables, the D-optimal design is usually unbalanced, but completely balanced and partially balanced designs tend to have high efficiency. Optimal designs for a logistic model with a single quantitative independent variable can be found in terms of the response probabilities. We also discussed optimal designs for two independent quantitative variables using an example and we explained the difference between bounded and unbounded design regions.

6

Designs for multilevel models

6.1   Design problem for multilevel models

Multilevel studies have a design where the factors (effects) are hierarchically related and are assumed to be either fixed or random. A classical example of such a hierarchically structured design is the split-plot design. In contrast to studies that assume the outcomes of different subjects to be independent, a multilevel study assumes that the outcomes of subjects depend on the particular groups that contain these subjects. In cases where subjects are nested within groups or clusters, it is plausible to assume that subjects within a group are more alike. For example, in a study that is set up to evaluate the effect of an instruction programme to improve the arithmetic performances of students within schools, the outcomes of the students within a school may be influenced by the school environment and the policy of the school towards arithmetic education. Moreover, their relation with one another may lead to correlated outcomes. In a study on the development of pressure ulcers of patients within nursing homes, the risk of developing pressure ulcers was found to depend on the differences in the rules and regulations that these nursing homes maintain to prevent the development of pressure ulcers. In the first example, the students are said to be nested within schools and in the pressure ulcer example, the patients are nested within nursing homes. In this chapter, we treat the words *cluster* and *group* as synonyms and the word *subjects* as a synonym for units within the groups. For simplicity, we also restrict our discussion to *two levels of the multilevel structure*.

The statistical analysis of nested or hierarchically structured data is known as multilevel *regression analysis*. See Goldstein (1995), Snijders and Bosker (1999), Hox (2002) and Raudenbush and Bryk (2002), among others, for details.

An Introduction to Optimal Designs for Social and Biomedical Research   M. P. F. Berger & W. K. Wong
© 2009, John Wiley & Sons, Ltd

Multilevel regression analysis assumes that both the groups in the study and the subjects within groups are random samples from a certain population. The more dependent the outcomes within a group or cluster are, the more the results of a multilevel analysis will differ from the traditional regression analysis that assumes independent outcomes.

Multilevel studies usually require large samples of subjects and groups and are therefore often expensive. The sampling of the data has to be done at different levels, that is, at the group level and at the subject level. This means that the design of a multilevel study is characterized by the levels of the independent variables and by the sampling of units at the different levels of the study. For the arithmetic instruction study, not only the students within schools but also the schools themselves are sampled. Likewise, in the pressure ulcer study, the nursing homes as well as the patients within the nursing homes are sampled. This makes the design of multilevel studies more complicated, especially if the costs of sampling subjects and the costs of sampling groups or clusters are taken into account.

6.1.1   The design

Randomization of a treatment can, in principle, be done at any level of the multilevel design. If a design consists of patients within nursing homes, then a treatment can be assigned in two different ways.

First, the treatment can be randomly assigned to the patients within each nursing home. Assuming that we want a balanced design, about half of the patients in each home will receive the treatment and the rest will not receive the treatment. Such a randomization at the patient level is referred to as *subject randomization* and the design is often called a *multi-center trial*. The second way is that the treatment is randomly assigned to the nursing homes as a whole, where all patients within a home receive the same treatment. In the latter case, all patients in about half of the nursing homes are treated, whereas all patients in the rest of the nursing homes are not treated. Such a design is usually called a *cluster randomized trial*, because the treatment is randomized at the cluster or group level.

A schematic representation of these two designs is given in Figure 6.1, where we assume that we have four groups of nursing homes. The shaded areas indicate the data obtained under the treatment condition and the blank areas indicate the data obtained under the control condition. The two designs in Figure 6.1 distinguish the treatment and the control groups by having two distinct design points $d_1(k)$ and $d_2(k)$ for $k = 1, \ldots, 4$ groups. The two values of $d_1(k)$ and $d_2(k)$ are coded as $x_{ik} = 1$ for treatment and $x_{ik} = -1$ for control. Here, x_{ik} is the value of the independent variable in the model. We assume that the total sample of subjects for the study is N and we plan to sample n_k subjects from the cluster k, $k = 1, 2, \ldots, K$. This means $N = \sum_k^K n_k$. The two types of trials of interest in this chapter can now be described succinctly as follows.

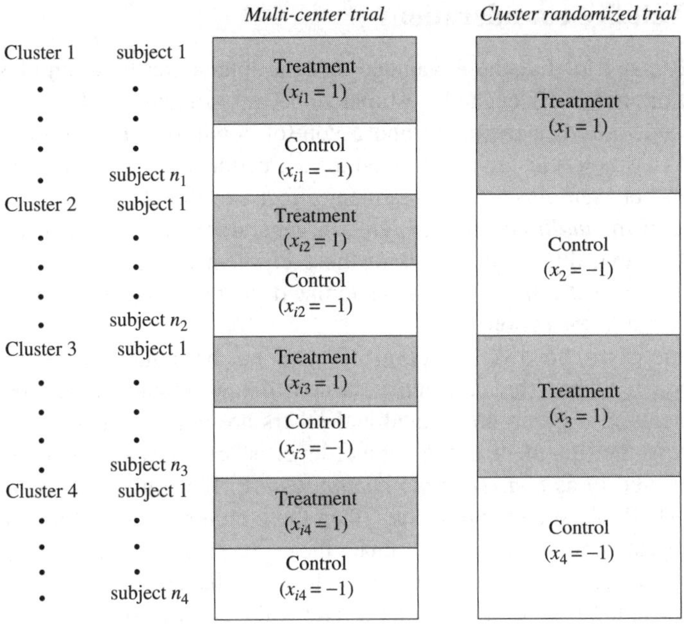

Figure 6.1 Schematic representation of subject and cluster level randomization of treatment.

The exact design for the *multi-center trial* in Figure 6.1 can be denoted by

$$\xi_N = \left\{ \begin{array}{cccccccc} d_1(1) & d_2(1) & d_1(2) & d_2(2) & d_1(3) & d_2(3) & d_1(4) & d_2(4) \\ \dfrac{n_1}{2N} & \dfrac{n_1}{2N} & \dfrac{n_2}{2N} & \dfrac{n_2}{2N} & \dfrac{n_3}{2N} & \dfrac{n_3}{2N} & \dfrac{n_4}{2N} & \dfrac{n_4}{2N} \end{array} \right\}. \quad (6.1)$$

The two design points $d_1(k)$ and $d_2(k)$ in cluster k have $n_k/2$ subjects in each group and the corresponding weight is $n_k/2N$.

On the other hand, the exact design for the *cluster randomized trial* can be denoted by

$$\xi_N = \left\{ \begin{array}{cccccccc} d_1(1) & d_2(1) & d_1(2) & d_2(2) & d_1(3) & d_2(3) & d_1(4) & d_2(4) \\ \dfrac{n_1}{N} & 0 & 0 & \dfrac{n_2}{N} & \dfrac{n_3}{N} & 0 & 0 & \dfrac{n_4}{N} \end{array} \right\}. \quad (6.2)$$

Cluster randomization of the treatment is visualized by the fact that only one design point in cluster k has weight n_k/N, while the other design point has zero weight. That is, all n_k subjects in cluster k are connected to only one of the two design points $d_1(k)$ or $d_2(k)$.

In the next section, we describe the potential advantages and disadvantages of multi-center and cluster randomized designs.

6.1.2   Validity considerations

A *multi-center trial* has the advantage that the interaction between clusters and treatment or control effect can be estimated, because subjects within every cluster are assigned to both a treatment and a control condition. In a *cluster randomized trial*, a whole cluster receives either a treatment or no treatment, and no interaction between cluster and treatment effect can thus be estimated. Another advantage of the *multi-center trial*, and, in general, of randomization at the subject level, is that this design leads to more efficient estimates of the treatment effect than when the treatment is randomized at the group level. This will be discussed in a later section.

In some cases, however, randomization cannot be done at the subject level. One reason may be ethical, when it is considered ethically unacceptable that some patients in a group are treated and others are not. Another reason for not randomizing the treatment at the subject level is *treatment contamination*. This is also referred to as *control group contamination*. If we give a new instruction method to half of the pupils in a class, then these children may inform each other about the treatment and thus contaminate the corresponding effect. In those cases where the subjects within a group are able to exchange information about the treatment, contamination may be large and influence the size of the treatment effect estimate. A procedure to estimate contamination is given by Moerbeek (2005b), who also discusses cost considerations.

Sometimes, blinding procedures may prevent contamination of a treatment effect, but in those cases where this is not possible and serious contamination of the treatment is expected, group or cluster randomization may be preferred. Group randomization may also be preferred in those cases where it is logistically more efficient to give a treatment to whole groups. One can imagine that it may be less expensive to randomly select half of the schools/teachers to implement a new instruction method, than to have all schools/teachers implement the new method. Finally, it may sometimes not be very interesting for schools and nursing homes to participate in a study where they are assigned as a whole to a control condition and they do not expect to profit from a new treatment. In that case, they may decline participation in the study. It should be noted, however, that contamination of a treatment effect may also play a role in a cluster randomized design. This especially occurs when the groups (wards, schools) are located physically close to each other and the subjects within these groups are able to exchange information about the research project. Thus, cluster randomized designs may also suffer from potential contamination effects.

When choosing a design with cluster or subject randomization, a researcher has to weigh the advantages against potential disadvantages. If a cluster randomized design is chosen, the researcher must be aware of efficiency loss that the cluster randomized design will have. The researcher may then possibly increase the sample size to reach the desired power for finding real differences. In addition, potential threats towards internal validity, such as selection bias, should also be taken into account.

6.1.2.1   Potential internal validity threats for cluster randomized trials

Both types of designs are based on identifiable groups, where members share social, physical or geographical characteristics. This implies that both designs may have potential sources of bias that threaten the validity of the study. This is especially true for cluster randomized trials. For example, when the number of groups is small, randomly assigning a treatment to the groups often cannot ensure that potential sources of bias are evenly distributed across the groups. Possible sources of threats towards internal validity have already been briefly summarized in Chapter 1. Among them are selection, (differential) maturation, (differential) history and (differential) regression towards the mean and contamination.

Selection bias may arise in cluster randomized trials when a limited number of clusters is randomly assigned to a treatment. A difference in clusters may then be misinterpreted as a difference due to treatment. Intact groups or clusters may have a different historical background that may cause bias in the results. The same accounts for a differential regression towards the mean when one (treated) group is more extreme in response than the other (non-treated) group. A detailed description of these threats for cluster randomization is given by Murray (1998, Chapter 2). Cook and Campbell (1979) and Kish (1987) provide a general discussion on validity threats with possible solutions, and a recent review of design issues for cluster randomized trials is given by Murray, Varnell and Blitstein (2004).

6.2   The multilevel regression model

Suppose that a treatment is evaluated by means of a randomized trial with K clusters or groups ($k = 1, \ldots, K$) and let n_k people be nested within cluster or group k. The total sample size is then $N = \sum_k^K n_k$. The treatment can be either randomly assigned to the clusters or to the subjects within a cluster. Thus, as discussed before, two designs can be distinguished, namely, a *cluster randomized trial* and a *subject randomized design* or a *multi-center trial*.

6.2.1   Cluster randomization of treatment

In a cluster randomized trial, whole clusters or groups are randomly assigned to a treatment and a control condition. For example, a cluster randomized trial – to evaluate the effect of an instruction program to improve the arithmetic performances of students in schools – would randomly assign whole schools to the instruction program so that all the students in a treatment school would receive the same instruction and all students in a control school would not receive the instruction. A simple regression model can be applied to model the relation between the response y_{ik} of subject i in cluster k and the treatment variable x_k:

$$y_{ik} = a_{0k} + \beta_1 x_k + \varepsilon_{ik} = (\beta_0 + b_{0k}) + \beta_1 x_k + \varepsilon_{ik}, \qquad (6.3)$$

where $a_{0k} = (\beta_0 + b_{0k})$ represents the random variation of the K clusters ($k = 1, \ldots, K$). The treatment variable is coded as $x_k = +1$ if cluster k receives the treatment and it is coded as $x_k = -1$ if it did not receive the treatment. The parameter β_0 is the overall intercept and the parameter β_1 is equal to one half of the treatment effect. Again, we assumed a balanced design with $K/2$ of the K clusters being given the treatment and the rest of the $K/2$ clusters are control clusters.

The random cluster effect parameter b_{0k} and the random errors ε_{ik} are assumed to be independently distributed with variances σ_0^2 and σ_ε^2, respectively. This means that the variance of the responses y_{ik} can be expressed as the sum of the two variances σ_0^2 and σ_ε^2:

$$\text{var}(y_{ik}) = \sigma_0^2 + \sigma_\varepsilon^2. \tag{6.4}$$

A more detailed explanation of the multilevel model is given in Snijders and Bosker (1999) and Hox (2002), among others.

6.2.1.1   Estimation of parameters in a cluster randomized design

The multilevel model in Equation (6.3) has a random effect b_{0k} and random errors ε_{ik} with variance component parameters σ_0^2 and σ_ε^2, respectively. In total, four parameters need to be estimated, namely, two fixed parameters β_0 and β_1 and two variance components σ_0^2 and σ_ε^2.

The maximum likelihood (ML) method is appropriate to use for this model. If the variance components are known, the ML estimators of β_0 and β_1 are the generalized least squares estimators and the estimators for the fixed parameters become $\hat{\beta}_0 = (\bar{y}_1 + \bar{y}_2)/2$ and $\hat{\beta}_1 = (\bar{y}_1 - \bar{y}_2)/2$, with $\bar{y}_1$ and $\bar{y}_2$ being the weighted means of the scores under the treatment and the control condition, respectively. If, on the other hand, the variance components are unknown, then their estimates can be obtained by restricted maximum likelihood (REML). See Searle, Casella and McCulloch (1992) and van Breukelen, Candel and Berger (2007, 2008), among others, for more details. In the following we restrict ourselves to the case that the variance components are known.

When $K/2$ of the clusters receive the treatment and the rest of the $K/2$ clusters do not receive the treatment, that is, $K/2$ control groups, the variance–covariance matrix of the fixed parameter estimators, with $\text{var}(\hat{\beta}_0)$ and $\text{var}(\hat{\beta}_1)$ on the main diagonal, is

$$\text{Cov}(\hat{\beta}) = \frac{1}{4} \begin{bmatrix} [\text{var}(\bar{y}_1) + \text{var}(\bar{y}_2)] & 0 \\ 0 & [\text{var}(\bar{y}_1) + \text{var}(\bar{y}_2)] \end{bmatrix}, \tag{6.5}$$

with $\text{var}(\bar{y}_1) = \left[\sum_{k=1}^{K/2} \frac{n_k}{n_k \sigma_0^2 + \sigma_\varepsilon^2} \right]^{-1}$ and $\text{var}(\bar{y}_2) = \left[\sum_{k=K/2+1}^{K} \frac{n_k}{n_k \sigma_0^2 + \sigma_\varepsilon^2} \right]^{-1}$, where for the treatment condition the index k runs from $k = 1, \ldots, K/2$ and for the control condition the index runs from $k = K/2 + 1, \ldots, K$. When the sample size in all clusters is equal, that is $n_k = n$, for all k, the expression for the variances of $\hat{\beta}_0$ and $\hat{\beta}_1$ in a cluster randomized design becomes $\mathrm{var}(\hat{\beta}_0) = \mathrm{var}(\hat{\beta}_1) = \frac{n\sigma_0^2 + \sigma_\varepsilon^2}{Kn}$.

As indicated in Section 6.1, the outcomes within the cluster of a multilevel data structure are dependent and a well-known measure for this dependency is the intra-class correlation. It is also referred to as *intra-cluster correlation*:

$$\rho = \frac{\sigma_0^2}{\sigma_0^2 + \sigma_\varepsilon^2}. \tag{6.6}$$

This intra-class correlation represents the proportion of total variance that is accounted for by the groups or clusters. The intra-class correlation forms the basis for the so-called *design effect* $= 1 + (n - 1)\rho$, where n is the common sample size of the clusters.

The variance of the treatment effect can now be rewritten in terms of the design effect as follows:

$$\mathrm{var}(\hat{\beta}_1) = \frac{n\sigma_0^2 + \sigma_\varepsilon^2}{Kn} = \frac{\sigma_0^2 + \sigma_\varepsilon^2}{Kn}[1 + (n - 1)\rho]. \tag{6.7}$$

It can be seen that, as the sample size n increases, the variance of the treatment effect $\mathrm{var}(\hat{\beta}_1)$ will decrease, while as the intra-class correlation ρ increases, the $\mathrm{var}(\hat{\beta}_1)$ will increase. This means that the more dependent the observations within a cluster or group are, the larger $\mathrm{var}(\hat{\beta}_1)$ will become and the less efficient the treatment parameter can be estimated. A similar line of reasoning is given in the next section for a multi-center trial.

6.2.2 Subject randomization of treatment

When the treatment is randomly assigned to the subjects within a cluster, the model relates the response variable y_{ik} of subject i in cluster k to the predictor variable x_{ik} by means of the simple regression equation

$$y_{ik} = a_{0k} + a_{1k}x_{ik} + \varepsilon_{ij}, \tag{6.8}$$

where ε_{ik} is normally distributed with mean zero and variance σ_ε^2. The treatment variable has two x_{ik} values, one for the treatment and the other for the control. We have assumed that these values are $+1$ and -1, respectively. The regression parameter a_{0k} is the mean of the responses y_{ik} within cluster k and with this particular coding of x_{ik}, the treatment effect parameter a_{1k} is half of the difference in outcome variable within cluster k. The regression parameters a_{0k} and a_{1k} vary across the clusters so that each cluster has its own (distinct) regression model. If β_0 and β_1 are the overall mean and overall (half of the) treatment effect, respectively, then the regression parameters a_{0k} and a_{1k} can be written as

$$a_{0k} = \beta_0 + b_{0k}$$

and   (6.9)

$$a_{1k} = \beta_1 + b_{1k},$$

where b_{0k} and b_{1k} are random effects that are assumed to be independently and normally distributed with mean zero and with a 2×2 variance–covariance matrix D:

$$D = \begin{bmatrix} \sigma_0^2 & \sigma_{01} \\ \sigma_{01} & \sigma_1^2 \end{bmatrix}.$$ (6.10)

The parameters σ_0^2, σ_1^2 and σ_{01} are the variances of b_{0k} and b_{1k} with their covariance, respectively. If the random effects b_{0k} and b_{1k} are assumed to be independent of the random errors ε_{ik} in Model (6.8), the variance of the responses y_{ik} becomes

$$\text{var}(y_{ik}) = \sigma_0^2 + \sigma_1^2 x_{ik}^2 + 2\sigma_{01}x_{ik} + \sigma_\varepsilon^2.$$ (6.11)

This equation shows that the variance of the responses depends on the variances and the covariance of the random effects b_{0k} and b_{1k}, and on the variance of the random errors. The covariance σ_{01} causes the variance of the responses var(y_{ik}) between treatment and control to be heterogeneous.

6.2.2.1 Estimation of parameters and efficiency

Apart from the individual random effects b_{0k} and b_{1k} and the random errors ε_{ik}, the model in Equation (6.8) is based on a total of six parameters, namely, the two fixed regression parameters β_0 and β_1 and four (co)variance parameters, that is, σ_0^2 and σ_1^2 with covariance σ_{01} and σ_ε^2, respectively. These can be estimated by ML and by REML methods. Asymptotically, the ML estimators of β_0 and β_1 lead to generalized least squares (GLS) estimators. Technical details for this subsection are given in Searle, Casella and McCulloch (1992) and van Breukelen, Candel and Berger (2007, 2008), among others.

For the special case that the X-variable is coded as $x_{ik} = +1$ for treatment and $x_{ik} = -1$ for control and $\sigma_{01} = 0$, the GLS estimators $\hat{\beta} = [\hat{\beta}_0, \hat{\beta}_1]'$ become

$$\hat{\beta}_0 = \sum_k^K \left[\frac{n_k(\bar{y}_{1k} + \bar{y}_{2k})/2}{n_k\sigma_0^2 + \sigma_\varepsilon^2} \right] \Big/ \text{var}(\hat{\beta}_0)$$

and   (6.12)

$$\hat{\beta}_1 = \sum_k^K \left[\frac{n_k(\bar{y}_{1k} - \bar{y}_{2k})/2}{n_k\sigma_1^2 + \sigma_\varepsilon^2} \right] \Big/ \text{var}(\hat{\beta}_1),$$

where $\bar{y}_{1k}$ and $\bar{y}_{2k}$ are the estimated response means for the subjects that are treated and not treated in cluster k. The sample size of cluster k is n_k. The 2×2 variance–covariance matrix of $\hat{\beta}$ is

$$\text{Cov}(\hat{\beta}) = \begin{bmatrix} \text{var}(\hat{\beta}_0) & \text{cov}(\hat{\beta}_0, \hat{\beta}_1) \\ \text{cov}(\hat{\beta}_0, \hat{\beta}_1) & \text{var}(\hat{\beta}_1) \end{bmatrix}, \tag{6.13}$$

and each element contains information on the uncertainty of these parameters. If the covariance of the random effects $\sigma_{01} = 0$, the covariance between $\hat{\beta}_0$ and $\hat{\beta}_1$ will also be zero, that is, $\text{cov}(\hat{\beta}_0, \hat{\beta}_1) = 0$. The following expressions for the variances of both estimators have been shown by van Breukelen, Candel and Berger (2007) to hold

$$\text{var}(\hat{\beta}_0) = \left[\sum_{k=1}^{K} \frac{n_k}{n_k \sigma_0^2 + \sigma_\varepsilon^2} \right]^{-1}$$

and (6.14)

$$\text{var}(\hat{\beta}_1) = \left[\sum_{k=1}^{K} \frac{n_k}{n_k \sigma_1^2 + \sigma_\varepsilon^2} \right]^{-1}.$$

If the sample sizes are equal over all clusters (i.e. $n_k = n$, for all k), then these expressions can be simplified to

$$\text{var}(\hat{\beta}_0) = \frac{n \sigma_0^2 + \sigma_\varepsilon^2}{Kn}$$

and (6.15)

$$\text{var}(\hat{\beta}_1) = \frac{n \sigma_1^2 + \sigma_\varepsilon^2}{Kn},$$

where n is again the common sample size for the clusters and K is the number of clusters in the design.

6.3 Cluster versus subject randomization

Four combinations of cluster and subject randomization versus a fixed and random treatment effect (i.e. a fixed and random slope) can be distinguished. In Table 6.1, the variances of the parameter estimators $\hat{\beta} = (\hat{\beta}_0, \hat{\beta}_1)'$ for each of these four conditions are summarized, assuming that the K clusters have equal sample sizes n.

Table 6.1 shows that subject randomization of the treatment leads to a smaller variance $\text{var}(\hat{\beta}_1)$ than cluster level randomization. For a multilevel model with a random intercept and a fixed slope, the relative efficiency for cluster versus

Table 6.1   Variances of the estimators $\hat{\beta}_0$ and $\hat{\beta}_1$ for two levels of randomization.

	Random intercept Fixed slope	Random intercept Random slope
Subject randomization	$\mathrm{var}(\hat{\beta}_0) = \dfrac{n\sigma_0^2 + \sigma_\varepsilon^2}{Kn}$ $\mathrm{var}(\hat{\beta}_1) = \dfrac{\sigma_\varepsilon^2}{Kn}$	$\mathrm{var}(\hat{\beta}_0) = \dfrac{n\sigma_0^2 + \sigma_\varepsilon^2}{Kn}$ $\mathrm{var}(\hat{\beta}_1) = \dfrac{n\sigma_1^2 + \sigma_\varepsilon^2}{Kn}$
Cluster randomization	$\mathrm{var}(\hat{\beta}_0) = \dfrac{n\sigma_0^2 + \sigma_\varepsilon^2}{Kn}$ $\mathrm{var}(\hat{\beta}_1) = \dfrac{n\sigma_0^2 + \sigma_\varepsilon^2}{Kn}$	$\mathrm{var}(\hat{\beta}_0) = \dfrac{n\sigma_0^2 + \sigma_\varepsilon^2}{Kn}$ $\mathrm{var}(\hat{\beta}_1) = \dfrac{n\sigma^2 + \sigma_\varepsilon^2}{Kn}$

Note. For cluster randomization and random slope, only the sum of the variances $\sigma^2 = \sigma_0^2 + \sigma_1^2$ can be estimated because all subjects within a cluster are in the same treatment condition. It is assumed that the covariance $\sigma_{01} = 0$.

subject randomization is

$$\mathrm{RE}_{\hat{\beta}_1} = \frac{\sigma_\varepsilon^2}{n\sigma_0^2 + \sigma_\varepsilon^2} = \frac{1 - \rho}{1 + (n - 1)\rho}, \tag{6.16}$$

where ρ is the intra-class correlation. The relative efficiency is $\mathrm{RE}_{\hat{\beta}_1} \leq 1$, and compares the efficiency of $\hat{\beta}_1$ in a cluster randomized trial with that of a multi-center trial, for the same total number of observations Kn. Note that the $\mathrm{RE}_{\hat{\beta}_1}$ is independent of the number of clusters K.

In Figure 6.2, the relative efficiencies $\mathrm{RE}_{\hat{\beta}_1}$ are plotted for six different values of the intra-class correlation, namely, $\rho = 0.01$, 0.03, 0.05, 0.10, 0.15 and 0.20, and for sample sizes $0 < n \leq 45$. These plots show that the efficiency drops very fast as the intra-class correlation increases and as the sample sizes increase.

The inverse of the $\mathrm{RE}_{\hat{\beta}_1}$ indicates how many times the cluster randomized trial has to be replicated to become as efficient as the multi-center trial with randomization at the subject level. Figure 6.2 shows that if we are prepared to accept a maximum efficiency loss of $(1 - \mathrm{RE}_{\hat{\beta}_1}) = 0.20$ (horizontal dotted line in the figure), and we assume that $\rho = 0.01$, the cluster randomized trial with $n \times K = 25K$ observations will require about $\left(\frac{1}{0.8} - 1\right)\% = 25\%$ more clusters each with $n = 25$ subjects to become as efficient as a multi-center trial with $25K$ observations.

Figure 6.2 also shows that for a given $\mathrm{RE}_{\hat{\beta}_1}$, the sample size n decreases as the intra-class correlation ρ increases. This means that if the responses within the clusters become more correlated, it will not be very efficient to increase the number of subjects in the clusters. Instead one should try to increase the efficiency by increasing the number of clusters.

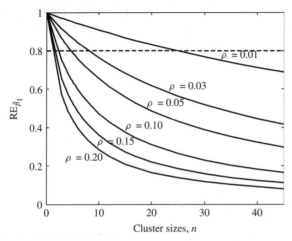

Figure 6.2 Relative efficiency for randomization at the cluster level versus subject level.

In conclusion, randomization of a treatment is always more efficient when it is done at the lowest level of a multilevel design. The drop of efficiency when randomization of the treatment is done at the cluster level is substantial, even for very small intra-class correlations and moderate sample sizes.

6.4   Cost function

In the previous section, we assumed that no special costs were connected to the sampling of clusters in a multilevel data structure. In practice, however, the costs of sampling clusters can be quite different from the costs of sampling subjects within clusters. In the following, simple cost function for a two-level design, the costs of recruitment of clusters (schools, nursing homes) are distinguished from the costs of sampling students or patients:

$$C \geq Knc_1 + Kc_2,$$

$$\text{where} \quad K \geq 2, \ c_1 > 0 \quad \text{and} \quad c_2 > 0. \tag{6.17}$$

The parameters c_1 and c_2 are the costs of sampling one subject and one cluster (nursing home, school), respectively. To maintain the multilevel data structure, it is assumed that the number of clusters will be $K \geq 2$. The total budget C is greater than or equal to the sum of the costs of sampling n subjects in K clusters, that is Knc_1, and the costs connected to sampling K clusters (homes, schools), namely, Kc_2. Because the total budget may include other costs, such as the costs of maintaining the research staff and the costs of analysing the final data set, we used the greater than or equal sign for the total budget C. The costs c_1 and c_2 may vary, and the optimal sample sizes may also vary a great deal for different costs.

The optimal design problem is now to find the minimum variance of the treatment effect estimator $\hat{\beta}_1$, given a maximum budget C. For equal cluster sizes, this problem can be formulated as

Minimize $\mathrm{var}(\hat{\beta}_1)$, subject to

$$C \geq Knc_1 + Kc_2, \quad \text{with} \quad K \geq 2, \ c_1 > 0 \quad \text{and} \quad c_2 > 0. \qquad (6.18)$$

The solution to this problem is presented in Table 6.2, where the minimum variance $\mathrm{var}(\hat{\beta}_1)^*$ is given together with the optimal sample sizes n^* and the optimal number of clusters K^*.

Table 6.2   Optimal $\mathrm{var}(\hat{\beta}_1)^*$, sample sizes n^* and clusters K^* as function of the costs c_1 and c_2.

	Random intercept Fixed slope	Random intercept Random slope
Subject randomization	$\mathrm{var}(\hat{\beta}_1)^* = \dfrac{c_1\sigma_\varepsilon^2}{C-2c_2}$	$\mathrm{var}(\hat{\beta}_1)^* = \dfrac{\left(\sqrt{c_2\sigma_1^2} + \sqrt{c_1\sigma_\varepsilon^2}\right)}{C}$
	$n^* = \dfrac{C-2c_2}{2c_1}$ $K^* = 2$	$n^* = \sqrt{\dfrac{c_2\sigma_\varepsilon^2}{c_1\sigma_1^2}}$ $K^* = \dfrac{C}{c_2 + \sqrt{\dfrac{c_1c_2\sigma_\varepsilon^2}{\sigma_1^2}}}$
Cluster randomization	$\mathrm{var}(\hat{\beta}_1)^* = \dfrac{\left(\sqrt{c_2\sigma_0^2} + \sqrt{c_1\sigma_\varepsilon^2}\right)^2}{C}$	$\mathrm{var}(\hat{\beta}_1)^* = \dfrac{\left(\sqrt{c_2\sigma^2} + \sqrt{c_1\sigma_\varepsilon^2}\right)^2}{C}$
	$n^* = \sqrt{\dfrac{c_2\sigma_\varepsilon^2}{c_1\sigma_0^2}}$ $K^* = \dfrac{C}{c_2 + \sqrt{\dfrac{c_1c_2\sigma_\varepsilon^2}{\sigma_0^2}}}$	$n^* = \sqrt{\dfrac{c_2\sigma_\varepsilon^2}{c_1\sigma^2}}$ $K^* = \dfrac{C}{c_2 + \sqrt{\dfrac{c_1c_2\sigma_\varepsilon^2}{\sigma^2}}}$

Notes. For cluster randomization and random slope, only the sum of the variances $\sigma^2 = \sigma_0^2 + \sigma_1^2$ can be estimated because all subjects within a cluster are in the same treatment condition. In practice the non-integer optimal sample sizes are rounded off to the nearest integers.

The optimal values for var$(\hat{\beta}_1)^*$ can be made smaller by increasing the total budget C. An increase of C for cluster randomization will lead to an increase of the optimal number of clusters and an increase of C for subject randomization will lead to an increase of the optimal number of subjects within a cluster for the model with a random intercept and a fixed slope.

A number of papers have been published on the design of multilevel studies. Snijders and Bosker (1993) and Cohen (1998) investigated the relationships of optimal sample sizes with cost constraints. Moerbeek, van Breukelen and Berger (2000) gave sample size formulae of multilevel designs with three levels of nesting and Mok (1995) compared different large sample designs via simulation. Donner, Birkett and Buck (1981) and Hsieh (1988) gave formulas of sample sizes for power calculations. Recent work include Liu (2003) who considered different costs for treatment and control and Moerbeek *et al.* (2003) and Moerbeek, van Breukelen and Berger (2008) who gave an accessible discussion on this topic, including a review of different approaches to the design problem in multilevel modelling.

In the next section, we illustrate how the optimal sample sizes and the optimal number of clusters are determined in practice using an example. We also evaluate the loss of efficiency when non-optimal values of n's and K's are used instead of the optimal ones.

6.5   Example: Nursing home study

Although very little information is available about the effect of massage on the prevention of pressure ulcers, massage is still one of the most often used preventive methods in nursing homes (Duimel-Peeters *et al.*, 2004). To evaluate the preventive effect of massage, two designs are possible, namely, a cluster randomized trial, similar to the one conducted by Duimel-Peeters *et al.* (2007) with a random assignment of the treatment (massage) to nursing homes and a multi-center trial, where the treatment is randomly assigned to the individual patients within a nursing home.

In Table 6.3, the optimal number of nursing homes K^* and optimal number of patients within the nursing homes n^*, together with the minimum variance of the treatment effect var$(\hat{\beta}_1)^*$, are presented for a number of different values of the costs c_1 and c_2, respectively. It is assumed that $C \geq Knc_1 + Kc_2$ and that the maximum budget is $C = €20\,000$ The costs c_1 vary from €100 to €500 and the costs c_2 vary from €100 to €400.

The optimal values are computed for both a cluster randomized trial and a multi-center trial. Throughout the following sections, we will assume that the multilevel model includes a random intercept and a fixed slope, that is, the model is $y_{ik} = \beta_0 + b_{0k} + \beta_1 x_{ik} + \varepsilon_{ik}$. In practice, the optimal values for K^* and n^* are rounded off to the nearest integers. To maintain a multilevel structure for subject randomization, the minimum value for K in Table 6.3 is set at $K = 2$. It should be noted that the optimal values for n^* under cluster randomization are sometimes

Table 6.3 Optimal cluster and optimal sample sizes for minimum $var(\hat{\beta}_1)^*$.

Cluster randomization of treatment

	c_2								
	100			200			400		
c_1	K^*	n^*	$var(\hat{\beta}_1)^*$	K^*	n^*	$var(\hat{\beta}_1)^*$	K^*	n^*	$var(\hat{\beta}_1)^*$
100	66.67	2.00	0.2250	41.42	2.83	0.2914	25.00	4.00	0.4000
200	52.24	1.41	0.3664	33.33	2.00	0.4500	20.71	2.83	0.5828
300	44.80	1.15	0.4982	28.99	1.63	0.5949	18.30	2.31	0.7464
400	40.00	1.00	0.6250	26.12	1.41	0.7328	16.67	2.00	0.9000
500	36.55	0.89	0.7486	24.03	1.26	0.8662	15.45	1.79	1.0472

Subject randomization of treatment

	c_2								
	100			200			400		
c_1	K^*	n^*	$var(\hat{\beta}_1)^*$	K^*	n^*	$var(\hat{\beta}_1)^*$	K^*	n^*	$var(\hat{\beta}_1)^*$
100	2.00	99.00	0.1010	2.00	98.00	0.1020	2	96.00	0.1042
200	2.00	49.50	0.2020	2.00	49.00	0.2041	2	48.00	0.2083
300	2.00	33.00	0.3030	2.00	32.67	0.3061	2	32.00	0.3125
400	2.00	24.75	0.4040	2.00	24.5	0.4082	2	24.00	0.4167
500	2.00	19.8	0.5050	2.00	19.6	0.5102	2	19.20	0.5208

Note. $\sigma_\varepsilon^2 = 20$, $\sigma_0^2 = 5$, $C = K^*n^*c_1 + K^*c_2 = \text{€}20\,000$.
The model includes only a random intercept: $y_{ik} = \beta_0 + b_{0k} + \beta_1 x_{ik} + \varepsilon_{ik}$.
In practice K^* and n^* are rounded off to the nearest integer.

$n^* < 2$. In this case, no estimate of the variance σ_ε^2 can be obtained from the data, but will have to be supplied from other sources of information. Because each nursing home is expected to have its own regime with rules and regulations to prevent patients from developing pressure ulcers, it is not unrealistic to assume that the intra-home correlation in this study is relatively high, that is, $\rho = 0.20$ and that $\sigma_\varepsilon^2 = 20$ and $\sigma_0^2 = 5$, respectively. In the following, it is helpful to note that $(1 - \rho)/\rho = \sigma_\varepsilon^2/\sigma_0^2$. This means that if we assume a value of $\sigma_0^2 = 5$ for the variance of the random intercept, the error variance becomes $\sigma_\varepsilon^2 = 20$.

6.5.1   Cluster randomization

For cluster randomization, it can be seen from Table 6.3 that the optimal number of homes K^* not only tends to become very large, but also decreases as the costs c_1 and c_2 increase. This is of course to be expected. From the formula in Table 6.2, it is seen that both c_1 and c_2 are in the denominator of the optimal number of clusters. If sampling nursing homes becomes more expensive, then this will lower the optimal number of nursing homes and if sampling patients becomes more expensive, then this will also lead to a reduction of the optimal number of nursing homes. It can also be seen that the optimal number of patients tends to become very small. However, to estimate the error variance σ_ε^2, the sample size n should be at least 2.

For example, consider the case that the cost of sampling a patient is $c_1 = €100$ and that the cost of sampling a nursing home is $c_2 = €400$. The optimal number of nursing homes for this case will be $K^* = 25$, with an optimal number of patients per home being equal to $n^* = 4$. In total, 100 patients will then be sampled with a total cost equal to $C = K^* n^* c_1 + K^* c_2 = €20\,000$. The optimal (minimal) variance is then $\text{var}(\hat{\beta}_1)^* = 0.4$.

Although these optimal numbers of nursing homes and patients within a home lead to a minimal variance $\text{var}(\hat{\beta}_1)^*$, this optimal design may not have sufficient power for finding a real treatment effect. In that case, the only thing that can be done is to increase the total budget C. The formulas in Table 6.2 show that a larger total budget C will directly lead to more clusters K^* but will not directly increase the n^*. But an increase of the number of nursing homes may become a problem in practice. Even the smallest optimal number of clusters $K^* = 15$ for the case that $c_1 = 500$ and $c_2 = 400$ (Table 6.3) may be too large in many practical cases. Reduction of the number of nursing homes together with an increase of the number of patients within nursing homes, however, will not be an efficient alternative in all cases.

Given the specific costs of sampling patients c_1 and nursing homes c_2 and the restriction that the total budget cannot exceed $C = 20\,000$, decreasing the number of nursing homes with an increase of the number of patients will lead to loss of efficiency. The relative efficiency of a cluster randomized design with alternative choices for K and n is a measure for efficiency loss:

$$RE_{\hat{\beta}_1} = \frac{var(\hat{\beta}_1)^*}{var(\hat{\beta}_1)}, \qquad (6.19)$$

where $var(\hat{\beta}_1)^*$ is the minimum variance of the parameter estimator for the optimal design with K^* and n^*.

In Figure 6.3, the relative efficiency $RE_{\hat{\beta}_1}$ is plotted as a function of K, for the same values of c_1 and c_2 as displayed in Table 6.3, namely, $c_1 = 100$, $200, 300, 400, 500$ and $c_2 = 100, 200, 400$. The restriction is $Knc_1 + Kc_2 = 20\,000$. The maximum relative efficiency is $RE_{\hat{\beta}_1} = 1$, and is, of course, connected to the optimal numbers K^* and n^* displayed in Table 6.3. In general, the $RE_{\hat{\beta}_1}$ increases rapidly with an increase of the number of clusters K. Figure 6.3 shows that for all the displayed cases, it will not be very efficient to design a cluster randomized trial with less than $K \leq 10$.

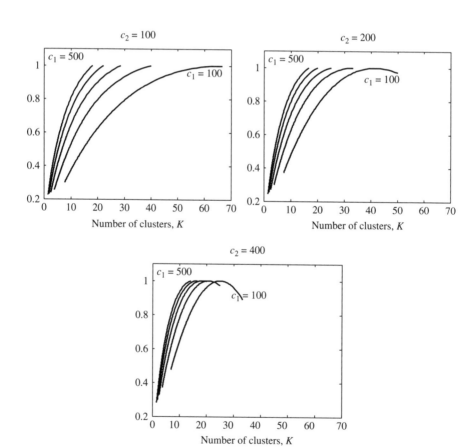

Figure 6.3 Relative efficiency $RE_{\hat{\beta}_1}$ for cluster randomized trials.

For $c_1 = 100$ and $c_2 = 400$, the figure shows that the optimal is $K^* = 25$, and the optimal number of patients can then be computed from $n^* = (20\,000 - 25c_2)/25c_1 = 4$, as is shown in Table 6.3. If it is not feasible to sample 25 nursing homes for our study, then given the same cost parameters, it may be possible to sample fewer nursing homes and more patients without too much efficiency loss. Figure 6.3 gives us an indication of how much efficiency is lost if only few nursing homes are sampled.

The third plot ($c_2 = 400$) in Figure 6.3 shows that for approximately $K = 14$, the relative efficiency will become

$$\mathrm{RE}_{\hat{\beta}_1} = \frac{\mathrm{var}(\hat{\beta}_1)^*}{\mathrm{var}(\hat{\beta}_1)} = \frac{0.4}{0.5} = 0.8, \tag{6.20}$$

with $n \approx (20\,000 - 14c_2)/14c_1 = 10$. The variance $\mathrm{var}(\hat{\beta}_1)$ can be obtained from $\mathrm{var}(\hat{\beta}_1) = (n\sigma_0^2 + \sigma_\varepsilon^2)/Kn = 0.5$. Therefore, instead of sampling 25 nursing homes with four patients in each, one can sample 14 nursing homes with 10 patients in each, with a moderate efficiency loss of 20%. Sampling less than 14 nursing homes and more patients will not be efficient because the relative efficiency will then drop rapidly.

Figure 6.3 also shows that with these cost parameters, the number of selected clusters should not become too large. It can be seen that for $c_1 = 100$ and $c_2 = 400$, the number of clusters can become $K > 25$ at the expense of the sample size n, but then the relative efficiency will drop a little.

6.5.2   Subject randomization

For randomization at the subject level, see Figure 6.4. Here, the (optimal) number of clusters tends to be small (minimum $K = 2$, to maintain the multilevel data structure) and the optimal cluster sizes tend to become large. The optimal number of patients n^* decreases as the costs of sampling patients c_1 increase, and does not decrease very much as the costs c_2 increase (Table 6.3). Of course, the total costs should not exceed $C \geq K^*n^*c_1 + K^*c_2$. In practice, K^* and n^* are rounded off.

In Figure 6.4, the relative efficiencies are plotted against the sample size n, again for $c_1 = 100, 200, 300, 400, 500$ and $c_2 = 100, 200, 400$. It can be seen that at first the $\mathrm{RE}_{\hat{\beta}_1}$ increases rapidly as the cluster size n increases, but that eventually the increase levels off.

Given a certain $\mathrm{RE}_{\hat{\beta}_1}$ value, Figure 6.4 shows that as the costs c_1 increase, the cluster size n will decrease, and that as the costs c_2 increase, one will need a larger sample n to maintain the same $\mathrm{RE}_{\hat{\beta}_1}$ level.

Now, consider the same case as before, where the cost parameters are $c_1 = 100$, $c_2 = 400$ and $C = 20\,000$. For these values, the optimal number of nursing homes and the optimal sample size are $K^* = 2$ and $n^* = 96$ (Table 6.3). In practice, it may, however, not be feasible to sample 96 patients from each nursing home because many patients and caregivers may not submit an informed consent or that the nursing homes simply do not have enough patients with pressure ulcer

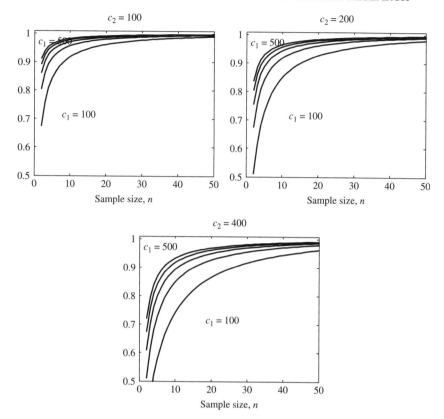

Figure 6.4 Relative efficiency $RE_{\hat{\beta}_1}$ *for subject randomized trials.*

problems. If only $n = 20$ patients are available per home, then given the cost parameters, the number of nursing homes will be $K = 20\,000/(2000 + 400) = 8.333$ (rounded off to 8). Figure 6.4 shows that for $n = 20$ (and $K \approx 8$), the relative efficiency remains relatively high, namely, $RE_{\hat{\beta}_1} > 0.8$.

Finally, Table 6.3 shows that the variance $\mathrm{var}(\hat{\beta}_1)^*$ for cluster randomization is much larger than for randomization at the patient level, given the same total cost $C = €20\,000$. This means that with the same resources, randomization at the patient level will always provide a more powerful test of the treatment effect hypothesis $H_0 : \beta_1 = 0$. Because power computations for such multilevel models are more complicated than for an ordinary regression model, the next section will explain power computations for multilevel models in more detail.

6.6   Optimal design and power

The power of a test is the probability that a null hypothesis is correctly rejected. In a multilevel design, where the total sample consists of K clusters each with

n subjects, the power of a test depends on the total sample size Kn, the treatment effect, the variance of the estimator of the treatment effect $\text{var}(\hat{\beta}_1)$ and the Type I error probability α. Indirectly, via $\text{var}(\hat{\beta}_1)$, the power also depends on the intra-class correlation ρ, representing the dependency of the observations within the clusters. For multilevel designs, variation of the number of clusters K and the number of observations n within each cluster makes computation of the power a little more complicated than in the usual independent sample case. Additionally, it also becomes more complicated when different costs of obtaining clusters and observations are taken into account.

Economically, a researcher who is planning to design a multilevel study would want to select a design that has maximum power for finding real differences, for a total maximum budget. This can actually be done in two steps:

1. Given a cost function with maximum total costs C, choose the optimal design with optimal n^* and K^* (Table 6.2), which will lead to a minimum variance $\text{var}(\hat{\beta}_1)^*$ of the treatment effect estimator. A minimum variance $\text{var}(\hat{\beta}_1)^*$ leads to maximum power for finding real effects.

2. If the computed power is not sufficiently high, then (if possible) the total budget is increased to obtain the desired power.

Suppose that the null hypothesis of no treatment effect $H_0 : \beta_1 = 0$ is tested against the alternative hypothesis $H_1 : \beta_1 \neq 0$. Given the null hypothesis and a known $\text{var}(\hat{\beta}_1)$, the test statistic $z = \hat{\beta}_1/\sqrt{\text{var}(\hat{\beta}_1)}$ is normally distributed. The power of this test can be expressed as the sum of two probabilities

$$\text{Power} = \left\{ \Phi\left(z_{\alpha/2} - \frac{\beta_1}{\sqrt{\text{var}(\hat{\beta}_1)}}\right) + \left[1 - \Phi\left(z_{1-\alpha/2} - \frac{\beta_1}{\sqrt{\text{var}(\hat{\beta}_1)}}\right)\right]\right\},$$
(6.21)

where $\Phi(.)$ is the cumulative standard normal distribution function with $z_{\alpha/2}$ and $z_{1-\alpha/2}$ being the $100\alpha/2$ and $100(1 - \alpha/2)$ standard normal percentile, respectively (note that $z_{\alpha/2} = -z_{1-\alpha/2}$). When the variance $\text{var}(\hat{\beta}_1)$ is unknown, then it will have to be estimated and the test statistic will then have a non-central t-distribution. In that case, the power will be a little lower because the variance $\text{var}(\hat{\beta}_1)$ has to be estimated as well. For large samples, however, the normal distribution will still provide an accurate approximation.

Because this is a two-sided test, the power is the sum of two areas under the standard normal distribution function, which are graphically displayed in Figure 6.5. The two probabilities summed in Equation (6.21) correspond to the two areas in Figure 6.5.

The left area in Figure 6.5 is extremely small, while the right area is larger than 0.5. This means that for this case, the first expression in Equation (6.21) can be ignored and the second expression is sufficient for accurate approximation of the power. However, if the effect β_1 is negative, then the first (left) part of the equation will be large, whereas the second (right) part will be ignorably small.

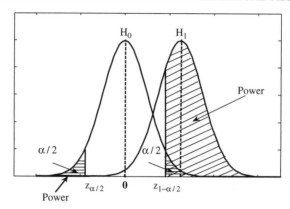

Figure 6.5 Graphical display of standard normal distribution under H_0 *and* H_1 *with power.*

The phenomenon that one of the two parts in Equation (6.21) can be ignored is especially encountered when the effect β_1 is relatively large with a relatively small variance var($\hat{\beta}_1$). Finally, for a one-sided test, the power can be computed from only one of the tails of the distribution with z_α (or $z_{1-\alpha}$) as percentile.

As can be seen from Figure 6.5, the power of the test increases as the effect β_1 increases, as the type I error probability α increases and as the variance var($\hat{\beta}_1$) decreases. This latter quantity decreases with an increasing sample size. The power of cluster randomized trials and multi-center trials can be computed with Equation (6.21) or by using the software program OPTDES (Raudenbush *et al.* 2004), which uses a non-central F-distribution because the variance var($\hat{\beta}_1$) has to be estimated.

6.6.1   Power for cluster randomized design

In Table 6.3 of the previous section, the optimal number of clusters and subjects within clusters were computed for different costs. For the nursing home example, it was shown that with a maximum budget of $C = €20\,000$ and with $c_1 = €100$ and $c_2 = €400$, the optimal number of clusters would be $K^* = 25$ together with the optimal number of patients within the clusters being equal to $n^* = 4$. The corresponding minimum variance is equal to var($\hat{\beta}_1$)$^* = 0.4$. It is assumed that the following parameters remain the same: $\sigma_\varepsilon^2 = 20$, $\sigma_0^2 = 5$ and $\rho = 0.2$, and that the computations are done for a multilevel model containing a random intercept and a fixed slope, that is, $y_{ik} = \beta_0 + b_{0k} + \beta_1 x_{ik} + \varepsilon_{ik}$. These optimal values $K^* = 25$ and $n^* = 4$ for $C = 20\,000$, guarantee that the power of the test of the hypothesis of no treatment effect will be as large as possible. To see how much power the test would actually have, the power function is computed.

In Figure 6.6 the power for $\alpha = 0.05$, $K^* = 25$, $n^* = 4$, and $C = 20\,000$ is plotted as a function of the effect β_1. Recall that the treatment effect is equal to half of the difference between weighted average outcomes in the treatment and

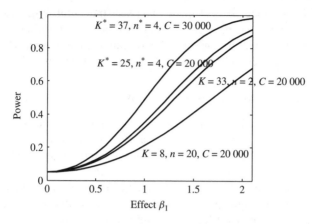

Figure 6.6 Power functions for cluster randomized trials.

control group. For example, if one expects the treatment effect to be 3, that is, $\beta_1 = 1.5$, the power will be approximately 0.65. Because this is the maximum power attainable with a budget of $C = 20\,000$, any small increase of the number of clusters together with a decrease of patients within nursing homes will lead to a drop of power. This can be seen in Figure 6.6 that also shows a power function for $K = 33, n = 2$, so that $C = Knc_1 + Kc_2 = 20\,000$. This drop of power corresponds to the small decrease of the RE_{β_1} in Figure 6.3.

For the nursing home example, it may be practically difficult or even impossible to find more than 8–10 nursing homes to participate in a study. The optimal number of homes $K^* = 25$ may really not be feasible. Figure 6.6 also shows a power function for $K = 8, n = 20$ and $C = 20\,000$. Although these values for the number of homes and the number of patients within a cluster seem more realistic in practice, extremely low power indicates that such a design is not recommendable.

In case that the power of 0.65 found for $\alpha = 0.05$, $K^* = 25, n^* = 4$, and $C = 20\,000$ and $\beta_1 = 1.5$, is considered to be too small, the only option a researcher has is to increase the total budget and to sample more homes. As indicated in the formulas in Table 6.2, it would not increase power if the sample size n were increased in a cluster randomized design. If the maximum budget is increased by 50% ($C = 30\,000$), the optimal design will have $K^* = 37$ homes and $n^* = 4$ patients within a home. The corresponding power function is also displayed in Figure 6.6. This plot indicates that for an effective size $\beta_1 = 1.5$, the power would become acceptably high at 0.80. But again such a large number of homes may not be feasible in practice. Alternatively, recruitment strategies in terms of costs for choosing between designs with a few large clusters and designs with many small clusters have been discussed by Flynn, Whitley and Peters (2002).

The above example illustrates a major problem with cluster randomized designs. To increase the power of a treatment effect, a larger number of clusters must be sampled. But in many practical cases, the number of clusters is limited and it may not be possible to increase the number of clusters in the design. Another potential problem with cluster randomized trials, not found in multi-center trials, is that they are not able to sample enough subjects from a cluster to guarantee sufficiently high power. This is explained in the following section.

6.6.2   Power for multi-center design

A much more powerful design is the multi-center trial, where the treatment is randomly assigned to subjects within a cluster. As has been shown in the previous section, the optimal number of clusters and subjects within a cluster for a multi-center trial with $c_1 = 100$, $c_2 = 400$ and $C = 20\,000$ are $K^* = 2$ and $n^* = 96$, respectively. The approximate power for these optimal values and $\alpha = 0.05$ is plotted in Figure 6.7 as a function of the treatment effect β_1. It can be seen that even for a smaller effect, for example, $\beta_1 = 1$, the power is almost 0.9.

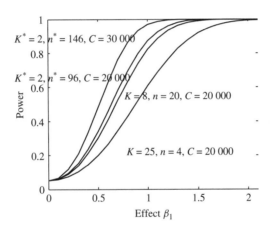

Figure 6.7 Power functions for multi-center trials.

If the optimal number of $n^* = 96$ patients within the homes is considered to be too large, and instead a reduced number of $n = 20$ is chosen, with an increased number of nursing homes equal to $K = 8$, then the power will drop moderately. For an effect of $\beta_1 = 1$, the power will be about 0.8.

For comparison reasons, the power is also plotted in Figure 6.7 for the optimal number of clusters and subjects in case the treatment was randomly assigned to the clusters, namely, $K^* = 25$ and $n^* = 4$. For a treatment effect equal to $\beta_1 = 1$, the power is now approximately 0.6. This is not high, but still even higher than the power for the same number of clusters and subjects in a cluster randomized trial (Figure 6.6), which is a little lower than 0.4. Finally, increase of the total budget will lead to an increase of the optimal number of subjects within a cluster. The optimal number of clusters for $C = 30\,000$ remains equal to $K^* = 2$, while the optimal number of subjects increases to $n^* = 146$.

In conclusion, in terms of efficiency and power, a multi-center trial, with a treatment randomly assigned to subjects within clusters or groups, will always be more efficient and more powerful in finding real effects than a cluster randomized trial. The optimal number of clusters in a cluster randomized trial may become unfeasibly large. But, even if such a large number of clusters could be included in a study, the random assignment of a treatment to subjects within clusters would be much more efficient than the random assignment of treatments to clusters as a whole.

6.6.3   Increase of efficiency and power by including covariates

We can increase the efficiency and power for finding real differences between two groups by including carefully selected covariates. For randomized trials, it has been shown by Allison *et al.* (1997) and Porter and Raudenbush (1987), among others, that it is often worthwhile to include a covariate to increase power when the correlation between a covariate and the outcome is high and when the costs of measuring the covariate are relatively low.

Inclusion of covariates to increase efficiency and power in a cluster randomized trial is more complicated, because their inclusion may change the variances of the errors in the model. Another difficulty is that covariates may vary at the subject level or at the cluster level and that covariates may differ in their costs of being measured. Moerbeek (2006) provided formulas to decide whether it is worthwhile to add covariates to increase efficiency and power in a cluster randomized trial. The conclusion of this study is that when the costs of measuring the covariate are small and the reduction of the variance components is large, it will be more efficient in terms of costs to include a covariate instead of increasing the number of clusters. The covariate should not be included if the correlation between the outcome and the covariate is not high enough.

We emphasize that covariates should only be included in the design of a cluster randomized trial after careful consideration of the relation between the covariate and the outcome. Adding a covariate to increase efficiency is likely to increase the probability of non-response or drop-out. This is especially true when the measurement of the covariate is a burden on the subjects. The complication that arises from the dropouts may further threaten the internal validity of the study.

6.6.4   Unequal sample sizes

In the previously presented computations, it was assumed that the sample sizes within the clusters were all equal. Formulas assuming equal sample sizes have been given by Raudenbush (1997), Raudenbush and Liu (2000), Moerbeek, van Breukelen and Berger (2000, 2001), Liu (2003) and Headrick and Zumbo (2005). Equal cluster sizes are optimal for estimating the treatment effect β_1 with minimum variance and for testing $\beta_1 = 0$ with maximum power (Ankerman, Aviles and Pinheiro, 2003).

In most cases, however, especially in quasi-experimental studies, the sample sizes are not equal due to non-response and drop out. These multilevel designs, with cluster randomization or subject randomization of the treatment, will therefore show loss of efficiency. The amount of efficiency loss depends on the variation of the sample sizes over the clusters and on the intra-class correlation, but not on the number of clusters K. van Breukelen, Candel and Berger (2007) showed numerically that the loss of relative efficiency in most commonly encountered designs with unequal sample sizes is usually less than 20% (i.e. RE $\geq$ 0.8), and in most cases, less than 10% (i.e. RE $\geq$ 0.9). This actually indicates that increasing the number of clusters by $(0.9^{-1} - 1)100\% = 11\%$ can compensate for such loss of efficiency and power due to unequal sample sizes. Similar conclusions were drawn by Candel *et al.* (2008) for small samples and different estimation methods for multilevel analysis.

In summary, the comparison in this chapter between a cluster randomized and a subject randomized design shows that in terms of efficiency and power, a subject randomized design or multi-center design should be preferred.

6.7   Design effect in multilevel surveys

Because social and economic data often have a multilevel structure, similar problems as those of the multilevel design discussed in previous sections are also encountered in these fields. For example, in the Los Angeles Family and Neighborhood Survey (L.A.FANS, Sastry *et al.*, 2006), which is designed to support analysis of child development, residential mobility and welfare reform, households are nested within neighbourhoods. The design of this study is a stratified random sample of 65 neighbourhoods, where an average of 41 households were randomly selected and interviewed within each neighbourhood. See Sastry *et al.* (2006) for more details on this sampling design. Although the actual design also included adults, children and caregivers, we will restrict ourselves to only two levels, namely, the neighbourhoods and the households.

In survey sampling, the optimal sampling design would be a sample that is completely randomly drawn from a specific population. Such a simple random sample (SRS) is, however, generally not very efficient in terms of costs, because it is generally very expensive to first completely list all households within the different neighbourhoods and then visit and interview the randomly selected households scattered all around the different neighbourhoods. To reduce the amount of effort and fieldwork, that is, to reduce the costs of sampling, survey researchers usually first sample a smaller number of primary sampling units (neighbourhoods) and then randomly select the secondary sampling units (households) via simple random sampling from these primary sampling units. Such two-stage sampling designs distinguish two levels and variables and characteristics are measured at each of these levels. The most suitable way to analyse these data is to apply multilevel analysis (Snijders and Bosker, 1999; Hox, 2002). The design problem for these large scale surveys is to find a design that will have sufficient power to estimate the fixed effects and the random effects with the lowest possible sampling costs.

Up to date, very little work has been done to find optimal survey designs. Owing to the different number of levels and variables measured at each level, this problem is often very complex, for which no general solution probably exists. A simple and heuristic method to estimate the effect of sampling relative to the costs of sampling is based on the so-called *design effect*. A *design effect* compares the variance of the effect of a sampling design with that of an SRS (Kish, 1965).

Suppose that we have obtained a simple random sample of N households, where households are selected with equal probability and where all possible sets of N distinct households from a population with size N_{pop} are equally likely to have been sampled. Then the variance of the sample mean $\bar{y}_{SRS}$ can be estimated by

$$\widehat{var}(\bar{y}_{SRS}) = \frac{s^2}{N},\qquad(6.22)$$

where $s^2 = \sum_i^N (y_i - \bar{y}_{SRS})^2/(N-1)$ is the unbiased estimator of the population variance of y. Note that we have assumed that the sampling fraction N/N_{pop} is very small, so that the difference between sampling with or without replacement can be ignored.

A *design effect* is the ratio of the variance of a parameter estimator $\hat{\beta}$ for a particular sampling design with the variance of that parameter estimator from an SRS design (Kish, 1965). In formula it is

$$Design\ effect = \frac{var(\hat{\beta}_{Design})}{var(\hat{\beta}_{SRS})}.\qquad(6.23)$$

Because the variance $var(\hat{\beta}_{SRS})$ is the smaller of the two, this *design effect* is larger than one. As such, it is inversely related to the relative efficiency measure, which lies between zero and one. But its interpretation in terms of sample size is similar. Using this *design effect*, the effective sample size can be formulated as the size of an SRS design that has the same variance of the estimator as the particular (clustered) sampling design with N households. This so-called effective sample size is given by

$$N_{eff} = \frac{N}{Design\ effect}.\qquad(6.24)$$

Now if we consider the two-stage sampling design, where first K neighbourhoods are randomly sampled and then within each neighbourhood a random sample of n households is drawn, and we wish to estimate the overall mean $\bar{y}_c = \sum_j^K \sum_i^n y_{ij}/N$, with $N = Kn$, then the variance of this estimator is estimated by

$$\widehat{var}(\bar{y}_c) = \frac{s^2}{N}[1 + (n-1)\rho],\qquad(6.25)$$

where $\widehat{\text{var}}(\bar{y}_{\text{SRS}}) = s^2/N$ and the corresponding *design effect* is equal to $[1 + (n - 1)\rho]$. The symbol ρ stands for the intra-class correlation coefficient that is a measure of the homogeneity of the y – scores in the neighbourhoods (clusters).

6.7.1   Values of intra-class correlation ρ

It is clear that the size of the *design effect* depends on two quantities, namely, the intra-class correlation and the size of the neighbourhoods (clusters). The difference in sample size N and effective sample size N_{eff} also depends on these two quantities. Because the intra-class correlation ρ is usually not known in advance, some kind of estimate from previous research must be used to estimate the *design effect*.

The values of ρ depend on the type of variable and lie between zero and one. Although estimates of ρ may become negative, this is rather unlikely for most variables in multilevel data structures. For natural variables, such as height, weight, age or gender, the dependency of their observations in a group may be very low and ρ can be assumed to be approximately equal to zero. However, for other kinds of variables, ρ may become quite high. For example, in a new neighbourhood with mainly young families with small children and with excellent health care and educational facilities, the opinions of the households about these facilities may be very homogeneous, resulting in a relatively high value of ρ. On the other hand, in an older neighbourhood, the opinions about these facilities may vary significantly because older households with no (small) children may differ quite a lot in their opinion about these facilities, thus causing the intra-class correlation to be relatively small. But even for relatively small ρ's, the design effect may become quite large. For example, if $\rho = 0.05$ or 0.10, and $n = 10$ households, the *design effect* will be 1.45 or 1.90, respectively.

6.7.2   Cluster randomized sampling versus simple random sampling

In Figure 6.8, the ratio N_{eff}/N, which is equal to the inverse of the *design effect*, is plotted against the number of neighbourhoods with a constant sample size of $N = Kn = 2000$. Four plots are displayed for $\rho = 0.001, 0.01, 0.05$ and 0.1. It can be seen that when the number of neighbourhoods K increases and the number of households n within the clusters decreases, the effective sample size N_{eff} will increase. For small values of ρ, the effective sample size N_{eff} approaches the sample size N.

We also observe that when the correlation ρ is larger than 0.05, we will need a lot more neighbourhoods in a cluster sampling design to even approach the information that is obtained from a simple random sample. In fact, if we have $K = 100$ neighbourhoods and $\rho = 0.05$, the effective sample size of a cluster sampling design is about $N_{\text{eff}} = 0.51 \times 2000 = 1025$ households, instead of the 2000 households that were actually sampled.

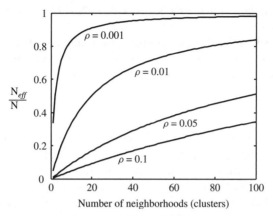

Figure 6.8 The ratio N_{eff}/N *by number of clusters with a total sample size* $N = 2000$.

These computations were done without any cost considerations. Now suppose that we consider the simple cost function $C = c_1 Kn + c_2 K$ from Section 6.4, with c_1 and c_2 being the costs of sampling a household and a neighbourhood respectively. Hansen, Hurwitz and Madow (1953, p. 172-173) developed a formula for the optimal number of households and neighbourhoods for estimating the means and Cohen (1998, 2005) provided an approximation as follows:

$$n^* = \sqrt{\frac{c_2(1 - \rho)}{c_1 \rho}}, \text{ and } K^* = \frac{C}{c_2 + c_1 n^*}. \tag{6.26}$$

We note that these formulas are equal to the formulas in Table 6.2 for the cluster randomization of the treatment condition. This is due to the fact that for the multilevel model in Equation (6.3), the intercept $\hat{\beta}_0$ is equal to the mean and $\text{var}(\hat{\beta}_0) = \text{var}(\hat{\beta}_1)$; see also Equation (6.5).

Finally, it is important to mention that estimation of several parameters is often the case in survey sampling and that the above simple situation will generally not hold when several parameters are to be estimated simultaneously. For this situation, Cohen (1998) proposed minimizing a linear combination of the variances of the several estimators, but the efficacy of this method has not been well studied. More research is needed on this topic. At present, the above formulas for determining the optimal number of households and neighbourhoods can only serve as a guide in planning a survey study.

6.8   Matrix formulation of the multilevel model

To facilitate the understanding of designs for the multilevel models for cluster randomized trials and subject randomized trials, this section uses matrix algebra.

Readers not familiar with matrix algebra can skip this section without much loss of information.

The multilevel models for cluster and subject randomization are special cases of the general linear mixed model described by Laird and Ware (1982), among others. These models can be written in matrix form as

$$y = X\beta + Zb + \varepsilon. \tag{6.27}$$

The $Kn \times 1$ vector y contains the n observations, stacked for the K clusters. Corresponding to the linear model with two fixed parameters in the 2×1 column vector $\beta = (\beta_0 \ \beta_1)'$, the design matrix X is of order $Kn \times 2$. The $Kn \times 2$ matrix Z is the design matrix for the two random parameters in the 2×1 vector b. The errors connected to each response in y are stacked in the same order in the vector ε. For general linear mixed models, it is assumed that the random parameters in b are normally distributed with mean 0 and variance–covariance matrix D and that the errors in ε are normally distributed with mean 0 and variance–covariance matrix Σ_ε. For the models in this chapter, the error variance–covariance matrix is assumed to be $\Sigma_\varepsilon = \sigma_\varepsilon^2 I$. The b's and ε's are independently distributed. It should be noted that this linear mixed model (Laird and Ware, 1982) is a member of the class of generalized linear mixed models as described by McCulloch and Searle (2001).

6.8.1   Cluster randomization of treatment

To illustrate how matrices are formed for the cluster randomized trial, consider the model for the cluster randomization of a treatment as given in Equation (6.3): $y_{ik} = \beta_0 + b_{0k} + \beta_1 x_k + \varepsilon_{ik}$. For illustrative purposes, we consider $K = 2$ clusters and $n = 4$ subjects within each cluster. This model can be written in matrix form as

$$
\begin{bmatrix} y_{11} \\ y_{21} \\ y_{31} \\ y_{41} \\ -- \\ y_{12} \\ y_{22} \\ y_{32} \\ y_{42} \end{bmatrix}
=
\begin{bmatrix} 1 & 1 \\ 1 & 1 \\ 1 & 1 \\ 1 & 1 \\ - & - \\ 1 & -1 \\ 1 & -1 \\ 1 & -1 \\ 1 & -1 \end{bmatrix}
\begin{bmatrix} \beta_0 \\ \beta_1 \end{bmatrix}
+
\begin{bmatrix} 1 & 0 \\ 1 & 0 \\ 1 & 0 \\ 1 & 0 \\ - & - \\ 0 & 1 \\ 0 & 1 \\ 0 & 1 \\ 0 & 1 \end{bmatrix}
\begin{bmatrix} b_{01} \\ b_{02} \end{bmatrix}
+
\begin{bmatrix} \varepsilon_{11} \\ \varepsilon_{21} \\ \varepsilon_{31} \\ \varepsilon_{41} \\ -- \\ \varepsilon_{12} \\ \varepsilon_{22} \\ \varepsilon_{32} \\ \varepsilon_{42} \end{bmatrix}.
\tag{6.28}
$$

In the second column of the design matrix X, the coding for all subjects in the treatment group is $x_k = +1$ and for all subjects in the control group is $x_k = -1$, respectively. The first column of X consists of all ones and corresponds to the intercept β_0. The design matrix Z for the random intercepts in the two groups is structured with zeros and ones, to ensure that the random intercept b_{01} is included in the model for the responses of the treatment group and b_{02} is included in the model for the control group. The horizontal dotted lines in Equation (6.28) separate the two groups. The variance–covariance matrix of the responses in y is given by

$$\text{Cov}(y) = (ZDZ' + \sigma_\varepsilon^2 I), \tag{6.29}$$

where $D = \sigma_0^2 I$, with a common variance of the random parameters b_{0k} on the main diagonal, $\sigma_\varepsilon^2 I$ is a diagonal matrix with error variances σ_ε^2 and I is an identity matrix. For the example with $K = 2$ and $n = 4$, the elements of $\text{Cov}(y)$ are

Cov(y)

$$= \begin{bmatrix}
\sigma_0^2+\sigma_\varepsilon^2 & \sigma_0^2 & \sigma_0^2 & \sigma_0^2 & 0 & 0 & 0 & 0 \\
\sigma_0^2 & \sigma_0^2+\sigma_\varepsilon^2 & \sigma_0^2 & \sigma_0^2 & 0 & 0 & 0 & 0 \\
\sigma_0^2 & \sigma_0^2 & \sigma_0^2+\sigma_\varepsilon^2 & \sigma_0^2 & 0 & 0 & 0 & 0 \\
\sigma_0^2 & \sigma_0^2 & \sigma_0^2 & \sigma_0^2+\sigma_\varepsilon^2 & 0 & 0 & 0 & 0 \\
0 & 0 & 0 & 0 & \sigma_0^2+\sigma_\varepsilon^2 & \sigma_0^2 & \sigma_0^2 & \sigma_0^2 \\
0 & 0 & 0 & 0 & \sigma_0^2 & \sigma_0^2+\sigma_\varepsilon^2 & \sigma_0^2 & \sigma_0^2 \\
0 & 0 & 0 & 0 & \sigma_0^2 & \sigma_0^2 & \sigma_0^2+\sigma_\varepsilon^2 & \sigma_0^2 \\
0 & 0 & 0 & 0 & \sigma_0^2 & \sigma_0^2 & \sigma_0^2 & \sigma_0^2+\sigma_\varepsilon^2
\end{bmatrix}. \tag{6.30}$$

This model is a random intercept model, which implies that the variances of the responses are constant and equal to $(\sigma_0^2 + \sigma_\varepsilon^2)$. The covariances among the responses within the clusters are also constant and equal to σ_0^2. This structure of the covariance matrix of the responses within each cluster is often referred to as *compound symmetry*, and the previously given intra-class correlation that defines the amount of correlation among the responses within the clusters is $\rho = \sigma_0^2/(\sigma_0^2 + \sigma_\varepsilon^2)$. The covariance matrix in Equation (6.30) also shows that the covariance of the responses between the two clusters is zero. This is of course to be expected because the responses come from subjects in different clusters and they are assumed to have nothing in common. The variance–covariance matrix of the parameter estimators $\text{Cov}(\hat{\beta})$ is now

$$\text{Cov}(\hat{\beta}) = \{X'[\text{Cov}(y)]^{-1}X\}^{-1} = \begin{bmatrix} \dfrac{4\sigma_0^2+\sigma_\varepsilon^2}{4\times 2} & 0 \\ 0 & \dfrac{4\sigma_0^2+\sigma_\varepsilon^2}{4\times 2} \end{bmatrix}. \tag{6.31}$$

It must be emphasized that this formula assumes known variance components σ_0^2 and σ_ε^2. As indicated in Subsection 6.2.1.1 estimation of these variance components is done by REML. Asymptotically, the estimators of the variance components are orthogonal to the fixed effects estimators $\hat{\beta}$ (Searle, Casella and McCulloch, 1992 and van Breukelen, Candel and Berger, 2007, 2008).

The diagonal elements of $\text{Cov}(\hat{\beta})$ are the variances $\text{var}(\hat{\beta}_0)$ and $\text{var}(\hat{\beta}_1)$, respectively, and it can be seen that they are the same as the formula $\text{var}(\hat{\beta}_0) = \text{var}(\hat{\beta}_1) = \frac{n\sigma_0^2 + \sigma_\varepsilon^2}{Kn}$ given in Subsection 6.2.1.1 for $K = 2$ and $n = 4$. The off-diagonal elements of $\text{Cov}(\hat{\beta})$ are zero, but this is due to the fact that we assume an equal number of subjects in the two clusters for this example. In general, the off-diagonal elements $\text{cov}(\hat{\beta}_0, \hat{\beta}_1)$ may not be zero. Finally, for this model, the variances of the two parameters are equal, that is, $\text{var}(\hat{\beta}_0) = \text{var}(\hat{\beta}_1)$. Therefore, we do not need to work with the D-, A- or E-optimality criteria to summarize variances and covariances of the two parameter estimators. Minimizing the sum or product of the main diagonal elements of $\text{Cov}(\hat{\beta})$ is equivalent to minimizing only one of these elements. But if other multilevel models are used, these optimality criteria can, of course, also be applied to optimize the multilevel design.

6.8.2   Subject randomization of treatment

The model for subject randomization as given in Section 6.2.2 is

$$y_{ik} = \beta_0 + b_{0k} + \beta_1 x_{ik} + b_{1k} x_{ik} + \varepsilon_{ik}. \tag{6.32}$$

Here, we distinguish two random parameters b_{0k} and b_{1k} with known variance components σ_0^2 and σ_1^2, respectively. Suppose again, that there are $K = 2$ clusters and $n = 4$ subjects within each cluster. Then this model can be written in matrix form as

$$
\begin{bmatrix}
y_{11} \\
y_{21} \\
y_{31} \\
y_{41} \\
-- \\
y_{12} \\
y_{22} \\
y_{32} \\
y_{42}
\end{bmatrix}
=
\begin{bmatrix}
1 & 1 \\
1 & 1 \\
1 & -1 \\
1 & -1 \\
- & - \\
1 & 1 \\
1 & 1 \\
1 & -1 \\
1 & -1
\end{bmatrix}
\begin{bmatrix}
\beta_0 \\
\beta_1
\end{bmatrix}
+
\begin{bmatrix}
1 & 1 & 0 & 0 \\
1 & 1 & 0 & 0 \\
1 & -1 & 0 & 0 \\
1 & -1 & 0 & 0 \\
- & - & - & - \\
0 & 0 & 1 & 1 \\
0 & 0 & 1 & 1 \\
0 & 0 & 1 & -1 \\
0 & 0 & 1 & -1
\end{bmatrix}
\begin{bmatrix}
b_{01} \\
b_{11} \\
b_{02} \\
b_{12}
\end{bmatrix}
+
\begin{bmatrix}
\varepsilon_{11} \\
\varepsilon_{21} \\
\varepsilon_{31} \\
\varepsilon_{41} \\
-- \\
\varepsilon_{12} \\
\varepsilon_{22} \\
\varepsilon_{32} \\
\varepsilon_{42}
\end{bmatrix}.
$$
$$\tag{6.33}$$

For subject randomization of the treatment, half of the subjects in each cluster will receive the treatment; therefore, $n/2 = 2$ subjects in each cluster will have the coding $x_k = +1$ and the other half will have the coding $x_k = -1$, as shown in the design matrix X. The design matrix Z corresponds to the two random parameters b_{0k} and b_{1k} for the two clusters. If we assume that these parameters are not correlated, that is, $\sigma_{01} = 0$, and that the parameters b_{0k} and b_{1k} have the same variances σ_0^2 and σ_1^2 in each cluster, then the variance–covariance matrix of the random parameters is a diagonal matrix

$$D = \begin{bmatrix} \sigma_0^2 & 0 & 0 & 0 \\ 0 & \sigma_1^2 & 0 & 0 \\ 0 & 0 & \sigma_0^2 & 0 \\ 0 & 0 & 0 & \sigma_1^2 \end{bmatrix}, \qquad (6.34)$$

with variances σ_0^2 and σ_1^2 on the main diagonal. The variance–covariance matrix of the responses $\mathrm{Cov}(y) = (ZDZ' + \sigma_\varepsilon^2 I)$ now becomes

$$\mathrm{Cov}(y) = \begin{bmatrix} A & 0 \\ 0 & A \end{bmatrix}$$

where

$$A = \begin{bmatrix} \sigma_0^2 + \sigma_1^2 + \sigma_\varepsilon^2 & \sigma_0^2 + \sigma_1^2 & \sigma_0^2 - \sigma_1^2 & \sigma_0^2 - \sigma_1^2 \\ \sigma_0^2 + \sigma_1^2 & \sigma_0^2 + \sigma_1^2 + \sigma_\varepsilon^2 & \sigma_0^2 - \sigma_1^2 & \sigma_0^2 - \sigma_1^2 \\ \sigma_0^2 - \sigma_1^2 & \sigma_0^2 - \sigma_1^2 & \sigma_0^2 + \sigma_1^2 + \sigma_\varepsilon^2 & \sigma_0^2 + \sigma_1^2 \\ \sigma_0^2 - \sigma_1^2 & \sigma_0^2 - \sigma_1^2 & \sigma_0^2 + \sigma_1^2 & \sigma_0^2 + \sigma_1^2 + \sigma_\varepsilon^2 \end{bmatrix}, \qquad (6.35)$$

This covariance structure is not simply compound symmetric, due to the inclusion of the random intercept b_{1k} in the two clusters, causing the covariances of the responses within the clusters to differ. Because of the specific coding $x_k = +1$ and $x_k = -1$, the covariances of the responses in the same clusters under the same conditions (treatment or control) are equal to $(\sigma_0^2 + \sigma_1^2)$, while the covariances of the responses in the same cluster under the two different conditions are equal to $(\sigma_0^2 - \sigma_1^2)$. The variance–covariance matrix of the fixed parameter estimators is given by

$$\mathrm{Cov}(\hat{\beta}) = \{X'[\mathrm{Cov}(y)]^{-1}X\}^{-1} = \begin{bmatrix} \dfrac{4\sigma_0^2 + \sigma_\varepsilon^2}{4 \times 2} & 0 \\ 0 & \dfrac{4\sigma_1^2 + \sigma_\varepsilon^2}{4 \times 2} \end{bmatrix}, \qquad (6.36)$$

and contain the variances $\mathrm{var}(\hat{\beta}_0)$ and $\mathrm{var}(\hat{\beta}_1)$ for the intercept and slope parameter estimator on the main diagonal. Their expressions are the same as the formulas given in Equation (6.15) with $K = 2$ and $n = 4$ substituted. Here again, the covariance of the fixed parameter estimators is zero, that is, $cov(\hat{\beta}_0, \hat{\beta}_1) = 0$.

It should be noted that if we assume a model with randomization at the subject level, but only assume the intercept to vary across clusters ($\sigma_0^2 > 0$) and the slope to be fixed ($\sigma_1^2 = 0$), then the variance of the fixed slope parameter in Equation (6.36) will become $\mathrm{var}(\hat{\beta}_1) = \sigma_\varepsilon^2 / Kn$; see also the corresponding entry in Table 6.1.

6.9 Summary

In this chapter, two different multilevel designs are described, namely, the multi-center trial with a treatment randomly assigned to subjects within clusters and a cluster randomized design with a treatment randomly assigned to the clusters. It is shown that in terms of efficiency, the multi-center trial is always more efficient than the cluster randomized trial, even if the intra-class correlation is small. This advantage of a multi-center trial over a cluster randomized design remains when different costs for sampling clusters and subjects within a cluster are taken into account. The example in this chapter also shows that power for testing a treatment effect is always higher in a multi-center trial than in a cluster randomized trial. Therefore, we recommend using a multi-center design unless the problem of treatment contamination cannot be adequately dealt with at the design stage by blinding procedures or otherwise. In that case, researchers are advised to use a cluster randomized design. In terms of efficiency increase, a multi-center trial will generally benefit more from an increase of the number of subjects than from an increase of the number of clusters. For a cluster randomized trial, the reverse is true.

Although inclusion of covariates in the multilevel model may sometimes lead to more powerful tests, such an inclusion is not recommended in general without verifying that there is a strong relation between the covariate and the outcome and without a check whether such an inclusion will not threaten the internal validity of the study. One of the problems that may arise due to inclusion of covariates is that the probability of non-response may rise. Results from the literature indicate that in most cases, unequal cluster sizes lead to an efficiency loss of about 10%. To compensate for such a moderate small efficiency loss, it is recommended to add an extra 11% of the number of clusters to the multilevel design.

7

Longitudinal designs for repeated measurement models

7.1 Design problem for repeated measurements

Repeated measurements are measurements that are obtained on an outcome or response variable from a unit or subject on a number of different occasions. Repeated measurements occur frequently in both experimental and observational studies. In this chapter we will distinguish repeated measurement designs from longitudinal designs.

A *repeated measurement design* is a design in which one or more variables are repeatedly measured at different occasions. Here, the occasion is *not* necessarily strictly time-structured. Different occasions may entail experimental conditions such as *treatment* and *no treatment* or different dosage levels of a drug. The ordering of the occasions in time may not be the same for the different units (subjects). In a crossover design, for example, a patient may first receive a treatment and then a placebo, or vice versa. The focus of repeated measurement studies is on inference about the differences among experimental conditions. Questions on treatment effects and differences in responses among dosage levels are usually of main interest in these studies. An often encountered repeated measurements design is the crossover design, which we will discuss separately in Chapter 8.

The term *longitudinal design* is usually used when the occasions at which measurements are taken are time-structured. On the basis of the ordering of the measurements in time, the main focus in longitudinal studies is on the modelling

An Introduction to Optimal Designs for Social and Biomedical Research M. P. F. Berger & W. K. Wong
© 2009, John Wiley & Sons, Ltd

of responses as a function of time. Research questions may be centred on the basic level of the responses or on the increase or decrease of the response function over time. In the latter case, the term *growth curves* is often used and the corresponding statistical models are generally called *growth curve models*. The generalized multivariate analysis of variance (MANOVA) model is an example of a growth curve model. See Kshirsagar and Smith (1995), among others, for a review. In this chapter we focus on longitudinal designs.

In longitudinal designs as well as in repeated measurement designs, the independent variables can be classified as *within-subject* and *between-subject* variables or factors. The variable *Time* is of course a within-subject variable. These variables distinguish different sources of variation in the study. For example, the variation of repeated measurements of a subject over time can be modelled as a function of the *within-subject* variable *Time*. If, on the other hand, a distinction is made between two groups of subjects, one receiving treatment *A* and the other receiving treatment *B*, then the treatment variable is a *between-subject* variable because it explains variation between the subjects in the two groups.

There are several textbooks on the analysis of longitudinal (repeated) measurements. Some examples are Lindsey (1993), Diggle, Liang and Zeger (1994), Kshirsagar and Smith (1995), Hand and Crowder (1996), Vonesh and Chinchilli (1997) and Verbeke and Molenberghs (2000). Monographs on this topic from a multilevel perspective include Goldstein (1995) and Snijders and Bosker (1999). However, research on design issues for longitudinal studies is relatively scarce perhaps because the issues are more complicated to address.

The problem of 'best' allocation of the repeated measurements in time is often circumvented by choosing equally spaced time points to measure the response variable. In some cases, however, unequally spaced time points may prove to be more efficient in terms of parameter estimation. It is sometimes recommended to cluster the time points at which measurements are taken in an interesting area of the time scale. For example, Matthews *et al.* (1990, p.235) proposed that extra observations could be taken in the area where a particular feature, such as a peak in the response, is known to occur. However, whether this would lead to extra information depends on the particular situation and on the method of analysis. Obtaining more observations at the point where the peak is likely to occur can be more informative for estimating the mean peak value if the location of the peak is roughly known. However, if we have no specific knowledge when the peak occurs in time, it is generally better to obtain observations in such a way that the variance of the time points at which the measurements are taken is as large as possible.

Clustering repeated measurements close to each other in time, that is, taking measurements with shorter time intervals between the measurements, will tend to increase the correlations among these measurements and in turn decrease efficiency. As an illustration, the effect of clustering repeated measurements on the variance of the estimator of a population mean is displayed in Figure 7.1. Let m be the number of repeated measurements and r be a correlation parameter. Let

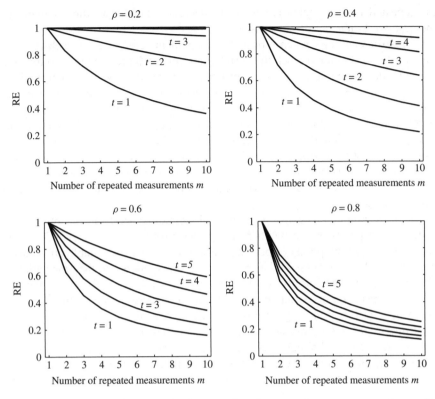

Figure 7.1 Relative efficiency plot for correlated repeated measurements ($r = \rho^t$).

var($\bar{y}_I$) and var($\bar{y}_C$) be the variances of the means $\bar{y}_I$ and $\bar{y}_C$ when we have independent observations and when we have correlated observations, respectively. These two variances can be compared using a relative efficiency (RE) measure defined by the inverse of the so-called *design effect*: $\text{RE} = \text{var}(\bar{y}_I)/\text{var}(\bar{y}_c) = 1/(1 + (m - 1)r)$. In Figure 7.1 the REs are plotted as a function of the number of repeated measurements m. The plots assume that the correlations among the repeated measurements are ruled by an auto-regressive process $r = \rho^t$, where t is the difference between two time points. For example, $t = 1$ for adjacent time points lying one unit apart on the time scale.

Figure 7.1 shows that it is not very efficient to take repeated (correlated) measurements with shorter time intervals between the measurements. As the observations are obtained closer to each other in time and their correlations increase (i.e. ρ^t increases), the RE values decrease. This decrease in efficiency becomes more prominent as the number of repeated measurements increases. Moreover, the RE plots decrease more sharply as the correlation ρ increases in value. Overall, Figure 7.1 suggests that it is desirable to spread out the repeated measurements as much as possible because in so doing, the correlations among the repeated measurements are likely to decrease and this in turn is likely to increase the efficiency of our estimates.

The efficiency loss when repeated measurements are closer to each other in time is also encountered when the correlations among the repeated measurements are not ruled by an auto-regressive process. For example, when all correlations are equal to $\rho = 0.6$, no matter how far apart they are measured in time, the RE plots will all be equal to the plot for which $t = 1$, showing sharp decrease in RE values as the number of repeated measurements increases. Winkens *et al.* (2006, Figure 2) found a similar observation in their longitudinal study to find linear divergent treatment effects. They showed that for a certain sample size, the addition of repeated measurements decreases the efficiency of a linear divergent treatment effect further as the correlations among the repeated measurements become higher.

Another important design issue for a longitudinal study is the number of distinct time points or the number of repeated measurements to include in the study. Too few time points (repeated measurements) will not enable an accurate description of the functional relation between the responses and the time variable. This may happen when only two time points are included, while the functional relation is based on more than two parameters. For these reasons, Willett, Singer and Martin (1998, p. 408) gave a rule of thumb and recommended including at least one more time point than the number of parameters in the model. Vickers (2003) argued that although increase of the number of repeated measurements from a single measurement to three or four measurements, will increase the power of a test, the benefit of an additional repeated measurement rapidly decreases as the number of measurements rises. His results support the conclusion that it is not very efficient to include too many repeated measurements (time points) in a study. Too many repeated measurements will be a waste of money and effort in many cases. Moreover, in clinical trials, especially where patients are exposed to strenuous measurement procedures, too many repeated measurements will also raise ethical objections.

On the other hand, increasing the number of time points (and measurements) will decrease the standard errors and increase reliability of the measurements. Especially when repeated measurements are obtained for a more precise estimate of an end point, it is often worthwhile from a statistical point of view, to obtain as many repeated measurements as feasible.

In deciding on the number of repeated measurements, the costs of data collection should also be taken into account. The costs of sampling subjects may be quite different from the costs of obtaining repeated measurements for each of these subjects. Usually, there is some kind of trade-off between the number of sampled subjects and the number of repeated measurements. Sometimes it may be more efficient in terms of parameter estimation and in terms of costs, to increase the sample of subjects at the expense of the number of repeated measurements.

A key feature of repeated measurements is that they are typically correlated. The actual correlation structure may be different for a longitudinal design and a repeated measurement design. A longitudinal design with time-structured measurements often has a serial- or auto-correlated covariance structure of the data, because correlations among repeated measurements tend to decrease when measurements are taken further apart in time. On the other hand, a repeated measurement (crossover) design may have a more compound symmetric structure, with more or less equal variances and equal covariances among the repeated measurements. In general, the efficiency of a design depends on the structure of the variance–covariance matrix of the responses and on the size of the correlations among the repeated measures. In particular, Berger and Tan (2004), Winkens *et al.* (2005) and Ortega-Azurduy, Tan and Berger (2009) studied the robustness of optimal designs against mis-specification of the parameters in the linear mixed effects model.

7.2   The design

We are interested in two types of designs: a longitudinal design, where the repeated measurements are obtained in a specific order of time, and, a repeated measurement design, where the measurements are not obtained in a fixed order for all subjects. This latter design is also referred to as a *crossover design* or *counter-balanced design* and will be discussed in Chapter 8. In this chapter we focus on the longitudinal design.

Consider a longitudinal design where m repeated measurements are obtained from a sample of n subjects. Such a longitudinal design can be schematically represented by the form:

$$\xi_N = \left\{ \begin{array}{ccccc} d_1 & d_2 & d_3 & \ldots & d_m \\ n/N & n/N & n/N & \ldots & n/N \end{array} \right\}. \tag{7.1}$$

The design points d_j (for $j = 1, \ldots, m$) are the time points at which the repeated measurements are taken with corresponding fractions n/N. This form represents an exact design where n is the (integer) sample size and N is the total number of observations, that is, $N = m \times n$. In some cases where for fixed values of N the sample size n turns out to be not an integer, integer approximation can be applied. See Pukelsheim and Rieder (1992) for rules on integer approximation.

Planners of repeated measurement studies usually start with the assumption that a design with repeated measurements is complete, that is, there are no missing data in the study. The design in Equation (7.1) assumes that all planned repeated measurements are actually available for analysis. Of course, this is not realistic in practice because patients are unlikely to show up at all occasions (time points) for examination, or they may drop out, or data can be simply lost or erroneously coded. Nevertheless, for simplicity, we will continue to assume that we have complete data at the end of the study for analysis. We refer the reader interested in analysing data with missing values to Little and Rubin (1987), where a full treatment of this topic is available.

7.3   Analysis techniques for repeated measures

Various statistical methods to analyse repeated measurements have been suggested. They range from relatively simple and easy to apply methods to more sophisticated statistical models. Here is an incomplete list of methods employed in other fields for analysing repeated measurements:

- Separate t tests of differences among groups at each occasion (repeated measurement).

- Use of a number of different summary measures to describe the response profile for each subject. Two often suggested summary measures are the mean of the repeated measurements per subject and the area under the curve (AUC).

- Analysis of the differences between mean changes from pre-treatment to the post-treatment measurements. This method is often referred to as the *analysis of change scores*.

- Analysis of covariance (ANCOVA).

- Analysis of variance (ANOVA) for repeated measurements and MANOVA.

- Linear mixed effects models.

Each of these methods has its own advantages and disadvantages. Everitt (1995) discussed the pros and cons of each method with illustrative examples.

Separate t tests for each of the repeated measurements are easy to apply, but do not take into account the multiple testing problem and do not use information from the correlations among the repeated measurements. For these reasons, the t-test approach is not recommended to analyse repeated measurements.

The use of summary measures as suggested by Matthews *et al.* (1990) has a potential disadvantage that summary measures do not use all information contained by the repeated measurements. Although a summary measure, such as the mean, may be more reliable than the individual repeated measurements, it may be quite difficult to select the most informative summary measure for a particular study. For example, the AUC of two response functions may not contain sufficient information to distinguish the response curves if they have different shapes.

In randomized studies, where pre-treatment observations are available, the ANCOVA is more efficient and powerful than the analysis of change scores and it is therefore usually recommended. Frison and Pocock (1992) showed that ANCOVA becomes increasingly more efficient than change score analysis as the mean correlation among the repeated measurements increases. However, in non-randomized trials, ANCOVA seems to be more biased than change score analysis (Van Breukelen, 2006).

The ANOVA for repeated measurements requires that the variance–covariance matrix of the repeated measures meets the so-called *circularity* or *sphericity* condition. This condition can be tested (e.g. Mauchly, 1940), and if the variance–covariance matrix departs from this assumption, the degrees of freedom for the test can be adjusted (Greenhouse and Geisser, 1959; Huynh and Feldt, 1976). A special and often considered case of *sphericity* is the *compound symmetric covariance matrix* with a 'common' variance on the main diagonal and a 'common' covariance as the off-diagonal element. Such a compound symmetric structure can be estimated by using the mean variance of the responses $\bar{\sigma}^2$ and mean covariance $\bar{\sigma}_{jj'}$ among the responses as estimates of the 'common' variance and 'common' covariance of the repeated measurements.

An alternative approach to circumvent the assumption about the sphericity structure of the variance–covariance matrix is to use MANOVA. The main disadvantage of MANOVA, however, is that for relatively small sample sizes, many repeated measurements and small departures from the sphericity assumption, this procedure will have less power than the ANOVA procedure (Rouanet and Lepine, 1970; Davidson, 1972).

There are many ways to analyse repeated measurements or longitudinal data; and while there is no agreement across disciplines on the best method, the linear mixed model is one of the most flexible methods to use for analysing such data. A key reason for this is that the linear mixed effects models are able to account for the different sources of variation from the fixed and random effects, and are also able to easily cope with unequally spaced time points and missing data.

7.4   The linear mixed effects model for repeated measurement data

Linear mixed effect models have been discussed by Diggle, Liang and Zeger (1994) and Verbeke and Molenberghs (2000), among others. The linear mixed effects model for repeated measurement data can be described in two steps. The measurement y_{ij} of the ith individual or subject at occasion j $(j = 1, \ldots, m)$ can be represented by the regression model:

$$y_{ij} = a_{0i}x_{0ij} + a_{1i}x_{1ij} + a_{2i}x_{2ij} + \cdots + a_{p-1,i}x_{p-1,ij} + \varepsilon_{ij}, \qquad (7.2)$$

where $x_{0ij}, x_{1ij}, x_{2ij}, \cdots, x_{p-1,ij}$ are p quantitative covariates that model the ith individual responses over the m occasions and $a_{0i}, a_{1i}, \cdots, a_{p-1,i}$ are the corresponding parameters. The errors ε_{ij} are assumed to be normally distributed with a mean of zero and a certain variance–covariance matrix. The second step relates the fixed parameters to the random parameters:

$$a_{0i} = \beta_0 + b_{0i}$$
$$a_{1i} = \beta_1 + b_{1i}$$
$$\ldots$$
$$\ldots$$
$$a_{q-1,i} = \beta_{q-1} + b_{q-1,i} \qquad (7.3)$$
$$a_{q,i} = \beta_q$$
$$\ldots$$
$$\ldots$$
$$a_{p-1,i} = \beta_{p-1}.$$

The q coefficients $b_{0i}, b_{1i}, \ldots, b_{q-1,i}$ are random (subject-specific) parameters, which describe the variation between the subjects. We assume that they are normally distributed with means equal to zero and that they are independent of the error terms ε_{ij}. The p parameters $\beta_0, \beta_1, \ldots, \beta_{p-1}$ are fixed parameters that describe the overall effects. Equation (7.3) displays the case where the number of fixed parameters is larger than or equal to the number of random parameters, that is, $p \geq q$. If the number of fixed parameters is equal to the number of random parameters, that is, if $p = q$, then the last $(p - q)$ fixed parameters $\beta_q, \beta_{q+1}, \ldots, \beta_{p-1}$ are zero. A direct substitution of Equation (7.3) into Equation (7.2) results in the so-called linear mixed effects model. In the next subsections, we focus on two special cases of the linear mixed effects model, namely, the random intercept (RI) model and, the RI and slope model.

7.4.1   Random intercept model

Consider, as an example, a longitudinal study (Lloyd *et al.*, 1993) to investigate the effect of daily calcium supplementation on bone gain in adolescent women over a 2-year period. The total body bone mineral density (TBBMD, gr/cm^2) was measured at $m = 5$ successive visits scheduled every 6 months. Although these visits were not exactly 6 months apart, we will assume that all visits were equally spaced in time and at 6-month intervals. Suppose that the specific profiles of the women and their mean profiles are all assumed to be linear and that the TBBMD profiles of the women are parallel and only vary in height. This means that we assume the variation of the TBBMD scores among women can be adequately described by a RI parameter $a_{0i} = \beta_0 + b_{0i}$. The slopes a_{1i} of the individual profiles are assumed to be equal, that is, $a_{1i} = \beta_1$, for all i. The model in Equations (7.2) and (7.3) now reduces to a so-called RI model which is given by:

$$y_{ij} = a_{0i}x_{0ij} + a_{1i}x_{1ij} + \varepsilon_{ij}$$

or

$$y_{ij} = (\beta_0 + b_{0i}) + \beta_1 t_j + \varepsilon_{ij}, \qquad (7.4)$$

where the covariate $x_{0ij} = 1$ for all (i, j), and the covariate $x_{1ij} = t_j$ describes the time schedule of the five visits and is numbered as $x_{1ij} = t_j = 1, 2, 3, 4$ or 5 (for all i). The interval between two adjacent time points $(t_j - t_{j-1})$ is 6 months (for $j = 2, 3, 4, 5$). The parameters β_0 and β_1 are the mean intercept and slope, respectively, while the parameter b_{0i} represents the deviation from the mean intercept for the ith woman.

Now suppose that the error terms are independently distributed with variance σ_ε^2. If the variance of the RI b_{0i} is denoted by σ_0^2, then the variance of the responses at each time point t_j is equal to $\text{var}(y_{ij}) = (\sigma_0^2 + \sigma_\varepsilon^2)$ and the covariance between measurements at time points t_j and $t_{j'}$ is $\text{cov}(y_{ij}, y_{ij'}) = \sigma_0^2$. This means that there is an equal correlation between any pair of repeated measurements over time:

$$\rho = \frac{\sigma_0^2}{\sigma_0^2 + \sigma_\varepsilon^2}. \qquad (7.5)$$

This quantity is known as the *intra-class* (intra-woman) *correlation coefficient*. The variance–covariance structure of the repeated measurements under this RI model is characterized by a common variance and a common covariance among the measurements over time. This variance–covariance structure is known as a *compound symmetry structure*.

7.4.2   Random intercept and slope model

An extension of the RI model is the RI and random slope model. If the profiles of the adolescent women are not parallel over time and each profile has its own slope, then the model in Equations (7.2) and (7.3) can be written as:

$$y_{ij} = (\beta_0 + b_{0i}) + (\beta_1 + b_{1i})t_j + \varepsilon_{ij}, \qquad (7.6)$$

where the RI, b_{0i} and random slope, b_{1i} represent deviations from the mean intercept and slope for the ith girl. The variance of the responses over time is now $\text{var}(y_{ij}) = (\sigma_0^2 + \sigma_1^2 t_j^2 + 2\sigma_{01} t_j + \sigma_\varepsilon^2)$, where σ_1^2 is the random slope variance and σ_{01} is the covariance between the RI and random slope. The covariance between measurements at distinct time points t_j and $t_{j'}$ is given by:

$$\text{cov}(y_{ij}, y_{ij'}) = (\sigma_0^2 + \sigma_1^2 t_j t_{j'} + \sigma_{01} t_j + \sigma_{01} t_{j'}). \qquad (7.7)$$

This formula indicates that the covariance structure of the responses over time consists of covariances that are different for different pairs of time points. When the five time points are coded as $t_j = 1, 2, 3, 4$ or 5, the covariance structure of the five responses in the vector y_i from the ith patient is

$\mathrm{Cov}(y_i)$

$$
= \begin{bmatrix}
\sigma_0^2 + 2\sigma_{01} + \sigma_1^2 + \sigma_\varepsilon^2 \\
\sigma_0^2 + 3\sigma_{01} + 2\sigma_1^2 & \sigma_0^2 + 4\sigma_{01} + 4\sigma_1^2 + \sigma_\varepsilon^2 \\
\sigma_0^2 + 4\sigma_{01} + 3\sigma_1^2 & \sigma_0^2 + 5\sigma_{01} + 6\sigma_1^2 & \sigma_0^2 + 6\sigma_{01} + 9\sigma_1^2 + \sigma_\varepsilon^2 \\
\sigma_0^2 + 5\sigma_{01} + 4\sigma_1^2 & \sigma_0^2 + 6\sigma_{01} + 8\sigma_1^2 & \sigma_0^2 + 7\sigma_{01} + 12\sigma_1^2 \\
\sigma_0^2 + 6\sigma_{01} + 5\sigma_1^2 & \sigma_0^2 + 7\sigma_{01} + 10\sigma_1^2 & \sigma_0^2 + 8\sigma_{01} + 15\sigma_1^2
\end{bmatrix}
$$

$$
\begin{matrix}
& \text{Symmetric} & \\
\sigma_0^2 + 8\sigma_{01} + 16\sigma_1^2 + \sigma_\varepsilon^2 & & \\
\sigma_0^2 + 9\sigma_{01} + 20\sigma_1^2 & \sigma_0^2 + 10\sigma_{01} + 25\sigma_1^2 + \sigma_\varepsilon^2 &
\end{matrix}
$$

$$(7.8)$$

This structure holds under the assumption that the errors are independently distributed. The linear mixed model with such a covariance matrix is sometimes called a *conditional independence model* because, conditional on the random parameters, the responses are independent (Verbeke and Molenberghs, 2000). The above two random effects models can be extended straightforwardly to describe a quadratic (curvilinear) fixed time effect, together with a quadratic function describing the subject-specific profiles. The structure of the variance–covariance matrix will be more complicated than the one presented in Equation (7.8).

7.5   Variance–covariance structures

The variance–covariance structure of repeated measurements for the linear mixed effects model in Equations (7.2) and (7.3) is composed of two sources of variation, namely, the within-subject or intra-subject variation and the between-subject or inter-subject variation. We note that measurement error can also be a third source of variation in a linear mixed effects model (Diggle, Liang and Zeger, 1994, p.80; Verbeke and Molenberghs, 2000, p.28), but for simplicity, we will restrict attention to only the first two sources of variation in this book.

These two sources of variation give rise to a variety of different variance–covariance structures for the repeated measurements. All are special cases of the general structure proposed by Diggle, Liang and Zeger (1994). See Jennrich and Schluchter (1986), Diggle (1988), Chi and Reinsel (1989) and Rochon (1992) for further discussion on covariance structures for repeated measurements.

7.5.1   Compound symmetry structure

A popular variance–covariance structure is the compound symmetry matrix, with equal variances and equal covariances. This structure arises for the RI model in Section 7.4.1. Its variances and covariances are $\text{var}(y_{ij}) = (\sigma_0^2 + \sigma_\varepsilon^2)$ and $\text{cov}(y_{ij}, y_{ij'}) = \sigma_0^2$, respectively. Here, σ_0^2 is the between-subject variance and σ_ε^2 is the within-subject variance of the independent errors.

Such a compound symmetry structure can also be found when the model is assumed to be a fixed effects (FE) model with errors having equal variances and equal covariances among the time points. These two models with a compound symmetric variance–covariance matrix of the repeated measurements are presented as illustration in Table 7.1 as Case I and Case II. It can be seen that these two models can be regarded as two parameterizations of the same model. As stated before, the compound symmetric structure is a special case of the more general sphericity variance–covariance structure (Greenhouse and Geisser, 1959; Huynh and Feldt, 1976), with equality of the variances of the differences between each pair of repeated measurements.

Table 7.1   Two models with a compound symmetric variance–covariance structure.

Case I: Random intercept model with independent errors

$y_{ij} = (\beta_0 + b_{0i}) + \beta_1 t_j + \varepsilon_{ij}$, with σ_ε^2 and $\text{cov}(\varepsilon_{ij}, \varepsilon_{ij'}) = 0$, for all $j \neq j'$

Covariance structure:

$$\text{Cov}(y_i) = \begin{bmatrix} \sigma_0^2 + \sigma_\varepsilon^2 & \sigma_0^2 & \cdots & \sigma_0^2 \\ \sigma_0^2 & \sigma_0^2 + \sigma_\varepsilon^2 & \cdots & \sigma_0^2 \\ \vdots & \vdots & \vdots & \vdots \\ \sigma_0^2 & \sigma_0^2 & \cdots & \sigma_0^2 + \sigma_\varepsilon^2 \end{bmatrix}$$

Case II: Fixed effects model with correlated errors ρ

$y_{ij} = \beta_0 + \beta_1 t_j + \varepsilon_{ij}$, with σ_ε^2 and $\text{cov}(\varepsilon_{ij}, \varepsilon_{ij'}) = \rho\sigma_\varepsilon^2$, for all $j \neq j'$

Covariance structure:

$$\text{Cov}(y_i) = \begin{bmatrix} \sigma_\varepsilon^2 & \rho\sigma_\varepsilon^2 & \cdots & \rho\sigma_\varepsilon^2 \\ \rho\sigma_\varepsilon^2 & \sigma_\varepsilon^2 & \cdots & \rho\sigma_\varepsilon^2 \\ \vdots & \vdots & \vdots & \vdots \\ \rho\sigma_\varepsilon^2 & \rho\sigma_\varepsilon^2 & \cdots & \sigma_\varepsilon^2 \end{bmatrix}$$

Note. ρ is the correlation among the errors.
σ_0^2 and σ_ε^2 are the variances of the random intercept and errors, respectively.
y_i is vector of responses for subject i.

7.5.2   Auto-correlation structure

In many practical situations where repeated measurements are time-structured, the assumption of independent errors is not realistic and a model with correlated errors is likely to provide a better fit to the data. Often times, one assumes that the errors follow a first-order auto-regressive, AR(1), correlation pattern. This pattern seems to be justified in many studies especially when the repeated measurements are obtained over a long period (Chi and Reinsel, 1989; Jones, 1990). Other relevant types of covariance matrices for repeated measurements include higher order auto-regressive variance–covariance structures studied by Jones and Boardi-Boteng (1991) and Rochon (1992) and the so-called Toeplitz matrices used by Jennrich and Schluchter (1986). Verbeke and Molenberghs (2000, p.99) reviewed different covariance structures and their uses in their monograph on linear mixed models.

The auto-correlation between errors from two time points t_j and $t_{j'}$ has the form $\rho^{|t_j - t_{j'}|}$, where ρ is a correlation parameter ($0 < \rho < 1$). Hence, the correlations between the errors depend on how far apart they are in time. The correlation between the errors decreases as the time points lie farther apart. For the time coding $t_j = 1, 2, 3, \ldots, m$, the auto-correlation matrix of the errors has the form:

$$
\text{Corr}(\varepsilon_{ij}, \varepsilon_{ij'}) =
\begin{bmatrix}
1 & \rho & \rho^2 & \ldots & \rho^{m-1} \\
\rho & 1 & \rho & \ldots & \rho^{m-2} \\
\rho^2 & \rho & 1 & \ldots & \ldots \\
\vdots & \vdots & \vdots & \vdots & \vdots \\
\rho^{m-1} & \rho^{m-2} & \ldots & \ldots & 1
\end{bmatrix}. \tag{7.9}
$$

Table 7.2 displays the variance–covariance matrices for the repeated measurements of the same two models in Table 7.1, but now with an AR(1) covariance structure for the errors. From the variance–covariance structure of the RI model (Case III), it can be seen that the variances of the responses are the sum of two elements, namely, the variance of the RI σ_0^2 and the variance of the errors σ_ε^2. The covariance between the responses also depends on these two variance components and on the correlation parameter ρ. It is equal to the sum of σ_0^2 and the covariance of the errors $\sigma_\varepsilon^2 \rho^{|t_j - t_{j'}|}$.

Changes in values of σ_0^2 (between-subject variance) and σ_ε^2 (within-subject variance) will change the form of the variance–covariance matrix $\text{Cov}(y_i)$ in Table 7.2 (Case III). If the variance σ_0^2 is very large relative of σ_ε^2, the auto-correlation pattern of the errors is less pronounced in $\text{Cov}(y_i)$. Likewise, if the variance σ_0^2 is small or near zero, the auto-correlation structure of the errors will become more pronounced in the covariance pattern of $\text{Cov}(y_i)$. The size of the auto-correlation parameter also influences the covariance pattern. If ρ is very small, the correlation among the errors will also be very small and so the covariance matrix is nearly diagonal. In other words, the covariance matrix becomes similar to the covariance pattern of uncorrelated errors. The FE model with auto-correlated errors is also presented in Table 7.2 as Case IV. Actually, the variance–covariance matrix of the responses of Case III is equal to the variance–covariance matrix of the responses of Case IV with the variance component σ_0^2 added to each element.

Table 7.2   Two models with first-order auto-regressive correlated errors.

Case III: Random intercept model with auto-correlated errors

$$y_{ij} = (\beta_0 + b_{0i}) + \beta_1 t_j + \varepsilon_{ij}, \text{ with } \sigma_\varepsilon^2 \text{ and } \text{cov}(\varepsilon_{ij}, \varepsilon_{ij'}) = \sigma_\varepsilon^2 \rho^{|t_j - t_{j'}|},$$
for all $j \neq j'$

Covariance structure:

$$\text{Cov}(y_i) = \begin{bmatrix} \sigma_0^2 + \sigma_\varepsilon^2 & \sigma_0^2 + \sigma_\varepsilon^2 \rho & \cdots & \sigma_0^2 + \sigma_\varepsilon^2 \rho^{m-1} \\ \sigma_0^2 + \sigma_\varepsilon^2 \rho & \sigma_0^2 + \sigma_\varepsilon^2 & \cdots & \sigma_0^2 + \sigma_\varepsilon^2 \rho^{m-2} \\ \vdots & \vdots & \vdots & \vdots \\ \sigma_0^2 + \sigma_\varepsilon^2 \rho^{m-1} & \sigma_0^2 + \sigma_\varepsilon^2 \rho^{m-2} & \cdots & \sigma_0^2 + \sigma_\varepsilon^2 \end{bmatrix}$$

Case IV: Fixed effects model with auto-correlated errors

$$y_{ij} = \beta_0 + \beta_1 t_j + \varepsilon_{ij}, \text{ with } \sigma_\varepsilon^2 \text{ and } \text{cov}(\varepsilon_{ij}, \varepsilon_{ij'}) = \sigma_\varepsilon^2 \rho^{|t_j - t_{j'}|}, \text{ for all } j \neq j'$$

Covariance structure:

$$\text{Cov}(y_i) = \begin{bmatrix} \sigma_\varepsilon^2 & \sigma_\varepsilon^2 \rho & \cdots & \sigma_\varepsilon^2 \rho^{m-1} \\ \sigma_\varepsilon^2 \rho & \sigma_\varepsilon^2 & \cdots & \sigma_\varepsilon^2 \rho^{m-2} \\ \vdots & \vdots & \vdots & \vdots \\ \sigma_\varepsilon^2 \rho^{m-1} & \sigma_\varepsilon^2 \rho^{m-2} & \cdots & \sigma_\varepsilon^2 \end{bmatrix}$$

Note. ρ is a correlation parameter.
σ_0^2 and σ_ε^2 are variances of the random intercept and errors, respectively.
y_i is vector of responses for subject i.

7.6   Estimation of parameters and efficiency

In the general linear mixed model there are two different kinds of parameters to be estimated. The fixed parameters are $\beta = (\beta_0, \beta_1, \ldots, \beta_{p-1})'$ and the variance components associated with the random parameters are $\sigma^2 = (\sigma_0^2, \sigma_1^2, \ldots, \sigma_{01}^2, \ldots, \sigma_{jk}^2, \sigma_\varepsilon^2)'$. For example, the RI model in Table 7.1 (Case I) has two fixed parameters $\beta = (\beta_0, \beta_1)'$, a variance component σ_0^2 for the RI and a variance component σ_ε^2 for the errors. In what is to follow, it is convenient to let θ be the vector containing all the parameters in the linear mixed model. For the simple linear model with an RI, we have $\theta = (\beta, \sigma^2)' = (\beta_0, \beta_1, \sigma_0^2, \sigma_\varepsilon^2)'$.

The parameters in the general linear mixed model can be estimated by maximizing the likelihood function based on all the fixed parameters and variance components in the model. The usual approach is to obtain estimators by maximizing the likelihood function $L(\theta)$ with respect to the parameters θ. For the case where the variance components are assumed to be known, the maximum likelihood (ML) estimators of the fixed parameters β are equivalent to the generalized least squares estimators given by Laird and Ware (1982). However, when the variance components are not known, suitable estimates of these variance components will be needed to obtain estimates of the fixed parameters β.

The variance components are usually estimated via ML or via restricted maximum likelihood (REML). The ML estimator for the variance components is obtained by maximizing the likelihood function after the fixed parameters β are substituted by their estimates $\hat{\beta}$. The REML estimators for the variance components are obtained by maximizing the likelihood function of a set of error contrasts; see Harville (1974), Searle, Casella and McCulloch (1992) and Verbeke and Molenberghs (2000), among others, for more details. Because ML estimation of the variance components produces (downward) biased estimates, the variances of the fixed effect parameter estimators will become smaller than those obtained by means of REML estimation. The differences in ML and REML estimates of the variance components will be more pronounced when the sample size is small. For large samples both methods will approximately lead to the same results.

For the RI model in Table 7.1 (Case I), the asymptotic variance–covariance matrix of the parameter estimators $\hat{\theta} = (\hat{\beta}_0, \hat{\beta}_1, \hat{\sigma}_0^2, \hat{\sigma}_\varepsilon^2)'$ can be schematically represented as:

$$
\text{Cov}(\hat{\theta}) = \begin{bmatrix} \text{var}(\hat{\beta}_0) & \text{cov}(\hat{\beta}_0, \hat{\beta}_1) & 0 & 0 \\ \text{cov}(\hat{\beta}_0, \hat{\beta}_1) & \text{var}(\hat{\beta}_1) & 0 & 0 \\ 0 & 0 & \text{var}(\hat{\sigma}_0^2) & 0 \\ 0 & 0 & 0 & \text{var}(\hat{\sigma}_\varepsilon^2) \end{bmatrix}. \tag{7.10}
$$

In the upper left part of this matrix the 2×2 variance–covariance matrix of the fixed parameters estimators $\hat{\beta} = (\hat{\beta}_0, \hat{\beta}_1)'$ is given. The remaining elements on the main diagonal are the variances of the variance components estimators $\hat{\sigma}_0^2$ and $\hat{\sigma}_\varepsilon^2$, respectively. The structure of $\text{Cov}(\hat{\theta})$ shows that the fixed parameter estimators $\hat{\beta} = (\hat{\beta}_0, \hat{\beta}_1)'$ and the variance component estimators $\hat{\sigma}^2 = (\hat{\sigma}_0^2, \hat{\sigma}_\varepsilon^2)'$ are asymptotically orthogonal (Ankerman, Aviles and Pinheiro, 2003; Searle, Casella and McCulloch, 1992).

Since the variances of the estimators of the fixed parameters β are a function of the variance–covariance matrix of the responses, that is, they depend on the specification of the variance components in σ^2, it is important to correctly specify the model and the form of the variance–covariance matrix $\text{Cov}(y_i)$. Inaccurate estimates of the variance components in σ^2, will in turn lead to inaccurate and biased estimates of the fixed parameters in $\hat{\beta}$. This will be illustrated in the following sections by means of an example.

7.6.1   Small sample behaviour of estimators

The ML and REML estimators of fixed and random parameters and the corresponding (likelihood ratio, Wald) tests in linear mixed models are all based on asymptotic theory and very little is known about their small sample properties. Crainiceanu, Ruppert and Vogelsang (2003) showed that asymptotic results can fail for the likelihood ratio tests. To reduce small sample bias, Kenward and

Rogers (1997) proposed a scaled Wald statistic with an adjusted estimator of the variance–covariance matrix, but Vallejo and Livacic-Rojas (2005) showed that compared to the method of Brown and Forsythe (1974), the Kenward and Rogers statistic did not perform overall best in small samples.

For the design problem of linear mixed models it is important to know whether the asymptotic variance–covariance matrix $\mathrm{Cov}(\hat{\beta})$ is a good approximation to the one that is obtained from a small sample with estimates of the variance components. Goos (2002, p. 98–101) gave some evidence that for linear mixed effects models with a compound symmetric variance–covariance structure, the determinants of the asymptotic $\mathrm{Cov}(\hat{\beta})$ and the corresponding small sample variance–covariance matrix estimator with REML estimates and educated guesses based on prior studies are not very different, but noted that the small sample determinants almost always are larger than the asymptotic determinants.

7.7   Bone mineral density example

Consider again, the study on the effect of calcium supplementation on the rate of bone gain in girls during early adolescence. Lloyd *et al.* (1993) gave results from a clinical trial in which adolescent women were given a daily calcium supplement. The TBBMD (gr/cm^2) was measured every 6 months over a 2-year period. Five repeated TBBMD measurements were collected in five successive visits, including a base line measurement. Further details of the study are available in Lloyd *et al.* (1993) and the data set is available in Vonesh and Chinchilli (1997, p. 229–230). For our purpose here, we ignore subjects with incomplete data and only use data from 44 women with complete records.

In Figure 7.2 the individual TBBMD profiles of the women who received the daily calcium supplement are displayed in the right plot. The mean profile is presented on the left plot, together with the 95% confidence intervals, which

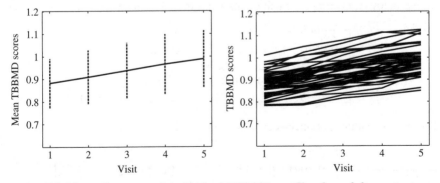

Figure 7.2 Plots of mean and individual TBBMD profiles for adolescent women with calcium supplementation (vertical dotted lines are 95% confidence intervals). (Data from: Vonesh and Chinchilli, 1997, p. 228–230).

are displayed as vertical dotted lines. These plots seem to indicate that a linear function would summarize both the individual development of bone mineral density in these women and their average profile very well. Moreover, the variation between the profiles of these women is mainly explained by the difference in height of TBBMD scores. The individual profiles are all rather parallel, and only differ in height.

Given these observations three different models can be selected to describe these data. An FE model with an unstructured covariance matrix for the error terms and two RI models, one assuming independent errors and the second assuming auto-correlated errors:

Model 1 : FE model with unstructured covariance matrix for errors :

$$y_{ij} = \beta_0 + \beta_1 t_j + \varepsilon_{ij}, \text{ with unknown } \sigma_\varepsilon^2 \text{ and unknown } \text{cov}(\varepsilon_{ij}, \varepsilon_{ij'}),$$
$$\text{for all } j \neq j';$$

Model 2 : RI model with independent errors :

$$y_{ij} = (\beta_0 + b_{0i}) + \beta_1 t_j + \varepsilon_{ij}, \text{ with } \sigma_\varepsilon^2 \text{ and } \text{cov}(\varepsilon_{ij}, \varepsilon_{ij'}) = 0,$$
$$\text{for all } j \neq j'; \tag{7.11}$$

Model 3 : RI model with auto-correlated errors :

$$y_{ij} = (\beta_0 + b_{0i}) + \beta_1 t_j + \varepsilon_{ij}, \text{ with } \sigma_\varepsilon^2 \text{ and } \text{cov}(\varepsilon_{ij}, \varepsilon_{ij'}) = \sigma_\varepsilon^2 \rho^{|t_j - t_{j'}|}$$
$$\text{for all } j \neq j'.$$

These three models contain a common slope parameter β_1. The two RI models have an additional term, the RI b_{0i} to describe the differences in height among the profiles. Note that *Model* 2 is the same as Case I in Table 7.1 and *Model* 3 is the same as Case III in Table 7.2.

Model 1 with only FE assumes an unstructured covariance matrix for the errors. The estimated correlation matrix for the responses from the five visits shown below was computed using data from the 44 women with complete data:

$$\text{Corr}(y_i) = \begin{bmatrix} 1.0000 & & & & \\ 0.9610 & 1.0000 & & \text{Symmetric} & \\ 0.9205 & 0.9632 & 1.0000 & & \\ 0.8929 & 0.9493 & 0.9731 & 1.0000 & \\ 0.8666 & 0.9294 & 0.9443 & 0.9625 & 1.0000 \end{bmatrix}. \tag{7.12}$$

The corresponding variance–covariance matrix is

$$\widehat{\text{Cov}}(y_i) = 10^{-2} \begin{bmatrix} 0.2978 & & & & \\ 0.3091 & 0.3473 & & \text{Symmetric} & \\ 0.3085 & 0.3486 & 0.3771 & & \\ 0.3240 & 0.3720 & 0.3974 & 0.4422 & \\ 0.2973 & 0.3443 & 0.3646 & 0.4024 & 0.3953 \end{bmatrix}. \tag{7.13}$$

The variances of the measurements at the five successive visits are on the main diagonal of $\widehat{\text{Cov}}(y_i)$. It can be seen that the correlations among the repeated measurements are quite high and that they tend to decrease as the visits lie farther apart in time. Moreover, the variances of the measurements seem to increase a little over time.

Model 1 is a FE model and hence all variation of the responses in Equation (7.13) is due to the variation of the errors in the model. Accordingly, $\text{Cov}(y_i) = \text{Cov}(\varepsilon_i)$. The variance–covariance matrix $\text{Cov}(y_i)$ is called *unstructured*, because its elements are computed from the data without assuming any specific structure. For the bone mineral example with five time points, there are $5(5 + 1)/2 = 15$ parameters (variance components of the errors) that need to be estimated. This rather unwieldy situation can be improved by assuming a structured variance–covariance matrix. In this way, the number of unknown variance components can be reduced, which in turn will generally increase efficiency of the fixed parameters and increase the power of statistical tests on these parameters.

Model 2 and *Model* 3 in Equation (7.11) assume a structured variance–covariance matrix of the errors and reduce the number of variance components to two, namely, the RI variance σ_0^2 and the error variance σ_ε^2. In most cases, we need a computer program to estimate these variance components via REML (Harville, 1977; Searle, Casella and McCulloch, 1992). However, our small sample size ($n = 44$) in this example is likely to be too small to meet the large sample requirements for ML or REML estimation. Therefore, in what is to follow, we will illustrate using heuristic estimators of the variance components. These estimators can be easily computed using a desk calculator.

Model 2 in Equation (7.11) only needs estimates for two variance components, namely, σ_0^2 and σ_ε^2 and is the same as Case I in Table 7.1. The variances of the responses (elements on the main diagonal) are assumed to be equal to $(\sigma_0^2 + \sigma_\varepsilon^2)$ and the covariances among the responses (elements off the main diagonal) are assumed to be equal to σ_0^2. We use the average covariance of the unstructured variance–covariance matrix in Equation (7.13) as a heuristic estimate of σ_0^2 and the average variance as an estimate of $(\sigma_0^2 + \sigma_\varepsilon^2)$. The heuristic estimates for the two variance components are $\bar{\sigma}_0^2 = 0.3468 \times 10^{-2}$ and $\bar{\sigma}_\varepsilon^2 = 0.0251 \times 10^{-2}$, respectively. Table 7.3 displays the resulting estimate of the variance–covariance matrix of the responses and also the variance–covariance matrix of estimates for the two fixed parameters.

Model 3 in Equation (7.11) is the same as the RI model with auto-correlated errors in Table 7.2 (Case III). The variance–covariance matrix of the responses has the special structure given in Table 7.2. For this model, there are two variance components σ_0^2 and σ_ε^2, and one correlation ρ to be estimated. As before, the estimates for σ_0^2 and σ_ε^2 are $\bar{\sigma}_0^2 = 0.3468 \times 10^{-2}$ and $\bar{\sigma}_\varepsilon^2 = 0.0251 \times 10^{-2}$. To estimate ρ, we first recall that the covariance of the auto-correlated errors is $\text{cov}(\varepsilon_{ij}, \varepsilon_{ij'}) = \sigma_\varepsilon^2 \rho^{|t_j - t_{j'}|}$. Now, let $\bar{r}_1$ be the average correlation of the responses y_j among adjacent time points (lag 1 correlations), that is, $\bar{r}_1 = (0.9610 + 0.9632 + 0.9731 + 0.9625)/4 = 0.9649$, and set this average correlation equal to $\bar{r}_1 = (\bar{\sigma}_0^2 + \bar{\sigma}_\varepsilon^2 \rho)/(\bar{\sigma}_0^2 + \bar{\sigma}_\varepsilon^2) = 0.9649$. Solving this last equation yields an estimate for the correlation parameter ρ, that is, $\hat{\rho} = [\bar{r}_1(\bar{\sigma}_0^2 + \bar{\sigma}_\varepsilon^2) - \bar{\sigma}_0^2]/\bar{\sigma}_\varepsilon^2 = 0.4845$.

We emphasize that we use heuristic estimators for estimating the variance components because they are simple and easy to compute. In most situations, it is of course easier to obtain ML or REML estimates with the help of a computer program. For this example with balanced data and no missing data or drop out, it turns out that these heuristic estimators are all very close to the ML and REML estimators, even though the sample size is small.

There is quite a large difference in values of the correlations between the responses and the correlation between the corresponding errors. The correlation of the errors associated with the measurements between two adjacent visits is $\hat{\rho} = 0.4845$ and the correlation of the errors incurred at the first and last visits is $\hat{\rho}^4 = 0.0551$. The latter correlation is very small, while the correlation of the responses themselves between the first and last visit is very high, namely, 0.8666. This difference between the correlation of the errors and the responses can be explained by the variance of the RI, which is relatively large compared to the variance of the errors.

Table 7.3 displays estimates for the two FE parameters β_0 and β_1 along with their estimated variance–covariance matrices $\widehat{\text{Cov}}(\hat{\beta})$ for the unstructured variance–covariance model and for two RI models, one with independent errors and the other with auto-correlated errors. It can be seen that the parameter estimates and their variances and covariances are different for these three models. Although the differences among the parameter estimates $\hat{\beta}_0$ and $\hat{\beta}_1$ seem small, there are clear differences among the variances and covariances from the three models. The efficiency of the fixed parameter estimators of these three models can be summarized using the determinant of the variance–covariance matrices of the FE estimators. We recall that estimators with a smaller determinant for the variance–covariance matrix are more efficient estimators. For this problem, the RI model with independent errors (*Model* 2) has the smallest determinant value $\text{Det}[\widehat{\text{Cov}}(\hat{\beta})] = 2.2099 \times 10^{-6}$. This implies that estimators from *Model 2* for the two FE parameters are the most efficient.

Estimators of the fixed parameters from the RI model with auto-correlated errors (*Model* 3) are less efficient and we also need to estimate the auto-correlation parameter ρ. Compared to the determinant value for the FE model (*Model* 1) with unstructured variance–covariance matrix, the estimators of the FE parameters from the RI model (*Model* 2) with uncorrelated errors are much more efficient. This improvement of efficiency can be quantified by the relative efficiency $\text{RE}_D = (2.2099/3.7895)^{1/2} = 0.76$.

The above result supports the conclusion that among the three models, the RI model with independent errors is *probably* the best fitting model. It should however be kept in mind, that it is generally not an easy task to decide on the choice of a best fitting model. Approximate statistical tests and likelihood ratio tests for the parameters and the fit of these models exist, and have been implemented in computer programs; see for example, Verbeke and Molenberghs (2000, Chapter 6). Moreover, one can use information criteria, like the Akaike's

Table 7.3 The covariance structures of three different models.

Model 1: Fixed effects model with unstructured covariance matrix

$y_{ij} = \beta_0 + \beta_1 t_j + \varepsilon_{ij}$, with unknown σ_ε^2 and unknown $cov(\varepsilon_{ij}, \varepsilon_{ij'})$, for all $j \neq j'$

Covariance structure:

$$\widehat{Cov}(y_i) = 10^{-2} \begin{bmatrix} 0.2978 \\ 0.3091 & 0.3473 \\ 0.3085 & 0.3486 & 0.3771 \\ 0.3240 & 0.3720 & 0.3974 & 0.4422 \\ 0.2973 & 0.3443 & 0.3646 & 0.4024 & 0.3953 \end{bmatrix} \text{Symmetric}$$

$$\widehat{Cov}(\hat\beta) = 10^{-1} \begin{bmatrix} 0.1395 & -0.0037 \\ -0.0037 & 0.0028 \end{bmatrix}$$
$$Det[\widehat{Cov}(\hat\beta)] = 3.7895 \times 10^{-6}$$
$$\hat\beta_0 = 0.8528, \hat\beta_1 = 0.0265$$

Model 2: Random intercept model with independent errors

$y_{ij} = (\beta_0 + b_{0i}) + \beta_1 t_j + \varepsilon_{ij}$, with σ_ε^2 and $cov(\varepsilon_{ij}, \varepsilon_{ij'}) = 0$, for all $j \neq j'$

Covariance structure:

$$\widehat{Cov}(y_i) = 10^{-2} \begin{bmatrix} 0.3720 \\ 0.3468 & 0.3720 \\ 0.3468 & 0.3468 & 0.3720 \\ 0.3468 & 0.3468 & 0.3468 & 0.3720 \\ 0.3468 & 0.3468 & 0.3468 & 0.3468 & 0.3720 \end{bmatrix} \text{Symmetric}$$

$$\widehat{Cov}(\hat\beta) = 10^{-1} \begin{bmatrix} 0.1872 & -0.0038 \\ -0.0038 & 0.0013 \end{bmatrix}$$
$$Det[\widehat{Cov}(\hat\beta)] = 2.2099 \times 10^{-6}$$
$$\hat\beta_0 = 0.8559, \hat\beta_1 = 0.0269$$

Model 3: Random intercept model with auto-correlated errors

$y_{ij} = (\beta_0 + b_{0i}) + \beta_1 t_j + \varepsilon_{ij}$, with σ_ε^2 and $cov(\varepsilon_{ij}, \varepsilon_{ij'}) = \sigma_\varepsilon^2 \rho^{|t_j - t_{j'}|}$, for all $j \neq j'$

Covariance structure:

$$\widehat{Cov}(y_i) = 10^{-2} \begin{bmatrix} 0.3720 \\ 0.3590 & 0.3720 \\ 0.3527 & 0.3590 & 0.3720 \\ 0.3497 & 0.3527 & 0.3590 & 0.3720 \\ 0.3482 & 0.3497 & 0.3527 & 0.3590 & 0.3720 \end{bmatrix} \text{Symmetric}$$

$$\widehat{Cov}(\hat\beta) = 10^{-1} \begin{bmatrix} 0.1918 & -0.0044 \\ -0.0044 & 0.0015 \end{bmatrix}$$
$$Det[\widehat{Cov}(\hat\beta)] = 2.6050 \times 10^{-6}$$
$$\hat\beta_0 = 0.8560, \hat\beta_1 = 0.0268$$

Note. $\bar\sigma_0^2 = 0.3468 \times 10^{-2}, \bar\sigma_\varepsilon^2 = 0.0251 \times 10^{-2}, \hat\rho = 0.4845$.

information criterion (Akaike, 1974; Bozdogan, 1987) to compare goodness of fit among models. These procedures, of course, are only valid asymptotically and hence we should have a large sample when we assess model adequacy.

7.7.1   Improvement of the longitudinal design

The design problem for repeated measurements focuses on the choice of the number of time points (design points) at which the measurements are taken and on the allocation of these time points. The original design for the bone mineral density example has five visits (time points). For simplicity we have assumed that the five visits are equally spaced in time and that the coding of the design points is 1, 2, 3, 4 and 5. The original exact design for a sample of $n = 44$ women can now be schematically represented as follows:

$$\xi_{220} = \left\{ \begin{array}{ccccc} d_1 = 1 & d_2 = 2 & d_3 = 3 & d_4 = 4 & d_5 = 5 \\ n/N & n/N & n/N & n/N & n/N \end{array} \right\}. \tag{7.14}$$

The fractions are all equal to $n/N = 0.2$ when the total number of observations is $N = 220$. This design is displayed in Table 7.4 as ξ_1.

The design used in the bone mineral density study can be improved from both the cost and efficiency perspectives. To show how the efficiency of ξ_1 can be increased, let us consider four other designs ξ_2, ξ_3, ξ_4 and ξ_5 also displayed in Table 7.4. We are going to study the performance of these designs for estimating

Table 7.4   The $\mathrm{Det}[\widehat{\mathrm{Cov}}(\hat{\beta})]$ (generalized variance) of five different designs for the three models in Table 7.3.

Designs	Model 1	Model 2	Model 3
$\xi_1 = \left\{ \begin{array}{ccccc} 1 & 2 & 3 & 4 & 5 \\ 0.2 & 0.2 & 0.2 & 0.2 & 0.2 \end{array} \right\}$	378.9471	220.9945	260.4963
$\xi_2 = \left\{ \begin{array}{ccccc} 1 & 2 & 3 & 4 & 5 \\ 0.25 & 0.25 & 0 & 0.25 & 0.25 \end{array} \right\}$	11.7039	11.9416	7.2606
$\xi_3 = \left\{ \begin{array}{ccccc} 1 & 2 & 3 & 4 & 5 \\ 0.3333 & 0 & 0.3333 & 0 & 0.3333 \end{array} \right\}$	2.11056	5.7945	1.7759
$\xi_4 = \left\{ \begin{array}{ccccc} 1 & 2 & 3 & 4 & 5 \\ 0.5 & 0 & 0 & 0 & 0.5 \end{array} \right\}$	1.4627	2.6052	1.2873
$\xi_5 = \left\{ \begin{array}{ccccc} 1 & 2 & 3 & 4 & 5 \\ 0.3333 & 0 & 0 & 0.3333 & 0.3333 \end{array} \right\}$	3.4729	5.3487	3.3933
	$(\mathrm{Det}[\widehat{\mathrm{Cov}}(\hat{\beta})]$ values $\times 10^{-8})$		

Note. *Model* 1: Fixed effects model with unstructured variance–covariance matrix.
Model 2: Random intercept model with independent errors.
Model 3: Random intercept model with auto-correlated errors.

the model parameters for the three models in Table 7.3. For this purpose, all five designs in Table 7.4 are assumed to have the same total observations, that is, $N = n \times m = 220$. This implies that any decrease in the number of repeated measurements m would require an increase of the number of subjects n to maintain the same total number of observations N, and vice versa. This necessitates us to consider approximate designs, because a fixed sample size N and a given fraction n/N may require a non-integer value of n. For example, designs ξ_3 and ξ_5 result in non-integer values for $n = 0.3333 \times 220 = 73.3333$. In practice, we round the non-integers to the nearest integer before we implement the design.

Table 7.4 lists the generalized variances of $\hat{\beta}$ from the five designs for the three models. The $\text{Det}[\widehat{\text{Cov}}(\hat{\beta})]$ values (generalized variances) of the fixed parameter estimators for the three models with different variance–covariance structures using design ξ_1 from Table 7.3 are reproduced in Table 7.4. The rest of the generalized variances of $\hat{\beta}$ in the table are computed using designs ξ_2, ξ_3, ξ_4 and ξ_5.

Table 7.4 shows that there is a clear increase in efficiencies of the estimators when the design has fewer design points. For instance, design ξ_1, with five equally weighted design points, has the largest value for $\text{Det}[\widehat{\text{Cov}}(\hat{\beta})]$ for all three models, while design ξ_4, with two equally weighted design points, has the smallest value. This seems to be in line with the fact that a D-optimal design for the simple homoscedastic linear model requires only two design points as far apart from each other as possible (Chapter 2). The computations in Table 7.4 suggest that such a design would also be D-optimal for a model with an RI and with auto-correlated or independent errors.

From Table 7.4, we also observe that design ξ_3 seems to be a second best choice for *Model* 1 and *Model* 3, and nearly so for *Model* 2 as well. Further, we notice that the $\text{Det}[\widehat{\text{Cov}}(\hat{\beta})]$ values for ξ_3 are not that much larger than the $\text{Det}[\widehat{\text{Cov}}(\hat{\beta})]$ values of design ξ_4, but ξ_3 has one more support point than ξ_4. Hence, even though ξ_4 is the 'best' design in Table 7.4 for estimating β, ξ_3 is a strong competitor because having an extra point allows model checks to be performed and is also likely to make ξ_3 potentially more robust than ξ_4 to model assumptions violation.

7.7.1.1 Correlated errors versus random effects

Repeated measurements frequently exhibit an auto-correlated structure in a longitudinal study. As the time points lie farther apart in time, the correlation between measurements at these time points decreases. Such a structure is also found in the correlation matrix of the responses of the bone mineral example given in Equation (7.12). For the FE model the variance–covariance structure of the responses is completely explained by the variances and covariances of the errors in the model. In many cases, however, when random parameters are added to the model, the variance–covariance matrix of the responses is not completely explained by the variance–covariance matrix of the errors. Parts of these variances and covariances are then explained by the random parameters.

Jones (1990) discussed the phenomenon that the auto-correlation is partially confounded with random between-subject effects in the model. The variance–covariance structure of the responses y_i in linear mixed effect models can be partially explained by the variance–covariance structure of the random effects and partially by the variance–covariance structure of the error terms in the model. The more (between subjects) random effects are added to the model, the more variance is explained by the random effects. In other words, the variances and covariances of the responses can be increasingly explained by adding random effects to the model, hence that relatively less remains unexplained.

A similar phenomenon is seen in the bone mineral example. We saw that more efficient parameter estimators $\hat{\beta}$ were obtained when an RI was added to the model (*Model* 2 and *Model* 3) and that *Model* 2 with independent errors gave the most efficient estimators. In this case the correlated variance–covariance structure of the responses can apparently be completely explained by an RI. This effect is often found in small samples, where there is simply not enough variation to be explained by more random parameters. In larger samples, however, more random parameters are often needed to fully describe all variance of the responses (Jones, 1990).

The trade-off between random effects and the correlation of the errors will also affect the optimality of a longitudinal design. If a model fits the data, the optimal design will usually need as many distinct design points m as fixed parameters p in the model, that is, $m = p$. If a model does not fit the data well and more parameters are needed to adequately describe the data, the number of distinct design points will probably have to be larger than the number of parameters, that is, $m > p$.

7.8   Cost function

The results in Table 7.4 assume an equal total number of observations for all five designs. If there is a one-to-one relation between the costs and the number of observations N, then these five designs will all have the same total costs of performing the study. In practice, the costs of sampling a subject in a longitudinal design may be quite different from the costs of obtaining a repeated measurement for that subject. It is therefore reasonable to make a distinction between these two costs and incorporate the cost structure in the study when we compare designs.

Suppose that the total costs of a study are denoted by C, and that the costs of an initial set-up of a study are c_0. The costs c_0 may include the salary of the research staff and other costs, which are not directly connected to the actual sampling of observations. Let the costs of sampling one subject be c_1 and let the costs of obtaining one repeated measurement per subject be c_2. The total costs of a longitudinal design can be written as the sum of the initial setup costs, the costs of sampling the n subjects, and the costs of obtaining a total of $n \times m$ repeated measurements:

$$C = c_0 + nc_1 + nmc_2. \qquad (7.15)$$

When c_1 is zero, the cost function reduces to $C - c_0 = nmc_2$, and there is a one-to-one relation between the total costs C and the $N = n \times m$ observations.

Let us first consider the simple case when c_0 and c_1 are both zero in the bone mineral density example. The original design ξ_1 in Table 7.4 measures $n = 44$ women at five different visits. Hence, the costs of obtaining the $N_1 = n_1 \times m_1 = 44 \times 5 = 220$ observations are $C = (44 \times 5)c_2$. The other four designs in Table 7.4 contain fewer than five repeated measurements, but use a larger sample of women to keep the total number of observations the same. For instance, design ξ_3 has only $m_3 = 3$ repeated measurements and hence we have to increase the number of women in this design to $n_3 = 44 \times 0.3333/0.2 = 73.3333$, to ensure that the total number of observations remains the same, that is, $N = n_3 \times m_3 = 73.3333 \times 3 = 220$. We reiterate that the non-integer sample size $n_3 = 73.3333$ is not feasible in practice, and is only used here to enable a fair efficiency comparison among the designs. In practice such sample sizes are rounded to the nearest integer value; so $n_3 = 73$ in this case.

In most practical applications, the costs c_0 and c_1 are both not zero. An approach to account for the possibly unequal non-zero costs in c_0 and c_1 is to standardize or normalize the variance–covariance matrix of the fixed parameter estimators with respect to the sampling costs of a design. See Fedorov, Gagnon and Leonov (2002), Fedorov and Leonov (2005) and Gagnon and Leonov (2005) for details. Such a normalization was applied to find D-optimal cohort designs in Tekle, Tan and Berger (2008b).

If one wants to compare the sample size requirement for a design ξ_2 with a design ξ_1 having n_1 subjects using the same cost structure and total budget, it is easy to show that the desired sample size n_2 for design ξ_2 is

$$n_2 = \frac{n_1(c_1 + m_1 c_2)}{(c_1 + m_2 c_2)}. \tag{7.16}$$

Table 7.5 presents the sample sizes of the five designs for different combinations of c_1 and c_2, expressed by means of a cost ratio $f = c_1/c_2$, where $f = 0$, 1, 5 and 10. This cost ratio represents the relative costs of sampling a subject with respect to the costs of obtaining one repeated measurement per subject. All sample sizes are related to the sample size of the original design ξ_1 and assume that the costs of obtaining the observations are equal to those of the original design. In the column with heading $f = 0$, the sample sizes are given for the case where only the costs of obtaining repeated measurements are taken into account. These are actually the sample sizes that are used in Table 7.4.

It can be seen that as the cost ratio f increases, the relative costs of sampling subjects increases, and the differences between the sample sizes n_1 and n_2 becomes smaller. The loss of efficiency in designs with less repeated measurements can be compensated by increasing the sample of subjects. But as the costs of sampling subjects become relatively higher, this may not be a very efficient option anymore. It may then be more efficient to apply a design with many repeated measurements.

Table 7.4 shows design ξ_4 seems the most efficient design for estimating the fixed parameters in *Models* 1, 2 and 3. We therefore choose design ξ_4 as

Table 7.5   Sample sizes for five designs and different cost ratios under the condition that all designs have the same total costs.

Designs	$f = 0$	$f = 1$	$f = 5$	$f = 10$
ξ_1	$n_1 = n_1$	$n_1 = n_1$	$n_1 = n_1$	$n_1 = n_1$
ξ_2	$n_2 = \dfrac{5}{4}n_1$	$n_2 = \dfrac{6}{5}n_1$	$n_2 = \dfrac{10}{9}n_1$	$n_2 = \dfrac{15}{14}n_1$
ξ_3	$n_3 = \dfrac{5}{3}n_1$	$n_3 = \dfrac{6}{4}n_1$	$n_3 = \dfrac{10}{8}n_1$	$n_3 = \dfrac{15}{13}n_1$
ξ_4	$n_4 = \dfrac{5}{2}n_1$	$n_4 = \dfrac{6}{3}n_1$	$n_4 = \dfrac{10}{7}n_1$	$n_4 = \dfrac{15}{12}n_1$
ξ_5	$n_5 = \dfrac{5}{3}n_1$	$n_5 = \dfrac{6}{4}n_1$	$n_5 = \dfrac{10}{8}n_1$	$n_5 = \dfrac{15}{13}n_1$

Note. Sample sizes of five designs are n_1, n_2, n_3, n_4 and n_5, respectively.

reference design and evaluate efficiencies of the other four designs relative to ξ_4. Specifically, for each model of interest we define RE_D equal to the square root of the ratio given by dividing the generalized variance of ξ_4 with that of the comparison design. These RE_Ds are displayed in Figure 7.3 for the three models and four cost ratios $f = 0$, $f = 1$, $f = 5$ and $f = 10$.

The fourth design ξ_4 is the reference design and therefore has an RE_D value equal to one. It is the most efficient design for *Model* 2. It is, however, not always the most efficient design for *Model* 1 and *Model* 3. The plots for *Model* 1 and *Model* 3 show an RE_D value of design ξ_3 that is greater than 1 if the cost ratios are $f = 5$ and $f = 10$.

It seems that for *Models* 1 and 3, it is more efficient to use design ξ_3 with three design points instead of the two design points of design ξ_4 when the costs of sampling subjects becomes relatively high compared to the costs of obtaining repeated measurements. *Model* 2 seems to fit the data best and gives us the most efficient parameter estimators for the original design ξ_1 (Figure 7.3). Design ξ_4 is the D-optimal design for this model. The original design ξ_1 seems to produce less efficient parameter estimators in *Model* 1 and *Model* 3, in part because these models do not seem to fit the data well. In this case, it appears that we can improve the efficiency of the parameter estimators by including an extra design point (design ξ_3) when the cost ratio is high.

From this example, the following recommendations can be offered to design a longitudinal study for a linear mixed effects model with polynomial fixed and random parameters:

- Since design efficiency strongly depends on the specified model, it is important to know what model will fit the data best.

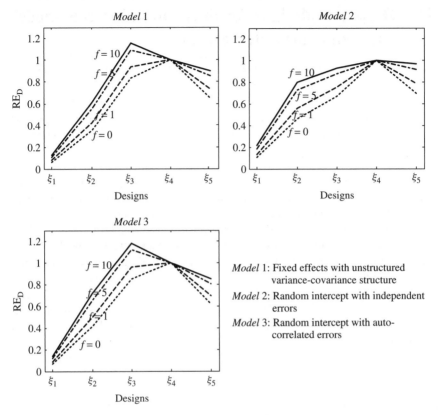

Figure 7.3 Relative efficiency RE$_D$ plots for five designs and three models.

- Choose a design with as less repeated measurements as possible. If one is sure that the model with p fixed polynomial parameters fits the data, a rule of thumb is to select no more repeated measurements m than there are fixed polynomial parameters in the model; thus the number of repeated measurements should be $m = p$. If, on the other hand, the model is suspected not to fit the data well, one or two extra design points should be added to cope with the unexplained variance, that is, the number of repeated measurements should be $m = p + 1$ or $m = p + 2$.

- It is surely unnecessary to design a study with a lot more repeated measurements than fixed parameters. This is not only inefficient in terms of parameter estimation but also in terms of costs and may also raise ethical objections.

- Money can be better spent and efficiency can be improved by sampling more subjects instead of obtaining more repeated measurements.

7.9   D-optimal designs for linear mixed effects models with auto-correlated errors

D-optimal designs for polynomial regression models with independent errors are well known. D-optimal design points for polynomial regression models with independent errors were first derived by Smith (1918) and Guest (1958), and these and other types of optimal designs are tabulated in optimal design texts, such as Pukelsheim (1993) and Atkinson and Donev (1992), among others (see also Table 3.5 in Chapter 3).

D-optimal designs in mixed effects models were studied in the literature. Abt *et al.* (1997, 1998) studied optimal designs for linear and quadratic growth curve models with auto-correlated errors, Atkins and Cheng (1999) investigated designs for an RI quadratic polynomial model with independent errors, Bischoff (1993) studied D-optimal designs for linear models with correlated errors, Cheng (1995) derived results on optimal designs for random block effects models, Mentré, Mallet and Baccar (1997) proposed an algorithm for finding D-optimal designs for random effects models and Tan and Berger (1999) presented balanced D-optimal designs for RI models under an AR(1) error correlation structure. These optimal designs are all constructed under the assumption that there are no missing data. We will also make this assumption in our discussion on optimal designs for linear mixed effects models.

Optimal designs for FE polynomial regression models have certain characteristics. From Table 3.5 (Chapter 3) we deduce that a D-optimal design depends on the degree of the polynomial. The optimal number of distinct design points is equal to the number of polynomial parameters in the model, that is $p = m$, and although the m design points are equally weighted, the optimal design points are not necessarily equally spaced. For example, D-optimal design points are not equally spaced when $p \geq 4$. Table 3.5 of Chapter 3 shows that A-optimal and E-optimal designs also have unequally spaced design points for $p \geq 4$, but in addition are also unequally weighted for $p \geq 3$.

A mixed effects model with polynomial description of longitudinal data can be derived from the general form in Equations (7.2) and (7.3). The measurement y_{ij} of the ith subject at time point t_j $(j = 1, \ldots, m)$ can be represented by the following mixed effects model:

$$y_{ij} = \beta_0 + \beta_1 t_j + \beta_2 t_j^2 + \cdots + \beta_{p-1} t_j^{p-1}$$
$$+ b_{0i} + b_{1i} t_j + b_{2i} t_j^2 + \cdots + b_{q-1,i} t_j^{q-1} + \varepsilon_{ij}. \qquad (7.17)$$

The p parameters $\beta_0, \beta_1, \ldots, \beta_{p-1}$ are the fixed polynomial parameters. The q random (subject-specific) parameters $b_{0i}, b_{1i}, \ldots, b_{q-1,i}$ are assumed to be normally distributed with mean zero and a $q \times q$ variance–covariance matrix, which reflects the dependencies among these random parameters. The errors in Model 7.17 are assumed to be normally distributed with mean zero and variance–covariance structure $\sigma_\varepsilon^2 \Sigma$. For time-structured data the variance–covariance structure of the errors is often auto-correlated with correlation parameter ρ such that the correlation between the errors at two time points t_j and $t_{j'}$ is $\rho^{|t_j - t_{j'}|}$. Such a structure is often referred to as AR(1) (see also Equation (7.9)). In addition, we assume that the errors and the random parameters are independent.

The model in Equation (7.17) includes the already described models in the previous subsections. For example, the RI model in Equation (7.4) can be obtained by assuming that there are $p = 2$ fixed parameters β_0 and β_1 and that there is only $q = 1$ RI b_{0i}. Similarly, the FE model with auto-correlated errors in Table 7.2 (Case IV) is also a special case of Model 7.17 with $p = 2$ fixed parameters β_0 and β_1, no random parameters and the errors have an auto-correlated variance–covariance structure.

D-optimal designs for mixed effects models with polynomial terms are not well known. These models not only vary in the number of parameters, they also make a distinction between fixed and random parameters and assume different covariance structures for the errors. D-optimal designs for linear mixed effects models not only depend on the degree of the polynomial, as was the case for a fixed effect model with independent errors but also depend on several other features in the model, such as:

- the number of fixed parameters p and the number of random parameters q $(q \leq p)$,

- the form of the covariance structure of the random parameters and the covariances among their estimates and

- the form of the covariance structure of the errors in the model and the correlations among the error terms.

Formally, a D-optimal design for a linear mixed effects model with p FE parameters $\beta = (\beta_0, \beta_1, \ldots, \beta_{p-1})'$ is a design that minimizes the generalized variance of the estimators of β. This is the same as finding a design that minimizes the determinant of the $p \times p$ variance–covariance matrix of parameter estimators, that is:

$$\min\{\text{Det}[\text{Cov}(\hat{\beta})]\}. \tag{7.18}$$

The minimization is over all designs on the design interval of interest. The D-optimal design is, strictly speaking, locally optimal because it depends on the correlations among the distinct time points, which are unknown. Hence, nominal values for all non-FE parameters are required to compute the D-optimal design. These nominal values may be available from pilot studies or related studies. This situation is similar to the case when we want to find optimal designs for nonlinear models in Chapter 10.

Analytical descriptions of D-optimal designs for linear mixed models are very difficult and probably impossible to obtain (Hungerford, 1990, p. 363). The usual approach is to compute optimal designs for linear mixed models numerically using optimization routines. Figures 7.4, 7.5 and 7.6 display D-optimal designs for four different types of models defined on the time-scaled interval $[-1, 1]$:

- FE model, with no random parameters;

- RI model, with only RI parameters b_{0i};

- Random intercept-random slope (RI-RS) model, with RI parameter b_{0i} and random slope (RS) parameter b_{1i};

- Correlated random intercept-random slope (RI-RS$_C$) model, with RI parameter b_{0i} and RS parameter b_{1i}, which are correlated.

All four types of models assume that the errors have an auto-correlated (AR(1)) variance–covariance structure. The D-optimal design points are plotted for values of the correlation parameter within the interval $0 \leq \rho \leq 0.95$, and each plot represents a combination of the number of polynomial parameters p and the number of repeated measurements $m \geq p$, for $p = 2$, 3 and 4, and $m = 2$, 3, 4, 5, 6 and 7. The weights are all equal and sum up to one. Similar plots of D-optimal designs were presented in Tan and Berger (1999) and Ouwens, Tan and Berger (2002).

Figure 7.4 displays the D-optimal design points for linear models with $p = 2$ polynomial parameters. Figures 7.5 and 7.6, respectively, display similar plots for a quadratic model with $p = 3$ parameters and for cubic polynomial models with $p = 4$ parameters. The figures show that D-optimal design points for the same combination of values of p and m are very similar for the different models. D-optimal design points depend strongly on the correlation ρ. In general, they are *not* equally spaced and tend to be a bit more spread out as the correlation increases. Another observation is that as the correlation parameter ρ approaches zero, the optimal number of distinct design points approaches p. This is because D-optimal design points for $\rho = 0$ are equal to D-optimal design points for polynomial models with independent errors. The D-optimal design points are also not necessarily symmetrically placed on the time scale $[-1+1]$, especially when the correlation is smaller than $\rho \leq 0.2$. For example, consider the D-optimal design points for cubic polynomial models where $(p - 1) = 3$ and $m = 5$. If five D-optimal time points are $(-1 \;\; -t_2 \; t_3 \; t_4 \; 1)$, then the time points $(-1 \;\; -t_2 \;\; -t_3 \; t_4 \; 1)$ are also D-optimal.

In practice, many repeated measurement studies have equally spaced time points. These figures show that the optimal design points become more equally spaced over the time scale as the correlation parameter increases. The differences between the D-criterion values for designs with equally spaced time points and D-optimal designs is very small for correlations $\rho > 0.1$, and this difference becomes even smaller as the number of time points m increases compared to p. In other words, when the correlation is large, an equally spaced design with an appropriate number of design points should approach the D-optimal design.

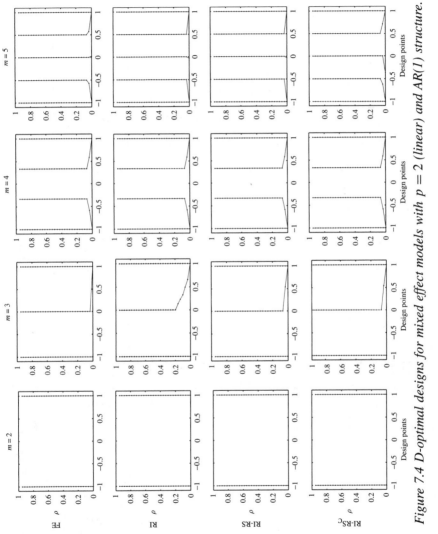

Figure 7.4 D-optimal designs for mixed effect models with p = 2 (linear) and AR(1) structure.

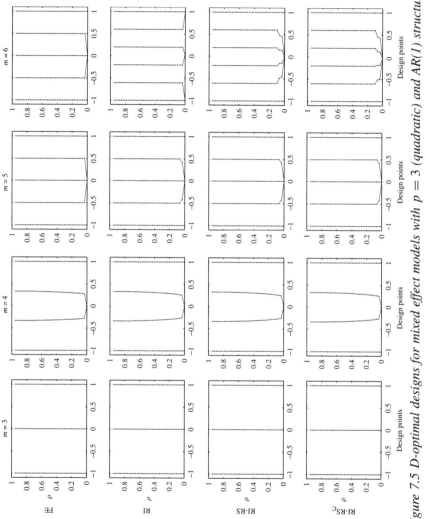

Figure 7.5 D-optimal designs for mixed effect models with p = 3 (quadratic) and AR(1) structure.

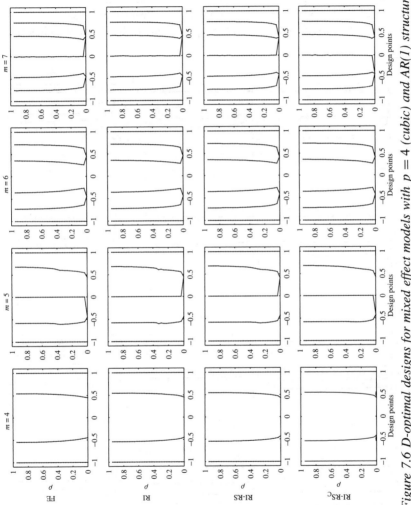

Figure 7.6 D-optimal designs for mixed effect models with $p = 4$ (cubic) and AR(1) structure.

The most striking effect of the D-optimal designs in these figures is their loss of efficiency as the number of distinct time points m increases. This effect is shown in Figure 7.7 where the relative efficiencies of D-optimal designs with $m > p$ distinct time points compared to the D-optimal design with $m = p$ distinct time points, are displayed as a function of the auto-correlation $0 \le \rho \le 0.95$. Plots are given for the four types of models with $p = 2$, $p = 3$ and $p = 4$ parameters, while maintaining the same sample size n in these plots.

We observe two other interesting observations from these figures. The first observation is that as the auto-correlation approaches 1, the RE approaches p/m.

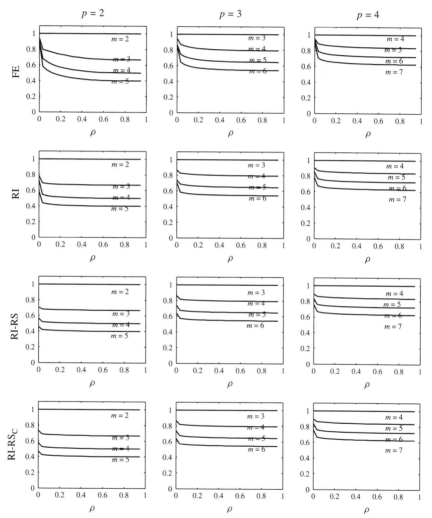

Figure 7.7 Relative efficiencies of D-optimal designs compared to the D-optimal design with $m = p$ distinct design points.

Actually this asymptotic result holds for any model (Tan and Berger, 1999). The ratio p/m can actually be seen as a lower bound for the efficiency of a D-optimal design with m distinct time points. The second observation is that as the auto-correlation approaches zero, the RE of a D-optimal design with m time points will approach 1. In other words, when the auto-correlation approaches zero, the D-optimal design tends to the D-optimal design with $m = p$ distinct time points. Figure 7.7 shows that the efficiency drop is largest in the range $0 \leq \rho \leq 0.1$. This is also the same set of ρ values for which the D-optimal time points are least equally spaced on the design interval.

Our last commentary for this subsection is that even when the auto-correlation ρ is small, the correlation between the error terms at two close time points can be quite large. For example, consider a small value of ρ, say $\rho = 0.01$ and a small absolute difference between two time points: $|t_j - t_{j'}| = 0.1$. Then the correlation between the errors at these two time points is $\rho^{|t_j - t_{j'}|} = 0.01^{0.1} = 0.6309$. And if the model is an FE model, then the correlation between the measurements at the two closely located time points is also 0.6309. For a model with random parameters, the correlation between the measurements at the two closely located time points is greater than 0.6309.

In summary, a design with equally spaced time points approaches the D-optimal design very well for $\rho \geq 0.2$. In addition, taking observations at many time points will result in loss of efficiency. Generally, it is preferable to use a design with relatively few time points from the ethical, cost and statistical considerations. From the sole perspective of parameter estimation, the number of time points should be $p + 1$ or $p + 2$ where p is the number of FE polynomial parameters in the model.

7.10 Miscellanea

7.10.1 Homoscedasticity

In the previous section, the optimal designs for linear mixed effects models were found when the correlation structure of the errors is AR(1) with homogeneous variances of the errors. In practice, this assumption may not hold and the variances of the errors at different time points may be quite different. The effect of heterogeneity of the error variances on the optimality of a design has been studied for linear models with independent errors by Wong and Cook (1993), Wong (1995), Atkinson and Cook (1995) and Montepiedra and Wong (2001), among others. Heterogeneity of the error variances for linear mixed models has been studied by Ortega-Azurduy, Tan and Berger (2008a). Their preliminary results suggest that the D-optimal time points for models with an AR(1) structure and heterogeneous error variances show moderate deviations from the D-optimal time points with homogeneous error variances. The time points tend to shift in the direction of the time scale where the errors have smaller variances. They also found that an incorrect specification of the error variances in the model is likely to overestimate the optimal sample size.

7.10.2   Uninformative dropout

Another problem encountered in a longitudinal study is missing information and dropouts. This happens frequently when the design has many repeated measurements and the total time interval for the study is large. The reason for this is more patients tend to miss scheduled visits or drop out altogether in long-term clinical trials. The effect of dropout on the optimality of designs for linear models with independent errors has been studied by Imhof, Song and Wong (2002, 2004). Galbraith and Marschner (2002) discussed unanticipated dropout for linear mixed models. Dropout did not seem to affect power for a first-degree polynomial mixed effects model. A framework for power calculation in the presence of dropout is given by Moerbeek (2008) and Ortega-Azurduy, Tan, and Berger (2008b) showed for different dropout functions with about 70% dropout at the last time point, that the D-optimal design points for linear mixed effects models tended to be displaced in the direction of the time scale where the probability of encountering dropout is smallest. Moreover, for the dropout functions that were studied, the maximum efficiency loss due to dropout is smaller than 15%. These results seem to suggest that for estimating parameter purposes, dropout is not a very serious problem in a repeated measurement study and that design efficiency can be maintained by increasing the planned sample size in advance by 15%.

7.11   Matrix formulation of the linear mixed effects model

In matrix notation, the linear mixed effects model can be formulated as:

$$y_i = X_i \beta + Z_i b_i + \varepsilon_i, \tag{7.19}$$

where the $m \times 1$ vector y_i contains the m repeated measurements for the ith subject, the $p \times 1$ column vector β contains the fixed parameters and the $q \times 1$ vector b_i contains the individual random parameters. The matrices X_i and Z_i are the known $m \times p$ and $m \times q$ design matrices for the fixed and random effects and the errors ε_i's are normally distributed each with mean zero and a common positive-definite variance–covariance matrix $\sigma_\varepsilon^2 \Sigma$. The matrix Σ can have any form, including the independent error case and the auto-correlated error case.

The random effects parameter vector b_i describes the deviation of the individual responses from the overall average response pattern and it is assumed that b_i is normally distributed with mean zero and $q \times q$ variance–covariance matrix D. The random effects b_i and the errors ε_i are also assumed to be independent.

The variances and covariances of the repeated measurements are the weighted sum of the variances and covariances of the random effects parameters and variances and covariances of the errors. Thus,

$$\mathrm{Cov}(y_i) = Z_i D Z_i' + \sigma_\varepsilon^2 \Sigma. \tag{7.20}$$

Let the vector $\theta = (\beta, \sigma^2)'$ contain all the fixed parameters β and variance components σ^2 of the linear mixed model. Then ML estimates can be obtained by maximizing the likelihood function $L(\theta)$ or by minimizing -2 times the logarithm of the likelihood, that is:

$$-2\ln(L(\theta)) = \sum_i^n \left\{ m\ln(2\pi) + \ln(\mathrm{Det}[\mathrm{Cov}(y_i)]) \right.$$

$$\left. + (y_i - X_i\beta)'\mathrm{Cov}(y_i)^{-1}(y_i - X_i\beta) \right\}. \tag{7.21}$$

The expression $\ln(.)$ stands for the natural logarithm, and $\mathrm{Det}[\mathrm{Cov}(y_i)]$ is the determinant of $\mathrm{Cov}(y_i)$. When the variance components are known, the ML estimators of β are equal to the generalized least squares estimators for the fixed parameters:

$$\hat{\beta} = \left(\sum_i^n X_i'[\mathrm{Cov}(y_i)]^{-1}X_i \right)^{-1} \sum_i^n X_i'[\mathrm{Cov}(y_i)]^{-1}y_i, \tag{7.22}$$

When the variance components are not known but estimates are available, Equation (7.22) can still be used as an estimator of β. ML estimates of the variance components are obtained by minimizing $-2\ln(L(\theta))$ in Equation (7.21) with respect to the variance components, after β is replaced by $\hat{\beta}$. REML estimates of the variance components are obtained by minimizing a similar function with respect to a set of error contrasts. See Verbeke and Molenberghs (2000, Chapter 5) and Diggle *et al.* (2002, Chapter 4) for more details on ML and REML estimation of the parameters.

The asymptotic variance–covariance matrix of the estimator $\hat{\beta}$ is equal to

$$\mathrm{Cov}(\hat{\beta}) = \left(\sum_i^n X_i' \, \mathrm{Cov}(y_i)^{-1}X_i \right)^{-1}, \tag{7.23}$$

and by means of the fixed parameter estimators $\hat{\beta}$, the estimator of σ_ε^2 can be obtained from

$$\hat{\sigma}_\varepsilon^2 = \frac{1}{N} \sum_i^n \left[(y_i - X_i\hat{\beta})'\mathrm{Cov}(y_i)^{-1}(y_i - X_i\hat{\beta}) \right]. \tag{7.24}$$

To illustrate the analysis of the bone mineral example, the RI model in Equation (7.4) (Table 7.1, Case I) has the form:

$$\begin{bmatrix} y_{i1} \\ y_{i2} \\ y_{i3} \\ y_{i4} \\ y_{i5} \end{bmatrix} = \begin{bmatrix} 1 & t_1 \\ 1 & t_2 \\ 1 & t_3 \\ 1 & t_4 \\ 1 & t_5 \end{bmatrix} \begin{bmatrix} \beta_0 \\ \beta_1 \end{bmatrix} + \begin{bmatrix} 1 \\ 1 \\ 1 \\ 1 \\ 1 \end{bmatrix} [b_{0i}] + \begin{bmatrix} \varepsilon_{i1} \\ \varepsilon_{i2} \\ \varepsilon_{i3} \\ \varepsilon_{i4} \\ \varepsilon_{i5} \end{bmatrix} \tag{7.25}$$

For this model, the vector b_i reduces to a scalar, that is, an RI parameter b_{0i}, having variance σ_0^2. The variance–covariance matrix of the responses for this model has a compound symmetric structure and is

$$\mathrm{Cov}(y_i) = Z_i D Z_i{}' + \sigma_\varepsilon^2 \Sigma$$

$$= \begin{bmatrix} \sigma_0^2 & \sigma_0^2 & \sigma_0^2 & \sigma_0^2 & \sigma_0^2 \\ \sigma_0^2 & \sigma_0^2 & \sigma_0^2 & \sigma_0^2 & \sigma_0^2 \\ \sigma_0^2 & \sigma_0^2 & \sigma_0^2 & \sigma_0^2 & \sigma_0^2 \\ \sigma_0^2 & \sigma_0^2 & \sigma_0^2 & \sigma_0^2 & \sigma_0^2 \\ \sigma_0^2 & \sigma_0^2 & \sigma_0^2 & \sigma_0^2 & \sigma_0^2 \end{bmatrix} + \begin{bmatrix} \sigma_\varepsilon^2 & 0 & 0 & 0 & 0 \\ 0 & \sigma_\varepsilon^2 & 0 & 0 & 0 \\ 0 & 0 & \sigma_\varepsilon^2 & 0 & 0 \\ 0 & 0 & 0 & \sigma_\varepsilon^2 & 0 \\ 0 & 0 & 0 & 0 & \sigma_\varepsilon^2 \end{bmatrix}.$$

$$(7.26)$$

The inverse of this variance–covariance matrix is

$$\mathrm{Cov}(y_i)^{-1}$$

$$= \frac{1}{\sigma_\varepsilon^2(5\sigma_0^2 + \sigma_\varepsilon^2)} \begin{bmatrix} 4\sigma_0^2 + \sigma_\varepsilon^2 & -\sigma_0^2 & -\sigma_0^2 & -\sigma_0^2 & -\sigma_0^2 \\ -\sigma_0^2 & 4\sigma_0^2 + \sigma_\varepsilon^2 & -\sigma_0^2 & -\sigma_0^2 & -\sigma_0^2 \\ -\sigma_0^2 & -\sigma_0^2 & 4\sigma_0^2 + \sigma_\varepsilon^2 & -\sigma_0^2 & -\sigma_0^2 \\ -\sigma_0^2 & -\sigma_0^2 & -\sigma_0^2 & 4\sigma_0^2 + \sigma_\varepsilon^2 & -\sigma_0^2 \\ -\sigma_0^2 & -\sigma_0^2 & -\sigma_0^2 & -\sigma_0^2 & 4\sigma_0^2 + \sigma_\varepsilon^2 \end{bmatrix}.$$

$$(7.27)$$

The variance–covariance matrix of the FE parameters is now obtained by pre- and post-multiplication of the $\mathrm{Cov}(y_i)^{-1}$ by $X_i{}'$ and X_i, respectively, and this product is weighted by n/N. Finally, after inverting the weighted product, we obtain the 2×2 variance–covariance matrix of the FE parameters:

$$\mathrm{Cov}(\hat{\beta}) = \left(\frac{n}{N}\right)^{-1} \begin{bmatrix} \sigma_0^2 + \frac{11}{10}\sigma_\varepsilon^2 & -\frac{3}{10}\sigma_\varepsilon^2 \\ -\frac{3}{10}\sigma_\varepsilon^2 & \frac{1}{10}\sigma_\varepsilon^2 \end{bmatrix}.$$

$$(7.28)$$

Substituting $n/N = 0.2$, and the heuristic estimates $\bar{\sigma}_0^2 = 0.3468 \times 10^{-2}$ and $\bar{\sigma}_\varepsilon^2 = 0.0251 \times 10^{-2}$ into Equation (7.28) results in

$$\widehat{\mathrm{Cov}}(\hat{\beta}) = 10^{-1} \begin{bmatrix} 0.1872 & -0.0038 \\ -0.0038 & 0.0013 \end{bmatrix}.$$

$$(7.29)$$

A straightforward calculation shows the determinant criterion value is $\mathrm{Det}[\widehat{\mathrm{Cov}}(\hat{\beta})] = 2.2099 \times 10^{-6}$. To compute the generalized least squares estimator $\hat{\beta}$, we now use Equation (7.22), so that $\hat{\beta}' = [0.8559, 0.0269]$ (see also Table 7.3).

7.12   Summary

The design of longitudinal studies is more complicated than the design of cross-sectional studies, because the efficiency of longitudinal designs depends on the correlation among the repeated measures. One of the most often applied and most flexible statistical models to analyse longitudinal data, is the linear mixed effects model. This model is flexible, because it can easily incorporate dropout and include measurements from subjects taken at different time points. Of course, optimal designs for longitudinal studies still require that the assumed linear mixed effects model provides an adequate fit for the data.

In general, efficiency of polynomial parameter estimators increases more by increasing the sample size than by increasing the number of repeated measurements. It is not very efficient to measure the outcome variable at too many different time points. A rule of thumb is to select no more repeated measurements than there are fixed polynomial parameters in the model, that is, the number of repeated measurements should be $m = p$. If, however, the linear mixed effects model with p polynomial parameters is suspected not to adequately fit the data, then one of two additional time points should be added to the design, and the design should have $m = p + 1$ or $m = p + 2$ distinct time points. It is surely not very efficient to design a longitudinal study with a lot more time points than fixed polynomial parameters in the model. It should be noted that this rule of thumb only applies to polynomial parameters in the model.

Unnecessary inclusion of extra time points would not only lead to a large loss of efficiency, but also to spending money, which could have been better used to increase the sample size. Moreover, if the measurement procedure is stressful or painful, or the treatment is controversial, then ethical objections may be raised as more repeated measurements are taken. Finally, inclusion of many repeated measurements may increase the probability of dropout. Even if non-informative or missing at random (MAR) dropout is encountered, the efficiency of a design will decrease as the number of repeated measurements increases. But, in general, the efficiency will show a small loss of about 15% for dropout patterns having up to 70% dropout at the last time point, and such a moderate efficiency loss can easily be compensated by increasing the sample size by about 15%.

8

Two-treatment crossover designs

8.1 Design problem for crossover studies

A *crossover* (*CO*) *design* is a repeated measures design in which different treatments are applied to the same subjects or experimental units in different time periods. We will use the term *experimental units* in general and subjects or patients for special examples. The term 'crossover' is used to emphasize that experimental units are crossed over from one treatment to another during the study. CO designs have been applied in psychological, biomedical and industrial experimental research, and are described in Ratkowsky, Evans and Alldredge (1993), Senn (2002) and Jones and Kenward (2003), among others. The CO design can be seen as a form of the traditional row–column design when the experimental units are considered to be blocks of units that are assigned the same sequences of treatments. Bailey (2008) provides a review of row–column design.

CO designs have been proven to be particularly suitable for medical studies investigating chronic diseases, where a treatment is not expected to cure a disease but is intended to diminish symptoms. CO designs are also extensively used in pharmacokinetics. For example, Soria *et al.* (2002) performed a bioequivalence study to compare the effectiveness of two different types of inhalers for asthma patients using a CO design. The reason for this was that very young or elderly asthma patients are often not able to use a *press-and-breath inhaler*, because the inhaler requires simultaneous inhalation and pressing action. For these patients, a *breath actuated inhaler* may be easier. To investigate whether the medication

An Introduction to Optimal Designs for Social and Biomedical Research   M. P. F. Berger & W. K. Wong
© 2009, John Wiley & Sons, Ltd

delivery of the two inhalers is equivalent, Soria *et al.* (2002) employed a CO design in which patients administered their own treatment using the two different inhalers during four separate periods of time.

CO designs have also been applied in social science research. For example, Parkes (1982) used a CO design in an experimental study on occupational stress. In a training programme, student nurses were randomly assigned to four sequences of ward types A, B, C and D. These sequences were AB, BA, CD and DC over two training periods. Similarly, Jensen, Watanabe and Richters (1999) used a CO design to study the effect of the order of presentation of two modules A and B of a structured interview schedule for use in the diagnosis of child and adolescent psychopathology. Families were randomly assigned to two sequences AB and BA of the modules.

An alternative design to the CO design is the *parallel-group (PG) design*, where subjects are randomly assigned to sequences with a single treatment instead of sequences with different treatments. The main advantage of CO designs over PG designs is that CO designs require fewer subjects (patients) to achieve the same level of efficiency and statistical power. The reason for this is that in a CO design, treatments are compared *within subjects* and not *between subjects* as is the case in PG designs. Since every subject in a CO design receives all the treatments successively, the variance between the subjects does not play a role in the treatment comparison. This implies that the variance between the subjects can be separated from the variance of the estimated difference between treatments, resulting in less noise in the comparison. Consequently, less resources are required in CO designs to perform as well as the PG designs.

There are drawbacks with CO designs, and assumptions in the CO designs must be met before their benefits can be realized. CO designs produce the so-called *carry-over effects* as nuisance effects. We speak of a carry-over effect when the effect of a treatment on the outcome from a specific period of time extends into the subsequent time periods. If two treatments are given in two successive periods of time and treatment A is offered before treatment B, the effect of treatment A on the outcome may continue to be present during the second period where treatment B starts to take effect. Thus, any variation in outcome during the second period could be a direct effect of treatment B, and also an indirect carry-over effect related to treatment A. The simultaneous occurrence of the effects of treatments A and B in the second period will complicate analysis and interpretation.

Carry-over effects are generally classified according to the number of time periods involved. The so-called *first-order carry-over effects* occur between two adjacent time periods. *Higher-order carry-over effects* refer to effects ranging across time periods that are more than one period apart. Higher-order carry-over effects can usually be neglected if there is no first-order carry-over effect and a non-significant test for first order carry-over effects is often sufficient to conclude that higher-order carry-over effects are not present. Even more severe problems can happen when the carry-over effects are different for different treatments; that is, when treatment A has a carry-over effect that differs from the carry-over effect of treatment B. Such carry-over effects are referred to as *differential carry-over effects*.

A carry-over effect may reveal a *period × treatment interaction* in the data analysis and this may in turn lead to biased conclusions about treatment effects. This means that CO designs cannot be applied routinely and certainly not in studies where treatments are known to change the 'state' that experimental units are in.

A drawback of CO designs is that their use can be limited either on ethical grounds or in terms of practical implementation. For example, CO designs may take longer to carry out because of the unknown duration of the carry-over effects or one of the treatments under investigation has to be administered immediately to save the patient when the patient becomes sick.

Another disadvantage of CO designs is that the effect of a particular treatment can cause experimental units to drop out of the study. In this case, the effects of other treatments cannot be measured for these experimental units and vital information will be missed. Dropout complicates the analysis of such designs. It can be argued that since treatment sequences are randomly assigned to experimental units, the dropout process of a unit due to a treatment effect may be thought to be missing at random (MAR) and so the usual methods of data analysis are still valid. However, one must always remain cautious, because the probability of dropout may be different in different treatment regimes. Of course, both carry-over effects and dropout may also be a problem in a PG design where the effect of one treatment is measured at different occasions. This is especially so when carry-over effects are different for different treatments and when dropout is caused by specific characteristics of a treatment. The end result is that estimators of treatment effects are biased and the internal validity of the study becomes questionable (Cook and Campbell, 1979).

These problems are well documented in the literature for the two-treatment and two-period CO design (AB and BA), and various solutions to overcome the problems have been proposed. Ebbutt (1984) and Laska and Meisner (1985), among others, suggested extending the number of periods from two to three or more periods. Earlier on, Wallenstein (1979), Kershner and Federer (1981), Laska, Meisner and Kushner (1983), Patel (1983), and Pocock (1983) proposed including baseline measurements in the design. Fleiss (1989), however, pointed out that these solutions have problems of their own and recommended that a two-treatment CO design be only used if equality or near equality of carry-over effects can a priori be assumed. Otherwise, a PG design, with each group of units receiving only one treatment, with possible baseline measurements as covariates, should be preferred. Cox (1958) already described CO designs under the specific assumption of no carry-over effects. We will discuss CO designs under the a priori assumption that carry-over effects are equal or nearly equal. Preferably, carry-over effects should be zero for a valid analysis of data from CO designs.

It should be clear by now that carry-over effects are a key issue with CO designs. An often applied method to decrease carry-over effects is the inclusion of a *washout period* between the successive occasions. Such a washout period is meant to make sure that the treatment effect of a previous period is no longer active in the current period. If a set washout period proves sufficient to guard against first-order carry-over effects, it can usually safely be assumed that higher-order carry-over effects are not present. However, it is difficult to determine how long a washout period should be. A long washout period may not be manageable or ethically justified, while if the washout period is too short it will miss the mark. In studies that involve medication, the life time of a drug often determines the length of the washout period. A drug's half-life is the time necessary for the drug levels in the blood to decrease by half in the population of interest. The length of a washout period is usually a multiple of a drug's half-life. In other fields, such as psychotherapy or education, there are no criteria to establish the length for an effective washout period. Although a washout period can be extended, there is no guarantee that a previous treatment does not still affect the experimental units in such a way that future responses remain influenced by that treatment.

Carry-over effects raise questions in data analysis as well. One of the most frequently applied procedures in a two-treatment and two-period design is the two-step procedure proposed by Grizzle (1965, 1974), which first tests for possible carry-over effects. If the test reveals significant carry-over effects, then only the data from the first period are analysed to test for treatment effects. If, however, the first test does not reveal significant carry-over effects, then the analysis is carried out on the complete data set. This procedure has been criticized by Freeman (1989), Senn (2002) and Senn and Lee (2004) as being potentially misleading because the first test and the final test in the procedure are highly dependent and the overall type I error rate of the procedure is not adequately controlled.

8.2   The design

A CO design can be denoted by the triplet (t, s, m), and is characterized by the number of different treatments (t), the sequences of treatments (s) and the number of occasions or periods in time (m), where the repeated measurements are taken (see also Hedayat and Afsarinejad, 1975). The location of a treatment can be symbolized as $d(j, k)$, where the index $j = 1, \ldots, m$ represents the periods in time and the index $k = 1, \ldots, s$ represents the different sequences. The total number of experimental units in an arbitrary CO design is $N = \sum_k^s n_k$, where n_k is the number of experimental units (subjects) that receive sequence k. The total number of observations is $N \times m$.

Apart from the *treatment effects*, which are of main interest for the study, a CO design also harbours the so-called *nuisance effects*, such as *sequence*, *period* and *carry-over effects*. Two features of a CO design can be used to distinguish treatment effects from nuisance effects. The first feature is *uniformity* and the second feature is *(strongly) balancedness* (Afsarinejad, 1990). A CO design (t, s, m) is said to be *uniform in periods* if each treatment is administered the same number of times in each period. A CO design (t, s, m) is said to be *uniform in sequences* if each treatment is administered the same number of times within each sequence. Such uniformity of a CO design enables us to disentangle period and sequence effects from treatment effects.

A *balanced* CO design with respect to (first-order) carry-over effects has each treatment preceding each other treatment the same number of times. A design is *strongly balanced* when each treatment precedes every other treatment, *including itself*, the same number of times. So, in a strongly balanced design with t treatments, all t^2 sequences of treatments occur the same number of times, while in a balanced design only the $t(t-1)$ sequences of different treatments arise. A strongly balanced CO design enables control of carry-over effects. In a later section, we explain how uniformity and balancing of CO designs affect the confounding of treatment effects with nuisance parameters.

In Table 8.1, three designs with two treatments A and B and two periods are displayed. The first design is a CO design $(2, 2, 2)$, with $s = 2$ sequences, each with $t = 2$ differently ordered treatments, namely AB and BA. The number of periods is $m = 2$. In this design, the total sample of units is randomly divided into two groups with n_k units, each receiving the treatments in a different order. The $(2, 2, 2)$ CO design is both uniform in periods and sequences and is also balanced because each treatment precedes the other treatment the same number of times (only once). The second design is a PG design. The PG design also has $s = 2$ sequences, $t = 2$ treatments and $m = 2$ period, but the treatments are not crossed over. Each group of experimental units receives only one of the two treatments and units are randomly assigned to each group. The PG design is uniform in periods but not uniform in sequences. Finally the third design combines a CO with a PG design. It distinguishes four sequences, namely AB, BA, AA and BB. The total sample of experimental units is randomly divided into four groups, each receiving the treatments in one of four-sequence orders. This design, called *Balaam (BM)'s design* (Balaam, 1968), is uniform in periods, but not within sequences and it is also strongly balanced, because each treatment precedes the other treatment, including itself, equally often. The treatment sequences AB, BA, AA and BB occur equally often (only once).

Table 8.1 Two-treatment and two-period designs.

	Cross-over design $(2, 2, 2)$		Parallel-group design $(2, 2, 2)$		Balaam's design $(2, 4, 2)$	
	Period		Period		Period	
	1	2	1	2	1	2
Sequence 1	A	B	A	A	A	B
Sequence 2	B	A	B	B	B	A
Sequence 3					A	A
Sequence 4					B	B

As indicated before, a CO design contains the problem that treatment effects cannot be separated from carry-over effects. Because of the problem of carry-over effects in (2, 2, 2) CO designs, Senn (2002), among others, questioned its usefulness and advised first investigating whether the PG design would be a better alternative.

The two-period designs in Table 8.1 can be extended with an extra period. For $t = 2$ treatments and $m = 3$ periods there are 2^3 possible sequences of treatments: ABB, BAA, AAB, BBA, ABA, BAB, AAA and BBB, and a three-period design can in principle consist of combinations of these sequences. In Table 8.2, three designs with $m = 3$ periods and $t = 2$ treatments are presented. Note that if the third period is deleted from each of these designs, they will reduce to one of the designs in Table 8.1.

Table 8.2   Two-treatment and three-period designs.

	Design (2, 2, 3)			Design (2, 4, 3)			Design (2, 4, 3)		
	Period			Period			Period		
	1	2	3	1	2	3	1	2	3
Sequence 1	A	B	B	A	B	A	A	B	B
Sequence 2	B	A	A	B	A	B	B	A	A
Sequence 3				A	A	A	A	A	B
Sequence 4				B	B	B	B	B	A

The three designs in Table 8.2 are not uniform within sequences because of the addition of an extra period, but they remain uniform within periods and they are balanced because each treatment precedes every other treatment the same number of times. Moreover, these designs are also *strongly balanced*, that is, each treatment precedes the other treatment, including itself, the same number of times. This can be checked by verifying that the combinations AB, BA, AA and BB occur equally often in the sequences of each design.

8.3   Confounding treatment effects with nuisance effects

We use the commonly used analysis of variance parameterization for CO designs to explain confounding treatment effects with nuisance effects. We say that a treatment effect and a nuisance effect are *confounded* if they cannot be separated from each other. The practical implementation of confounding is that we have biased estimates for the treatment effect. To elaborate on this important concept, consider the two-treatment and two-period CO design presented in Table 8.3. As indicated before, this design distinguishes different effects. Let μ represent the overall population mean of the outcome scores and let τ_A and τ_B be the direct effects of two treatments A and B, respectively. The nuisance parameters π_1 and π_2 represent the period effects for periods 1 and 2 and ζ_1 and ζ_2 are the sequence effects for sequences AB and BA, respectively. The carry-over effect from treatment A to treatment B is λ_A and from B to A is λ_B. In Table 8.3, the expected values of an outcome variable of each of the four cells for a (2, 2, 2) CO design are now represented as the sum of direct treatment effects and nuisance parameters.

Table 8.3   Expected values of (2, 2, 2) cross-over design.

	Period 1	Period 2
Sequence AB	$\mu + \tau_A + \zeta_1 + \pi_1$	$\mu + \tau_B + \zeta_1 + \pi_2 + \lambda_A$
Sequence BA	$\mu + \tau_B + \zeta_2 + \pi_1$	$\mu + \tau_A + \zeta_2 + \pi_2 + \lambda_B$

The expected values of the treatment estimators $\hat{\tau}_A$ and $\hat{\tau}_B$ are obtained from the averages of the cells that contain τ_A and τ_B, respectively:

$$E(\hat{\tau}_A) = \tfrac{1}{2}[(\mu + \tau_A + \zeta_1 + \pi_1) + (\mu + \tau_A + \zeta_2 + \pi_2 + \lambda_B)]$$

$$= \mu + \tau_A + \tfrac{1}{2}(\zeta_1 + \zeta_2) + \tfrac{1}{2}(\pi_1 + \pi_2) + \tfrac{1}{2}\lambda_B$$

and

$$E(\hat{\tau}_B) = \tfrac{1}{2}[(\mu + \tau_B + \zeta_2 + \pi_1) + (\mu + \tau_B + \zeta_1 + \pi_2 + \lambda_A)]$$

$$= \mu + \tau_B + \tfrac{1}{2}(\zeta_1 + \zeta_2) + \tfrac{1}{2}(\pi_1 + \pi_2) + \tfrac{1}{2}\lambda_A. \tag{8.1}$$

The expected difference between these two treatment estimators is

$$E(\hat{\tau}_A - \hat{\tau}_B) = (\tau_A - \tau_B) - \tfrac{1}{2}(\lambda_A - \lambda_B). \tag{8.2}$$

It can be seen that the treatment differences are confounded with carry-over effects. The estimator of this treatment difference is biased and will only be unbiased when the two carry-over effects are equal, that is when $\lambda_A = \lambda_B$.

Although the parameterization in Table 8.3 is straightforward and easy to explain, all the parameters cannot be estimated independently. In total, nine parameters are distinguished, while there are only four cells. Therefore, some restrictions have to be imposed upon the parameters. Commonly applied restrictions in analysis of variance are $\sum_j \pi_j = 0$ and $\sum_k \zeta_k = 0$, so that the period effects in Table 8.3 are $\pi_2 = -\pi_1$ and the sequence effects are $\zeta_2 = -\zeta_1$. But these restrictions will not change the confounding of the expected difference of treatment estimators with carry-over effects in Equation (8.2). The fact that in the (2, 2, 2) CO design, sequence effects ζ_k and period effects π_j are not confounded with the difference of treatment estimators is due to the feature that this design is uniform in both periods and sequences.

A design that is not uniform in sequences is BM's design shown in Table 8.1. This design is, however, uniform in periods and is also strongly balanced. To see how these features affect the confounding with nuisance effects, the expected outcome values for BM's design are given in Table 8.4.

Table 8.4   Expected values of (2, 4, 2) cross-over design (Balaam's design).

	Period 1	Period 2
Sequence AB	$\mu + \tau_A + \zeta_1 + \pi_1$	$\mu + \tau_B + \zeta_1 + \pi_2 + \lambda_A$
Sequence BA	$\mu + \tau_B + \zeta_2 + \pi_1$	$\mu + \tau_A + \zeta_2 + \pi_2 + \lambda_B$
Sequence AA	$\mu + \tau_A + \zeta_3 + \pi_1$	$\mu + \tau_A + \zeta_3 + \pi_2 + \lambda_A$
Sequence BB	$\mu + \tau_B + \zeta_4 + \pi_1$	$\mu + \tau_B + \zeta_4 + \pi_2 + \lambda_B$

The parameterization in Table 8.4 is the same as that of Table 8.3. In addition it is assumed that the carry-over effect λ_A of A onto B (sequence AB, period 2) is equal to the carry-over effect of A onto A (in sequence AA, period 2). Likewise, the carry-over effect λ_B of B onto A (in sequence BA, period 2) is equal to the carry-over effect of B onto B (sequence BB, period 2). The expected values of treatment estimators for A and B are the averages of the cells that contain the corresponding treatment effect:

$$E(\hat{\tau}_A) = \mu + \tau_A + \tfrac{1}{4}(\zeta_1 + \zeta_2 + 2\zeta_3) + \tfrac{1}{4}(2\pi_1 + 2\pi_2) + \tfrac{1}{4}(\lambda_A + \lambda_B)$$

and   (8.3)

$$E(\hat{\tau}_B) = \mu + \tau_B + \tfrac{1}{4}(\zeta_1 + \zeta_2 + 2\zeta_4) + \tfrac{1}{4}(2\pi_1 + 2\pi_2) + \tfrac{1}{4}(\lambda_A + \lambda_B).$$

For this design, the expected difference between the treatment estimators reduces to

$$E(\hat{\tau}_A - \hat{\tau}_B) = (\tau_A - \tau_B) + \tfrac{1}{2}(\zeta_3 - \zeta_4). \qquad (8.4)$$

From Equation (8.4), it is seen that the difference between the treatment effect estimators is not confounded with carry-over effects. This is due to the fact that the design is strongly balanced. Since the design is not uniform within sequences, the difference in treatment effect estimators is confounded with sequence effects ζ_k. This confounding will only disappear if $\zeta_3 = \zeta_4$ or $\zeta_3 = \zeta_4 = 0$.

The three designs in Table 8.2 have the same features as BM's design. They are uniform in periods, but not uniform in sequences, and they are strongly balanced, meaning that they do not suffer from confounding of treatment differences with carry-over effects. Because the difference of treatment estimators is confounded with sequence effects, it is important to assume that these sequence effects are negligible. Although random assignment of experimental units to sequences is a method to increase the validity of this assumption, it may not completely guarantee zero-sequence effects for small samples.

In summary, here are a few conclusions about uniformity and balancing of CO designs. Afsarinejad (1990) and Vonesh and Chinchilli (1997) explained that a uniform and strongly balanced CO design is the best safeguard against confounding of treatment differences with nuisance parameters:

- A CO design that is uniform in periods will not have treatment differences confounded with period effects. This is important, because period effects are quite often influenced by natural or physical (daily, seasonal) rhythms of subjects. For example, learning processes of students or adaptation of patients to a treatment in later periods can cause the results in later periods to increase or decrease.

- A CO design that is uniform in sequences will not have treatment differences confounded with sequence effects. Random assignment of experimental units to sequences may protect against confounding with sequence effects.

- A CO design that is strongly balanced will not have treatment differences confounded with first-order carry-over effects. Protection against confounding of higher-order carry-over effects is more complicated.

8.4 The linear model for crossover designs

In a CO design, treatment effects are distinguished from sequence and period effects. Throughout it is assumed that only one measurement per period is obtained from an experimental unit. Although actually more measurements per period can be obtained, we assume that these measurements are adequately represented by just one score per period per experimental unit. The standard linear model for a measurement y_{ijk} of the ith experimental unit in period j receiving the kth sequence of treatments is

$$y_{ijk} = \mu + \tau_{d(j,k)} + \xi_{ik} + \pi_j + \lambda_{d(j-1,k)} + \varepsilon_{ijk}, \qquad (8.5)$$

where μ is the overall mean, $\tau_{d(j,k)}$ is the effect of treatment $d(j,k)$ in period j ($j = 1, \ldots, m$) and sequence k ($k = 1, \ldots, s$) and ξ_{ik} is the effect of the ith experimental unit or subject in sequence k, with $i = 1, \ldots, n_k$, where n_k is the number of experimental units receiving the treatments according to sequence k. The period effect is π_j and $\lambda_{d(j-1,k)}$ is the residual or first-order carry-over effect of treatment $d(j - 1, k)$ in the $(j - 1)$th period. For period j, the notation $d(j - 1, k)$ refers to the treatment in the preceding period, that is the $(j - 1)$th period. If $(j - 1) = 0$, the expression $d(j - 1, k)$ indicates no treatment and $\lambda_{d(j-1,k)} = 0$. Finally, the errors ε_{ijk} are usually assumed to be independently and normally distributed as $N(0, \sigma_\varepsilon^2)$.

There are two assumptions for this model. The first assumption is that the subject-specific parameters ξ_{ik} are fixed parameters. In this case, the model is referred to as a *fixed effects model* because all (treatment and nuisance) parameters are fixed. The second assumption is that the subject-specific parameters ξ_{ik} are random parameters, which are normally distributed as $N(\zeta_k, \sigma_\xi^2)$. It is often convenient but not necessary to assume that $E(\xi_{ik}) = \zeta_k = 0$. Under the assumption of random subject-specific parameters ξ_{ik}, the model is referred to as a *linear mixed-effects model*, with both random subject parameters and fixed parameters. Currently, analyses for CO designs typically assume a linear mixed model with random subject parameters.

For the random subject model, the variance of the responses in Model (8.5) is equal to the sum of the variances between subjects σ_ξ^2 and within subjects σ_ε^2, that is $\text{var}(y_{ijk}) = \sigma_\xi^2 + \sigma_\varepsilon^2$, while the covariance of the responses between period j and j' is $\text{cov}(y_{ijk}, y_{ij'k}) = \sigma_\xi^2$. This means that the variance–covariance matrix of responses has a compound symmetric structure:

$$\text{Cov}(y_i) = \begin{bmatrix} \sigma_\xi^2 + \sigma_\varepsilon^2 & \sigma_\xi^2 & \cdots & \sigma_\xi^2 \\ \sigma_\xi^2 & \sigma_\xi^2 + \sigma_\varepsilon^2 & \cdots & \sigma_\xi^2 \\ \vdots & \vdots & \vdots & \vdots \\ \sigma_\xi^2 & \sigma_\xi^2 & \cdots & \sigma_\xi^2 + \sigma_\varepsilon^2 \end{bmatrix}. \tag{8.6}$$

Let us return to the simple $(2, 2, 2)$ CO design. In Table 8.5, the model in Equation (8.5) describes the observations in the four cells of a $(2, 2, 2)$ CO design, with $s = m = 2$ and two treatments A and B, that is $d(1, 1) = d(2, 2) = A$, and $d(1, 2) = d(2, 1) = B$. We note that the expected values of the random subject parameter ξ_{ik} in the kth treatment sequence are equal to the sequence effect, that is $E(\xi_{ik}) = \zeta_k$, where ζ_k is the kth sequence effect.

Table 8.5   Model for $(2, 2, 2)$ cross-over design.

	Period 1	Period 2
Sequence AB	$y_{i11} = \mu + \tau_A + \xi_{i1}$ $+ \pi_1 + \varepsilon_{i11}$	$y_{i21} = \mu + \tau_B + \xi_{i1} + \pi_2$ $+ \lambda_A + \varepsilon_{i21}$
Sequence BA	$y_{i12} = \mu + \tau_B + \xi_{i2}$ $+ \pi_1 + \varepsilon_{i12}$	$y_{i22} = \mu + \tau_A + \xi_{i2} + \pi_2$ $+ \lambda_B + \varepsilon_{i22}$

To be able to estimate these parameters, we impose restrictions. In total, we have seven parameters and four independent cells. This means that three restrictions must be imposed to obtain a saturated model where the number of estimated parameters is equal to the number of cells.

Several different restrictions can be imposed to estimate the parameters. If we assume that there are carry-over effects such that $\lambda = \lambda_A = -\lambda_B$, and that the period effects are $\pi_2 = -\pi_1$, but that there are no sequence effects (expected subject parameters within a sequence), that is $\zeta_1 = \zeta_2 = 0$, then four remaining parameters can be estimated independently, namely $(\mu + \tau_A)$, $(\mu + \tau_B)$, π_1 and λ. It should be noted that under this set of restrictions, the carry-over effect λ is confounded with the difference between treatment estimators, that is $(\hat{\tau}_A - \hat{\tau}_B)$ $- \lambda = \frac{1}{2}(\bar{y}_{11} + \bar{y}_{22} - \bar{y}_{12} - \bar{y}_{21})$, where $\bar{y}_{jk}$ is the mean of observations in the cell (jk).

On the other hand, we can assume that there are no carry-over effects, that is $\lambda_A = \lambda_B = 0$, and that the period effects are $\pi_2 = -\pi_1$. If we would also want to distinguish sequence parameters as $\zeta_2 = -\zeta_1$, then there are four parameters to be estimated independently, namely two treatment effects $(\mu + \tau_A)$ and $(\mu + \tau_B)$ and two nuisance parameters π_1 and ζ_1. An unbiased estimator of the treatment difference in Table 8.5 is then $(\hat{\tau}_A - \hat{\tau}_B) = \frac{1}{2}(\bar{y}_{11} + \bar{y}_{22} - \bar{y}_{12} - \bar{y}_{21})$.

8.5 Estimation of parameters and efficiency

The model in Equation (8.5) has fixed (treatment) parameters and fixed nuisance parameters for periods, sequences and carry-over effects. It also includes subject-specific parameters that are often assumed to be random. The usual way of estimating the fixed (treatment, period and carry-over) parameters along with the random (subject-specific) parameters or their variance components (between- and within-subject variances) is by means of restricted maximum likelihood (REML). This method consists of two steps. In the first step, we eliminate the fixed parameters from the likelihood function. The resulting marginal likelihood function does not depend on the fixed parameters but only on the variance components σ_ξ^2 and σ_ε^2. The estimates of these variance components are found by maximizing the marginal likelihood. In the second step, generalized least squares (GLS) is applied with an estimate of the variance–covariance matrix $\text{Cov}(y_i)$ based on the REML estimates of σ_ξ^2 and σ_ε^2. See Harville (1974), Searle, Casella and McCulloch (1992), Diggle, Liang and Zeger (1994) and Verbeke and Molenberghs (2000), among others, for details on maximum likelihood (ML) and REML estimation of the fixed and random parameters in a mixed effects model. Chinchilli and Esinhart (1996) gave details on ML and REML estimation for CO experiments, and showed that closed form expressions for ML and REML estimators of the variance components exist when the CO design is uniform in terms of sequences. REML estimates are usually less biased downwards than ML estimates, but the differences disappear when sample sizes become large.

8.6 Cost and efficiency of the crossover design

The main argument for using a CO design is that treatment effects can be estimated more precisely than using a completely randomized (CR) design or a PG design. We demonstrate this by studying the efficiency and cost efficiency of a (2, 2, 2) CO design with two treatment sequences AB and BA in comparison to three alternative designs; namely, a CR design with two groups, each receiving a single treatment (A or B) only once, a PG design with two treatment sequences AA and BB and BM's design with four treatment sequences AB, BA, AA and BB. Note that the CR design has no repeated measurements, while the

CO, PG and BM designs have two repeated measurements. Also note that BM has four different treatment sequences, while the other three designs only have two treatment sequences (groups).

In order to make a valid comparison, we must assume that the carry-over effects in the CO design are zero or at least equal for both treatments. Equation (8.2) shows that for the CO design, the difference of treatment effects estimators and carry-over effects are confounded. For the PG design, a similar confounding of treatment differences with carry-over effects is encountered, but in this case the carry-over effects appear in the second period where the same treatment is given (i.e. within the same treatment) while in CO designs the carry-over effects emerge between different treatments. We will assume that the carry-over effect λ_A from A to B into the second period is the same as the carry-over effect λ_A from A to A in the second period. Likewise, the carry-over effect λ_B from B to A in the second period is identical to the carry-over effect λ_B from B to B in the second period. This assumption is especially important in BM's design, where it is assumed that the carry-over effect of A in the second period is equal for the sequences AB and AA, and the carry-over effect of B in the second period is identical for the sequences BA and BB. Finally, we note that the expected estimate of the treatment difference in a CR design does not contain carry-over effects.

Table 8.6 shows the expected values of the differences between treatment effect estimators for the four designs and their corresponding variances. The total number of subjects is divided randomly and equally over the groups. For the CO, CR, PG and BM designs the total number of subjects is denoted as N_{CO}, N_{CR}, N_{PG} and N_{BM}, respectively. Each group (sequence of treatments) in the CO, CR and PG designs contains $N_{CO}/2$, $N_{CR}/2$ and $N_{PG}/2$ subjects, respectively. The total number of subjects N_{BM} for BM's design is equally divided over four treatment sequence groups, that is, each sequence group contains $N_{BM}/4$ subjects. Because of the random assignment of subjects to sequences, we will assume that the sequence effects are zero.

Table 8.6   Expected differences of treatment estimators and their variances.

Designs	Expectation	Variance
CO	$E(\hat{\tau}_A - \hat{\tau}_B) = (\tau_A - \tau_B)$ $- \frac{1}{2}(\lambda_A - \lambda_B)$	$\text{var}_{CO}(\hat{\tau}_A - \hat{\tau}_B) = \dfrac{2\sigma_\varepsilon^2}{N_{CO}}$
CR	$E(\hat{\tau}_A - \hat{\tau}_B) = (\tau_A - \tau_B)$	$\text{var}_{CR}(\hat{\tau}_A - \hat{\tau}_B) = \dfrac{4(\sigma_\xi^2 + \sigma_\varepsilon^2)}{N_{CR}}$
PG	$E(\hat{\tau}_A - \hat{\tau}_B) = (\tau_A - \tau_B)$ $+ \frac{1}{2}(\lambda_A - \lambda_B)$	$\text{var}_{PG}(\hat{\tau}_A - \hat{\tau}_B) = \dfrac{2(2\sigma_\xi^2 + \sigma_\varepsilon^2)}{N_{PG}}$
BM	$E(\hat{\tau}_A - \hat{\tau}_B) = (\tau_A - \tau_B)$	$\text{var}_{BM}(\hat{\tau}_A - \hat{\tau}_B) = \dfrac{2(\sigma_\xi^2 + \sigma_\varepsilon^2)}{N_{BM}}$

Note. (i) Within-treatment and between-treatment carry-over effects are equal; (ii) the variances assume equal carry-over effects $\lambda_A = \lambda_B$; (iii) all sequence effects are assumed to be zero.

The expected values of the differences between treatment effect estimators in Table 8.6 include distinct carry-over effects λ_A and λ_B, but the variances of these differences in treatment effect estimators assume the two carry-over effects to be equal. For example, the difference of the treatment estimators for the CO design is unbiased only if $\lambda_A = \lambda_B$. The corresponding variance of the difference of the treatment effect estimators is then

$$\text{var}_{CO}(\hat{\tau}_A - \hat{\tau}_B) = \frac{\sigma_\varepsilon^2}{2}\left(\frac{2}{N_{CO}} + \frac{2}{N_{CO}}\right) = \frac{2\sigma_\varepsilon^2}{N_{CO}}, \qquad (8.7)$$

where σ_ε^2 is the variance of the error term.

From Table 8.6, we observe that if the total sample sizes of the four designs are equal, that is if $N_{CO} = N_{CR} = N_{PG} = N_{BM}$, the variances of the difference of treatment effect estimators are related as follows:

$$\text{var}_{CO}(\hat{\tau}_A - \hat{\tau}_B) \leq \text{var}_{BM}(\hat{\tau}_A - \hat{\tau}_B) \leq \text{var}_{PG}(\hat{\tau}_A - \hat{\tau}_B) < \text{var}_{CR}(\hat{\tau}_A - \hat{\tau}_B). \quad (8.8)$$

Given equal sample sizes, the CO design is the most efficient design with the smallest variance of the estimator of a treatment difference and the CR design is the least efficient with the largest variance. The variance $\text{var}_{CO}(\hat{\tau}_A - \hat{\tau}_B)$ is only equal to the variance $\text{var}_{BM}(\hat{\tau}_A - \hat{\tau}_B)$ or the variance $\text{var}_{PG}(\hat{\tau}_A - \hat{\tau}_B)$ if the between-subject or between-unit variance $\sigma_\xi^2 = 0$, that is if the model is a fixed effects model. The reason why the CR design is the least efficient is because the CR design has only $N = N_{CR}$ observations, while the other designs all have twice as many observations, that is $2 \times N$.

The relative efficiency (RE) of a design compared to another design is the ratio of the two variances of the estimators of the treatment effect differences. Table 8.7 shows the six REs obtained from the pair-wise comparison of the CO, CR, PG and BM designs. If the between-subject variance $\sigma_\xi^2 = 0$ (fixed effects model), the formulas show that the REs depend only on the sample sizes. However, if the between-subject variance is greater than zero, that is, if $\sigma_\xi^2 > 0$ (random effects model), the RE can be formulated as a function of the corresponding sample sizes and the between-subject variance σ_ξ^2 together with the within-subject variance σ_ε^2.

These REs indicate how many observations are needed in one design to be equally efficient as the other design. Table 8.7 also shows the relationships among the sample sizes for any two of the CO, CR, PG and BM designs to be equally efficient. These relationships may or may not depend on the variances σ_ξ^2 and σ_ε^2. For example, a two-group CR design with a total of N_{CR} experimental units will require at least twice as many experimental units to have the same efficiency as the CO design with a total of N_{CO} units: $N_{CR} = 2N_{CO}/(1 - \rho)$. If the between-subject variance $\sigma_\xi^2 = 0$ and the intra-class correlation $\rho = 0$, then $N_{CR} = 2N_{CO}$. Further discussion on this relationship is given in Chassan (1970) and Brown (1980). Likewise, the two-group CR design will need at least twice as many experimental units as needed in the BM's design where the total of N_{BM} units are evenly divided over the four groups, that is $N_{CR} = 2N_{BM}$.

Table 8.7   Relative efficiency and sample sizes under equal efficiency.

Design pairs	Relative efficiency	Sample size under equal efficiency	
CO/CR $\mathrm{RE_{CO/CR}}$	$= \dfrac{N_{CR}\sigma_\varepsilon^2}{2N_{CO}(\sigma_\xi^2 + \sigma_\varepsilon^2)}$	$N_{CR} = \dfrac{2N_{CO}(\sigma_\xi^2 + \sigma_\varepsilon^2)}{\sigma_\varepsilon^2}$	$= \dfrac{2N_{CO}}{(1-\rho)}$
CO/PG $\mathrm{RE_{CO/PG}}$	$= \dfrac{N_{PG}\sigma_\varepsilon^2}{N_{CO}(2\sigma_\xi^2 + \sigma_\varepsilon^2)}$	$N_{PG} = \dfrac{N_{CO}(2\sigma_\xi^2 + \sigma_\varepsilon^2)}{\sigma_\varepsilon^2}$	$= \dfrac{N_{CO}(1+\rho)}{(1-\rho)}$
PG/CR $\mathrm{RE_{PG/CR}}$	$= \dfrac{N_{CR}(2\sigma_\xi^2 + \sigma_\varepsilon^2)}{2N_{PG}(\sigma_\xi^2 + \sigma_\varepsilon^2)}$	$N_{CR} = \dfrac{2N_{PG}(\sigma_\xi^2 + \sigma_\varepsilon^2)}{(2\sigma_\xi^2 + \sigma_\varepsilon^2)}$	$= \dfrac{2N_{PG}}{(1+\rho)}$
CO/BM $\mathrm{RE_{CO/BM}}$	$= \dfrac{N_{BA}\sigma_\varepsilon^2}{N_{CO}(\sigma_\xi^2 + \sigma_\varepsilon^2)}$	$N_{BM} = \dfrac{N_{CO}(\sigma_\xi^2 + \sigma_\varepsilon^2)}{\sigma_\varepsilon^2}$	$= \dfrac{N_{CO}}{(1-\rho)}$
BM/PG $\mathrm{RE_{BM/PG}}$	$= \dfrac{N_{PG}(\sigma_\xi^2 + \sigma_\varepsilon^2)}{N_{BM}(2\sigma_\xi^2 + \sigma_\varepsilon^2)}$	$N_{PG} = \dfrac{N_{BM}(2\sigma_\xi^2 + \sigma_\varepsilon^2)}{(\sigma_\xi^2 + \sigma_\varepsilon^2)}$	$= N_{BM}(1+\rho)$
BM/CR $\mathrm{RE_{BM/CR}}$	$= \dfrac{N_{CR}(\sigma_\xi^2 + \sigma_\varepsilon^2)}{2N_{BM}(\sigma_\xi^2 + \sigma_\varepsilon^2)}$	$N_{CR} = 2N_{BM}$	

Note. Intra-class correlation: $\rho = \sigma_\xi^2/(\sigma_\xi^2 + \sigma_\varepsilon^2)$.

In this pair-wise comparison of efficiency of designs, we assume that there is no distinction between the costs of sampling subjects and the costs of treating and measuring them. However, if the cost of recruiting subjects and the cost of measuring and treating subjects are different, we should account for the cost differential when we compare the four types of designs. This is the topic of the next section.

8.6.1 Cost function

Consider the cost function that was used in previous chapters. Let c_0 represent the initial cost for setting up a study and let c_1 and c_2 be the costs of sampling an experimental unit and the costs of treating and measuring that unit, respectively. Then the total cost of performing a study with each of the designs is the sum of the costs for the initial set-up of the study, the costs of sampling all experimental units and the costs of treating and measuring these units:

$$\text{CO design: } C_{CO} = c_0 + N_{CO}c_1 + 2N_{CO}c_2,$$
$$\text{CR design: } C_{CR} = c_0 + N_{CR}c_1 + N_{CR}c_2, \qquad (8.9)$$
$$\text{PG design: } C_{PG} = c_0 + N_{PG}c_1 + 2N_{PG}c_2 \text{ and}$$
$$\text{BM design: } C_{BM} = c_0 + N_{BM}c_1 + 2N_{BM}c_2.$$

Here, we assume that c_0, c_1 and c_2 are the same for all designs. When two designs are equally efficient, a relative cost function R can be used to compare the costs of two designs. For example, when we compare the CO design with the CR design, the relative cost is

$$R_{CO/CR} = \frac{C_{CO} - c_0}{C_{CR} - c_0} = \frac{N_{CO}c_1 + 2N_{CO}c_2}{N_{CR}c_1 + N_{CR}c_2} = \frac{N_{CO}(f + 2)}{N_{CR}(f + 1)}, \tag{8.10}$$

where $f = c_1/c_2$. The relative cost $R_{CO/CR}$ is a function of c_1, c_2, N_{CO} and N_{CR}. Because the RE of two designs is unity, the sample size formula $N_{CR} = 2N_{CO}/(1 - \rho)$ from Table 8.7 can be substituted into the relative cost function $R_{CO/CR}$ in Equation (8.10), giving $R_{CO/CR} = ((1 - \rho)(f + 2))/(2(f + 1))$. Table 8.8 shows the relative cost function together with the relative cost functions of the other pairs of designs. Additionally, the table shows that when any two of the four designs are equally efficient, the relative cost is a function of the intra-class correlation ρ or the function $f = c_1/c_2$. Figure 8.1 shows the relative costs $R_{CO/CR}$, $R_{PG/CR}$ and $R_{BM/CR}$ as functions of $f = c_1/c_2$, where $0 < f < 10$ and $\rho > 0$.

Table 8.8   Relative cost for paired CO, CR, PG and BM designs when the paired designs are equally efficient.

Design pairs	Relative costs
CO/CR	$R_{CO/CR} = \dfrac{(1 - \rho)}{2} \dfrac{(f + 2)}{(f + 1)}$
CO/PG	$R_{CO/PG} = \dfrac{(1 - \rho)}{(1 + \rho)}$
PG/CR	$R_{PG/CR} = \dfrac{(1 + \rho)}{2} \dfrac{(f + 2)}{(f + 1)}$
CO/BM	$R_{CO/BM} = (1 - \rho)$
BM/PG	$R_{BM/PG} = \dfrac{1}{(1 + \rho)}$
BM/CR	$R_{BM/CR} = \dfrac{1}{2} \dfrac{(f + 2)}{(f + 1)}$

Note. Intra-class correlation $\rho = \sigma_\xi^2/(\sigma_\xi^2 + \sigma_\varepsilon^2)$, $f = c_1/c_2$ (for $c_2 > 0$).

The relative cost function $R_{CO/CR}$ compares the CO design with the CR design and it approaches unity when the ratio f and the intra-class correlation ρ become increasingly small, that is $R_{CO/CR} \to 1$ as $f \to 0$ and $\rho \to 0$. See also Brown (1980) for a slightly different form of $R_{CO/CR}$. When both CO and CR designs are equally efficient, the CO design will always cost less than the CR design. The cost is the same for both designs only when we can recruit subjects freely (i.e. $f = 0$) and the variance between subjects is zero (or $\rho = 0$). Another way of interpreting this comparison is that the CR design will become more expensive than the CO design as the costs of recruiting units relative to the costs of treating and measuring them increase, and as the correlation ρ becomes larger.

The relative cost function $R_{PG/CR}$ compares the PG design with the CR design, and it shows that as $f \to 0$ and $\rho \to 0$, the relative cost $R_{PG/CR} \to 1$. This effect is plotted in Figure 8.1b. Unlike the $R_{CO/CR}$ function, the $R_{PG/CR}$ may or may not exceed unity implying that the CR design is sometimes more and sometimes less expensive than the PG design. For high ρ values, the CR design is generally less costly, while for low ρ values the PG is generally less costly; the latter is especially true when $f = c_1/c_2$ is high.

The relative cost $R_{BM/CR}$ compares the BM design with the CR design and it does not depend on the intra-class correlation ρ. It is only a function of f, and the plot in Figure 8.1c shows that $R_{BM/CR} \to 1$ as $f \to 0$. Figure 8.1d shows the relative cost functions $R_{CO/PG}$, $R_{CO/BM}$ and $R_{BM/PG}$ versus values of

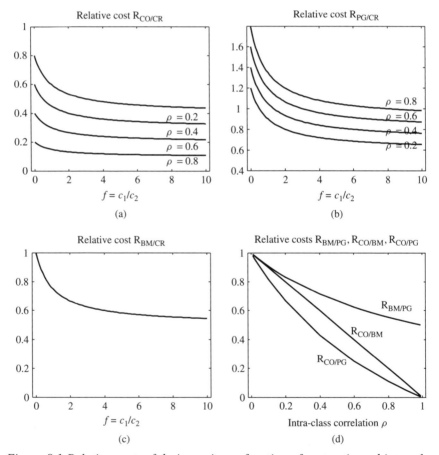

Figure 8.1 Relative costs of design pairs as function of cost ratio and intra-class correlation.

the intra-class correlation. We observe that as $\rho \to 0$, these relative costs all approach 1.

In summary, this cost–efficiency comparison shows that resources can be saved by proper choice of the design. When all four designs are equally efficient, the CO design is always preferable because it is also the cheapest of them all. Figure 8.1d shows that savings of more than 50% can be obtained for moderate and high ρ values by using the CO design instead of the PG design. However, deciding whether a CR or PG design is more cost efficient is less straightforward. The CR design is cheaper than the PG design for high ρ values and low f values, but for low ρ values and high f values the PG design is cheaper.

Our cost–efficiency comparison is restricted to the CO, CR, PG and BM designs. This procedure can be applied to other designs as well, as long as we are willing to assume equal carry-over effects for the treatments and assume that the model has independent errors. Kunert (1991) described two-treatment CO designs with correlated errors and Carriere and Huang (2000) investigated the impact of auto-correlated errors for a variety of CO designs. They compared the cost savings of CO designs relative to CR designs and found that the CR design was only better than the CO design in terms of cost, if the model has very high (positive and negative) auto-correlated errors and small cost ratios and small intra-class correlations ($f < 1, \rho < 0.5$).

8.7   Optimal crossover designs for two treatments

Hedayat and Afsarinejad (1978) were probably the first to consider optimal CO designs. Since then, there has been extensive work on this topic by Cheng and Wu (1980), Laska, Meisner and Kushner (1983) and Kushner (1997, 1998), among others. Additional references are given in Matthews (1988), Afsarinejad (1990) and Stufken (1996).

Two popular models for analysing a CO design with two treatments are the so-called fixed effects model and the random subject model. The fixed effects model distinguishes treatment effects from fixed nuisance parameters, and the random subject (unit) model in addition to the fixed nuisance parameters assumes random effects for the subjects or experimental units. Recall from Equation (8.5) that the basic model of interest here is given by

$$y_{ijk} = \mu + \tau_{d(j,k)} + \xi_{ik} + \pi_j + \lambda_{d(j-1,k)} + \varepsilon_{ijk}, \tag{8.11}$$

where, as before, y_{ijk} are the responses of the ith experimental unit in period j receiving the kth sequence of treatments. The symbols $\mu, \tau_{d(j,k)}, \pi_j, \lambda_{d(j-1,k)}$ and ε_{ijk} are the overall mean, the effect of treatment $d(j, k)$, the period effect, the first-order carry-over effect and the random error, respectively.

The only difference between the fixed effects and the random subject model is that the subject parameters ξ_{ik} are assumed to be either fixed or random. In the latter case, ξ_{ik} is assumed to be normally distributed with a zero mean and variance σ_{ξ}^2. The variance–covariance matrices of the responses in the fixed effect and random subject models are compound symmetric, and it can be shown that the best linear unbiased estimators of the treatment effect differences and the difference between the carry-over effects of the two treatments are the same for the two models. Likewise, optimal designs with minimum variance for the estimators of both treatment differences and differences in carry-over effects are the same for both the fixed effects model and the random subject model. In general, designs that are 'good' when the subject effects ξ_{ik} are fixed, will also be 'good' when model contains random subject effects (Hedayat, Stufken and Yang, 2006).

Cheng and Wu (1980) and Laska, Meisner and Kushner (1983) found the following properties of optimal CO designs. The results do not depend on whether or not carry-over effects exist or whether the design includes baseline measurements.

- Among all designs with s sequences, m periods and t treatments, a *strongly balanced and uniform (in periods and sequences) design* is optimal for estimating treatment effects and carry-over effects.

- A design that is *balanced and uniform* (in periods and sequences) in the first $(m - 1)$ periods and that is given the same treatment in the mth period as in the $(m - 1)$th period for every sequence is optimal for estimating treatment effects and carry-over effects.

These two statements can be used to check the optimality of the CO designs with two treatments A and B shown in Table 8.9. The optimality of these designs was established by Cheng and Wu (1980) and Laska, Meisner and Kushner (1983). Matthews (1990) gave a method for constructing the optimal designs. The optimal designs in Table 8.9 assume that an equal number of experimental units is randomly assigned to each of the sequences of treatments. So, for the optimal designs with two distinct sequences and a total sample of N experimental units, $N/2$ units are assigned to each sequence, while for a four-sequence design, $N/4$ units are assigned to each sequence. All four designs are strongly balanced designs and uniform within periods. Design 3, however, is not only strongly balanced but also uniform both in periods and sequences.

Table 8.9   Optimal cross-over designs for two treatments.

	Design 1 (2, 4, 2)		Design 2 (2, 2, 3)			Design 3 (2, 4, 4)				Design 4 (2, 2, 5)				
	Period		Period			Period				Period				
Sequence	1	2	1	2	3	1	2	3	4	1	2	3	4	5
1	A	B	A	B	B	A	B	B	A	A	B	B	A	A
2	B	A	B	A	A	B	A	A	B	B	A	A	B	B
3	A	A				A	A	B	B					
4	B	B				B	B	A	A					

Note. The triplet (t, s, m) represents a cross-over design with s sequences, m periods and t treatments.

8.7.1   Some further observations

We provide here further remarks on the optimality of the designs in Table 8.9. Design 1 with triplet (2, 4, 2) was proposed by Balaam (1968) and combines a simple two-treatment CO design (2, 2, 2) with a PG design. Design 1 is optimal for estimating both treatment and carry-over effects, but it is not optimal when carry-over effects are not present. In this case, the (2, 2, 2) CO design is optimal. A potential problem with the AA and BB sequences in Design 1 is that there may be clinical and ethical objections to giving patients the same treatment twice, especially if the efficacy of a treatment is unknown or a treatment is burdensome or stressful. These objections can be addressed by relaxing the rule of assigning equal numbers of patients to each of the four sequences. Carriere and Reinsel (1992) studied the unequal assignment of patients to the four treatment sequences of the (2, 4, 2) CO design (Design 1) under different model assumptions. They recommend using the (2, 4, 2) CO design (Design 1), with or without baseline measurements, and allocating 80% of patients to the AB and BA sequences and 20% of patients to the AA and BB sequences. Moderate efficiency can then be maintained under different model assumptions.

A second remark concerns the increase of the number of periods. One way to deal with the carry-over effect problem of the (2, 2, 2) CO design, and avoid the clinical and ethical objections against giving only one treatment to the same patients twice as in the (2, 4, 2) design, is to increase the number of periods. The (2, 2, 2) CO design can be expanded to include more periods such as in a (2, 2, 3) design or (2, 2, 5) design. This increase has the advantage that carry-over effects can be estimated more efficiently. A cost–efficiency comparison by Yuan and Zhou (2005) showed that among a number of $m = 3$ and $m = 4$ period CO designs, the (2, 2, 3) and (2, 4, 4) CO designs in Table 8.9 perform the best in terms of cost efficiencies. These designs are also optimal within the class of same period designs without cost constraint (Cheng and Wu, 1980; Laska, Meisner and Kushner, 1983). The (2, 4, 2) design (BM's design) performed overall worst in terms of cost efficiency. However, a disadvantage of including extra periods is that the trial will last longer, which in turn increases the probability of dropout and the occurrence of a significant treatment × period interaction, both of which can hinder interpretation (Matthews, 1988).

Carriere (1994) studied the gain in efficiency of using $m = 3$ period CO designs with $t = 2$ treatments. She found that high efficiency is maintained for these designs as long as the optimal sequences ABB and BAA are included. Moreover, under the assumption that dropout is random, her results showed that the (2, 2, 3) design was still more efficient than the (2, 2, 2) design, even if there was a high level (80%) of missing data in the third period. Carriere and Reinsel (1993) and Carriere (1995) also studied $m = 2$ period CO designs with $t \geq m$ treatments and showed that for a model with random subject parameters, balanced designs perform almost as well as strongly balanced designs.

In general, optimal designs are strongly model dependent and a design chosen on the basis of one model is likely not to be efficient for other models. The optimal designs in Table 8.9 are based on the model in Equation (8.5). Fleiss (1988) and Senn and Hildebrand (1991) argued that the model in Equation (8.5) does not make a distinction between different carry-over effects from A on to B and from B on to B in a sequence ABB, and that a treatment-by-carry-over effect interaction is likely to occur. Another deficiency of the model is that it assumes independent errors, which means that the variance–covariance matrix of the responses is compound symmetric. Since these designs are repeated measurement designs, this assumption may not be valid. These designs are not optimal in cases where the compound symmetry assumption does not hold. The assumption of auto-correlated errors may seem more realistic in designs with more than two periods. Matthews (1987) considered auto-correlated errors for two-treatment CO designs in the presence of carry-over effects and fixed subject effects, and showed that the three- and four-period designs in Table 8.9 are also highly efficient when the errors in the model are auto-correlated. Matthews (1987) concluded that the best all-round $m = 3$ period design is one that allocates equal numbers of experimental units to the sequences AAB, BBA, ABB and BAA. Among the $m = 4$ period CO designs, Design 4 with sequences ABBA, BAAB, AABB and BBAA, from Table 8.9, is perhaps the most efficient for the estimation of treatments effects and carry-over effects. Matthews (1987) found that the efficiency of these two designs is at least 82% across the range of auto-correlated values $-0.8 < \rho < 0.8$.

8.8 Matrix formulation of the mixed model for crossover designs

The random subject model that is commonly applied to analyse data from CO designs can be formulated in matrix form as

$$ y_{ik} = X_{ik}\beta + Z_{ik}\xi_{ik} + \varepsilon_{ik}, \tag{8.12} $$

where the $m \times 1$ vector y_{ik} contains repeated measurements in m periods for subject i in a kth treatment sequence. The $p \times 1$ vector β contains fixed parameters. The random parameter for subject i in sequence k is ξ_{ik} and Z_{ik} is an $m \times 1$ vector of ones. The design matrix X_{ik} is of the order $m \times p$. The errors ε_{ik}'s are assumed to be normally distributed each with mean zero and an arbitrary positive-definite variance–covariance matrix $\sigma_\varepsilon^2 \Sigma$. For simplicity, we assume to have independently distributed errors with the same constant, that is $\Sigma = I$, where I is an identity matrix.

The random effects parameter ξ_{ik} describes the deviation of the response of the ith subject from the overall average response, and it is assumed that ξ_{ik} is normally distributed as $N(0, \sigma_\xi^2)$. Finally, the subject parameters ξ_{ik} and the errors are assumed to be independently distributed.

The variances and covariances of the repeated measurements are the weighted sum of the variances and covariances of the random effects parameters and variances and covariances of the errors. Thus,

$$\text{Cov}(y_{ik}) = \sigma_\xi^2 Z_{ik} Z_{ik}' + \sigma_\varepsilon^2 I. \tag{8.13}$$

When the variance components σ_ξ^2 and σ_ε^2 are known and recalling that n_k is the number of subjects in the kth sequence, $k = 1, 2, \ldots, s$, the ML estimators of β are equal to the GLS estimators of the fixed parameters:

$$\hat{\beta} = \left(\sum_k^s \sum_i^{n_k} X_{ik}' [\text{Cov}(y_{ik})]^{-1} X_{ik} \right)^{-1} \sum_k^s \sum_i^{n_k} X_{ik}' [\text{Cov}(y_{ik})]^{-1} y_{ik}. \tag{8.14}$$

If the variance components are not known but their ML or REML estimates are available, these estimates can be substituted in Equation (8.13) instead (Verbeke and Molenberghs, 2000, Chapter 5; Diggle *et al.*, 2002, Chapter 4). The asymptotic variance–covariance matrix of the estimator $\hat{\beta}$ is equal to

$$\text{Cov}(\hat{\beta}) = \left(\sum_k^s \sum_i^{n_k} X_{ik}' \text{Cov}(y_{ik})^{-1} X_{ik} \right)^{-1}. \tag{8.15}$$

To illustrate the analysis of a CO design, consider the parameterization for the expected cell values in Table 8.3, where there are nine parameters in β, that is $\beta = (\mu, \tau_A, \tau_B, \zeta_1, \zeta_2, \pi_1, \pi_2, \lambda_A, \lambda_B)$. To be able to estimate these parameters, we impose the following restrictions. Let the two treatment effect parameters be $(\mu + \tau_A)$ and $(\mu + \tau_B)$, and let the period effects be related as $\pi = \pi_1 = -\pi_2$. Further, the sequence effects are $\zeta_1 = \zeta_2 = 0$ and the carry-over effects are $\lambda = \lambda_A = -\lambda_B$. The upper half of Table 8.10 presents the expected values for the cells with these restrictions. Notice that the difference between treatment

Table 8.10   Expected values of (2, 2, 2) cross-over design for two different parameterizations.

	Period 1	Period 2
Sequence AB	$(\mu + \tau_A) + \pi$	$(\mu + \tau_B) - \pi + \lambda$
Sequence BA	$(\mu + \tau_B) + \pi$	$(\mu + \tau_A) - \pi - \lambda$

$$E(\hat{\tau}_A - \hat{\tau}_B) = (\tau_A - \tau_B) - \lambda$$

	Period 1	Period 2
Sequence AB	$(\mu + \tau_A) + \pi + \zeta$	$(\mu + \tau_B) - \pi + \zeta$
Sequence BA	$(\mu + \tau_B) + \pi - \zeta$	$(\mu + \tau_A) - \pi - \zeta$

$$E(\hat{\tau}_A - \hat{\tau}_B) = (\tau_A - \tau_B)$$

parameter estimators is biased, that is $E(\hat{\tau}_A - \hat{\tau}_B) = (\tau_A - \tau_B) - \lambda$, because the two carry-over effects are not equal (Equation 8.2).

The 4×1 column vector of fixed parameters with four parameters is $\beta = (\mu + \tau_A, \mu + \tau_B, \pi, \lambda)'$, and the model for the ith subject in two-treatment sequences AB and BA can now be written as

AB ($k = 1$):

$$
\begin{bmatrix} y_{i11} \\ y_{i21} \end{bmatrix} = \begin{bmatrix} 1 & 0 & 1 & 0 \\ 0 & 1 & -1 & 1 \end{bmatrix} \begin{bmatrix} \mu + \tau_A \\ \mu + \tau_B \\ \pi \\ \lambda \end{bmatrix} + \begin{bmatrix} 1 \\ 1 \end{bmatrix} [\xi_{i1}] + \begin{bmatrix} \varepsilon_{i11} \\ \varepsilon_{i21} \end{bmatrix}
$$

and   (8.16)

BA ($k = 2$):

$$
\begin{bmatrix} y_{i12} \\ y_{i22} \end{bmatrix} = \begin{bmatrix} 0 & 1 & 1 & 0 \\ 1 & 0 & -1 & -1 \end{bmatrix} \begin{bmatrix} \mu + \tau_A \\ \mu + \tau_B \\ \pi \\ \lambda \end{bmatrix} + \begin{bmatrix} 1 \\ 1 \end{bmatrix} [\xi_{i2}] + \begin{bmatrix} \varepsilon_{i12} \\ \varepsilon_{i22} \end{bmatrix}.
$$

These equations show that the variance within subjects (units, patients) is due to the variance σ_ε^2 between the errors at different periods. The variance of responses to treatments is the sum of the between-subject variance σ_ξ^2 and the within-subject variance σ_ε^2.

A second set of restrictions leads to the parameterization as shown in the lower half of Table 8.10. This time it is assumed that there are no carry-over effects, that is $\lambda_A = \lambda_B = 0$, so that the difference between treatment effect estimators $(\hat{\tau}_A - \hat{\tau}_B)$ will be unbiased (Equation 8.2). Further, it is assumed that period effects and sequence effects are restricted by $\pi = \pi_1 = -\pi_2$ and $\zeta = \zeta_1 = -\zeta_2$, respectively. The corresponding models for a subject i in sequence k are

AB ($k = 1$):

$$
\begin{bmatrix} y_{i11} \\ y_{i21} \end{bmatrix} = \begin{bmatrix} 1 & 0 & 1 & 1 \\ 0 & 1 & -1 & 1 \end{bmatrix} \begin{bmatrix} \mu + \tau_A \\ \mu + \tau_B \\ \pi \\ \zeta \end{bmatrix} + \begin{bmatrix} 1 \\ 1 \end{bmatrix} [\xi_{i1}] + \begin{bmatrix} \varepsilon_{i11} \\ \varepsilon_{i21} \end{bmatrix}
$$

and   (8.17)

BA ($k = 2$):

$$
\begin{bmatrix} y_{i12} \\ y_{i22} \end{bmatrix} = \begin{bmatrix} 0 & 1 & 1 & -1 \\ 1 & 0 & -1 & -1 \end{bmatrix} \begin{bmatrix} \mu + \tau_A \\ \mu + \tau_B \\ \pi \\ \zeta \end{bmatrix} + \begin{bmatrix} 1 \\ 1 \end{bmatrix} [\xi_{i2}] + \begin{bmatrix} \varepsilon_{i12} \\ \varepsilon_{i22} \end{bmatrix},
$$

and again the variance of the responses is $\text{var}(y_{ik}) = \sigma_\xi^2 + \sigma_\varepsilon^2$. The variance–covariance matrix of the responses of the ith subject for this model has a compound symmetric structure:

$$\text{Cov}(y_{ik}) = \sigma_\xi^2 Z_{ik} Z_{ik}' + \sigma_\varepsilon^2 I = \begin{bmatrix} \sigma_\xi^2 & \sigma_\xi^2 \\ \sigma_\xi^2 & \sigma_\xi^2 \end{bmatrix} + \begin{bmatrix} \sigma_\varepsilon^2 & 0 \\ 0 & \sigma_\varepsilon^2 \end{bmatrix}$$

$$= \begin{bmatrix} \sigma_\xi^2 + \sigma_\varepsilon^2 & \sigma_\xi^2 \\ \sigma_\xi^2 & \sigma_\xi^2 + \sigma_\varepsilon^2 \end{bmatrix}. \tag{8.18}$$

There are three options for estimating the fixed effect parameters β:

- Substitution of ML or REML estimators in place of the variance components σ_ξ^2 and σ_ε^2 in Equation (8.18) and the use of the estimate of $\text{Cov}(y_{ik})$ to estimate the fixed parameters β by means of Equation (8.14).

- Computation of the unstructured variance–covariance matrix of the responses and its use to estimate the fixed parameters β in Equation (8.14).

- Assume that the between-subject variance is $\sigma_\xi^2 = 0$, so that $\text{Cov}(y_{ik}) = \sigma_\varepsilon^2 I$, and estimate the fixed parameters in β by means of Equation (8.14) with $\text{Cov}(y_{ik}) = \sigma_\varepsilon^2 I$ substituted. This option actually reduces the GLS estimator of β to the ordinary least squares (OLS) estimator of β.

Since both models for the designs in Table 8.10 are saturated (four parameters and four cells), the above described options for estimating the fixed parameters in β will result in the same estimates $\hat{\beta}$. But the variances and covariances in $\text{Cov}(\hat{\beta})$ will, generally, not be the same.

8.9 Summary

This chapter discusses the two-treatment CO designs that are commonly used in the social and biomedical disciplines. A key advantage CO designs have over PG designs is that CO designs are more efficient for estimating treatment effects when there are no carry-over effects. Our efficiency and cost comparison shows that, among the four two-treatment designs, namely the CO, CR, PG and BM designs, the CO design is preferable in terms of cost and efficiency. In particular, savings up to 50% can be obtained by using a CO design instead of the PG design when the intra-class correlations are moderately high or high. However, we emphasize that the CO design has the strong assumption that there are no carry-over effects.

The CO design that is strongly balanced and uniform (in terms of periods and sequences) is optimal for estimating both treatment effects and carry-over effects.

This CO design assumes random assignment of an equal number of experimental units to the various treatment sequences. In this chapter, we have also considered a few two-treatment optimal CO designs with an extended number of periods. These optimal designs will require additional time to collect the data, but their main advantage is that they can estimate carry-over effects more efficiently than two-period CO designs.

9

Alternative optimal designs for linear models

9.1 Introduction

We recall from Chapters 2 and 3 that a D-optimal design minimizes the volume of the confidence ellipsoid for the model parameters and so such designs produce the most accurate estimators of all model parameters simultaneously. For example, in the simple linear regression model (Chapter 2), the confidence region for the intercept and slope parameter from a D-optimal design has the smallest area compared with those from any other designs. Consequently, D-optimal designs provide the most accurate estimators for both parameters simultaneously.

An appealing feature of D-optimal designs is that for many problems, they can be described analytically. This means that the optimal design can be described using a formula. This is highly desirable because for these design problems, no iterative methods are required to find the D-optimal designs and so we can easily study the properties of D-optimal designs. In particular, we can deduce how the D-optimal designs will change if certain design parameters are changed. For example, the formula usually will enable us to determine the new design points when the design space is changed. The formula may also allow us to deduce how the weights in a D-optimal design change when certain aspects of the design problem are changed.

However, there are many instances where D optimality is not applicable. The researcher may be interested in estimating only some of the parameters, and not all parameters, which is what D optimality is based on. The researcher may also be interested in predicting the response at some point or region, which may or

An Introduction to Optimal Designs for Social and Biomedical Research   M. P. F. Berger & W. K. Wong
© 2009, John Wiley & Sons, Ltd

may not be in the design space; in the latter case, we have an extrapolation design problem. Extrapolation design problems are common in dose–response studies where it is desired to model at dose levels that may be unsafe or untested.

In this chapter, we present alternative optimal designs for linear models. Many of the ideas and methods we employ here can be applied directly to design problems for nonlinear models, and we will elaborate upon them further in the next chapter. Throughout this chapter, we assume that all errors are independently and normally distributed with mean zero and constant variance. As before, the simple linear model that we will frequently refer to is the straight-line model with both the intercept and slope terms and the error terms are independent with zero mean and constant variance. Unlike design books currently available, we choose not to deal with technical details; instead, we present design issues and concepts for a few linear models and refer the readers to references for mathematical details.

9.2   Information matrix

In the previous chapters, we focused on the variances and covariances of the parameter estimators to explain the problems in finding optimal and highly efficient designs. In this and the following chapters, we use the information matrix to study and construct optimal designs. In general, the information matrix is inversely related to the asymptotic variance–covariance matrix of parameter estimators and as such is a convenient alternative form to describe the design problem.

Let us consider the linear model for a single response: $y = f'(x)\beta + \varepsilon$. This model contains the vector of parameters β and a single independent variable with values x. The function $f'(x)$ is the transpose of a function of values of the independent variable. For each observation at x_j, the information is given by the product $f(x_j)f'(x_j)$ under the normality assumption. When the observations are independent, the total information is found by adding the contribution from each level. More generally, the information matrix for any design ξ is given by

$$M(\xi) = \sum_j w_j f(x_j)f'(x_j), \tag{9.1}$$

where the weights w_j correspond to the distinct values x_j and $\sum_j w_j = 1$. For example, for the simple linear model $f'(x_j) = (1, x_j)$, the information matrix becomes

$$M(\xi) = \sum_j w_j f(x_j)f'(x_j) = \sum_j w_j \begin{bmatrix} 1 & x_j \\ x_j & x_j^2 \end{bmatrix}. \tag{9.2}$$

The inverse of the information matrix is proportional to the asymptotic variance–covariance matrix for the parameter estimators $\hat{\beta}$ using design ξ. A 'large' information matrix implies more efficient estimators for β. Here, 'large' can take several forms. For example, one popular measure of large is the magnitude of the determinant of the information matrix. Using calculus, one can show that a design with a large determinant for its information matrix always has a small volume of the confidence ellipsoid for β. Consequently, between two competing designs, we prefer the one with a larger determinant for its information matrix because it has a smaller volume for the confidence region between the two designs. Alternatively, in terms of the inverse of the information matrix, we want a design that has a small determinant for the inverse of the information matrix. This is so because the determinant of the inverse of a matrix is the reciprocal of the determinant of the matrix.

The diagonal elements of the inverse of the information matrix $M(\xi)$ are proportional to the asymptotic variance of parameter estimators $\hat{\beta}$. In particular, the ith diagonal element of the inverse of the information matrix is proportional to the variance of the ith component of $\hat{\beta}$. So, another way to find a good design for estimating all parameters in the model is to minimize the sum of the variances of all the $\hat{\beta}$'s. We discuss this option further in Section 9.5.

In the next few sections, we present other types of optimal designs for linear models that are useful for different purposes.

9.3 D_A- or D_s-optimal designs

Sometimes, not all the parameters β in the model are of interest. There are nuisance parameters in the model and research interest is on estimating only a couple of selected parameters or only a few linear combinations of parameters. When designing for such a study, it is therefore natural to allocate resources for estimating only parameters of interest. When only a subset of the model parameters is of interest, the D-optimality criterion can be straightforwardly modified to reflect the practitioner's interest. The names for optimal designs for estimating only a subset of the model parameters are variously called D_s optimal or D_A optimal. Similar to D-optimality, the D_A- or D_s-optimal design is found by minimizing the determinant of a matrix with components that are submatrices extracted from the information matrix $M(\xi)$, where σ_ε^2 is the usual variance of the independent errors in the model.

The variance–covariance matrix for $A'\beta$ is proportional to $[A'M(\xi)^{-1}A]$ and the D-optimality criterion becomes

$$D_A\text{-criterion} = \text{Det}[A'M(\xi)^{-1}A], \qquad (9.3)$$

where the symbol D_A indicates the dependence of the determinant on the matrix A. If the interest is only on a subset of s parameters in β, we have the D_s-criterion, where 's' stands for subset and the matrix A now contains only zeros and ones in the right places to capture the subset of β of interest. See also Section 3.5 in Chapter 3.

In the simple linear model with parameters $\beta = (\beta_0, \beta_1)'$, for example, the researcher may be interested in only the slope parameter β_1 and not the intercept β_0. Resources should therefore be used to estimate only the slope parameter. In terms of the linear combinations $A'\beta$, the parameter of interest is $\beta_1 = A'\beta$

where $A' = (0\ 1)$. The D_s-optimal design for this case is unique and is simply the design that takes equal observations at both ends of the design space. If the intercept term β_0 is the only parameter of interest, then in terms of $A'\beta$ we have the intercept $\beta_0 = A'\beta$ where $A' = (1\ 0)$. When the design interval is $[-1, 1]$, it is interesting to note that the optimal design for estimating the slope parameter is also an optimal design for estimating the intercept. However, the optimal design for estimating the intercept term is no longer unique. Any design on the design space, such that the average value of the design points is zero, is also optimal for estimating the intercept. So, there are infinitely many optimal designs for estimating the intercept. Some examples are the uniform design supported at $[-1, 0, 1]$ and the design equally supported at $[-1, -0.5, -0.5, 1, 1]$.

The same approach for estimating a subset of the model parameters can be applied directly to other models. For instance, Preitschopf and Pukelsheim (1987) provided optimal designs for estimating all possible subsets of the three parameters in the quadratic model. In particular, they showed that by setting $A' = \begin{pmatrix} 0 & 1 & 0 \\ 0 & 0 & 1 \end{pmatrix}$ for the model $y = \{1, x, x^2\}\beta + \varepsilon$, the optimal design for estimating the coefficients of the linear and quadratic terms is equally supported at $[-1\ 0\ 1]$.

We mention here two other examples of D_s-optimal designs given in Atkinson and Donev (1992, p. 110). They first considered the homoscedastic quadratic model on the design interval $[-1, 1]$ and found the D_s-optimal design for estimating the coefficient β_2 for the quadratic term. This coefficient reflects curvature in the model and is often included in the linear model, as a first step, when there is doubt about the linearity assumption. The optimal design is to take 25% of the

Table 9.1 D-, D_s- and extrapolation optimal designs for polynomial regression models.

Model	Estimation	Design interval	Optimal design
$y_i = \beta_0 + \beta_1 x_{1i} + \beta_2 x_{1i}^2 + \varepsilon_i$	$(\beta_0, \beta_1, \beta_2)$	$[-1, 1]$	$\xi = \left\{ \begin{matrix} -1 & 0 & 1 \\ 1/3 & 1/3 & 1/3 \end{matrix} \right\}$
$y_i = \beta_0 + \beta_1 x_{1i} + \beta_2 x_{1i}^2 + \varepsilon_i$	(β_2)	$[-1, 1]$	$\xi = \left\{ \begin{matrix} -1 & 0 & 1 \\ 1/4 & 1/2 & 1/4 \end{matrix} \right\}$
$y_i = \beta_1 x_{1i} + \beta_2 x_{1i}^2 + \varepsilon_i$	(β_2)	$[0, 1]$	$\xi = \left\{ \begin{matrix} \sqrt{2}-1 & 1 \\ 1/\sqrt{2} & (\sqrt{2}-1)/\sqrt{2} \end{matrix} \right\}$
$y_i = \beta_0 + \beta_1 x_{1i} + \varepsilon_i$	Response at $x_0 = 2$	$[-1, 1]$	$\xi = \left\{ \begin{matrix} -1 & 1 \\ 1/4 & 3/4 \end{matrix} \right\}$
$y_i = \beta_0 + \beta_1 x_{1i} + \beta_2 x_{1i}^2 + \varepsilon_i$	Response at $x_0 = 2$	$[-1, 1]$	$\xi = \left\{ \begin{matrix} -1 & 0 & 1 \\ 1/7 & 3/7 & 3/7 \end{matrix} \right\}$

observations at each of the extreme design points, that is $x = -1$ and $x = 1$ and the rest at $x = 0$. This design is different from the D optimal design in terms of simultaneous estimation of all the three parameters. Atkinson and Donev (1992, p. 94) also provided another example using the quadratic model without intercept on the design interval [0, 1], and again interest was only in estimating the coefficient β_2 for the quadratic term. The optimal design for estimating only this term is a two-point design and takes $100/\sqrt{2}$ percentage of the observations at $x = \sqrt{2} - 1$ and the rest of the observations at $x = 1$. Both these D_s designs are summarized in Table 9.1. We do not provide technical details here and refer the reader to the cited monograph. An interesting application of a D_s-optimal design to sequentially estimate treatment effects in a multi-group trial was given by Atkinson (1982). An example for formulating the D-, A- and E-optimality criteria for estimating a subset of the model parameters is also given in Chapter 3.

9.4   Extrapolation optimal design

Let us assume that our model is $y = f'(x)\beta + \varepsilon$, where y is the outcome and $f(x)$ is a known regression function defined on the given design interval $[x_{\min}, x_{\max}]$. Sometimes, we may want to estimate the mean or predicted response at some value x_0. This frequently happens in dose–response studies where the value x_0 is outside the typical dose range for the drug, that is $x_0 < x_{\min}$ or $x_0 > x_{\max}$.

The objective is to find a design that estimates $f'(x_0)\beta$ as accurately as possible. If x_0 is in the design interval, that is, $x_{\min} \leq x_0 \leq x_{\max}$, the design problem is trivial; one simply takes all observations at x_0 and this is the optimal design for estimating $f'(x_0)\beta$. Of course, this design is of very limited use other than just being able to estimate $f'(x_0)\beta$ with minimum variance. If x_0 is outside the design interval $[x_{\min}, x_{\max}]$, it is not possible to observe the outcome at x_0, and so the design problem is more difficult. A typical situation is in dose–response study where the design interval is the known safety dose limits of the drug or any subset of the safety dose limits of the drug and the researcher wants to model the outcome at a possibly unsafe dose; see for example, Krewski and Kovar (1982) and Gaylor, Chen and Kodell (1985) for extrapolation design issues in cancer research. The extrapolation optimal design is to find an allocation scheme on the design space so that the variance of the estimated $f'(x_0)\beta$ is minimized.

Some simple examples of extrapolation optimal designs for the linear and quadratic models are given in Wong and Lachenbruch (1996). They considered homoscedastic models on the prototype design interval $[-1, 1]$. We recall that such prototype intervals are obtained after rescaling the actual drug dose levels. These optimal designs minimize the variance of the estimated response at the extrapolated dose. For instance, consider finding a design to make inference on the response at $x_0 = 2$ for the linear and quadratic regression model when the design interval is $[-1, 1]$. It can be shown that the extrapolation optimal design for the simple linear model is to take one-fourth of the observations at $x = -1$ and the rest at $x = 1$. For the quadratic model, the extrapolation optimal design for the point $x_0 = 2$ is to take one-seventh of the observations at $x = -1$ and the rest of the observations at $x = 0$ and $x = 1$ in equal proportions. These extrapolation designs are presented in Table 9.1. The theory behind the construction of extrapolation optimal design is complex, and interestingly for many problems the weights of the support points of the optimal design depend on the extrapolated point of interest and the support points of the optimal design do not depend on the extrapolated point.

We emphasize that extrapolation design problems are particularly risky because the statistical assumptions of the model over the design space may no longer apply outside the design space. For example, the mean function may take on a different form outside the design space or errors may have varying variances outside the design space. Therefore, the researcher must proceed with great caution when making inference outside the design space.

9.5   L-optimal designs

A less well known but very flexible optimality criterion is the so-called L-optimality criterion. This criterion is defined as

$$\text{L-Criterion} = \text{Trace}[L\, M(\xi)^{-1}], \qquad (9.4)$$

where the matrix L is a user-selected matrix. A design that minimizes the L-optimality criterion is an L-optimal design. The L-optimality criterion is actually a class of criteria because it contains several other optimality criteria as special cases. How do we choose the matrix L? We choose L to reflect the researcher's interest in the study. For instance, if estimation of all model parameters is of interest, one may choose L to be the identity matrix. In this case, the L-optimality criterion becomes $\text{Trace}[M(\xi)^{-1}]$ and the L-optimal design minimizes the sum of the variances of all estimated parameters in the model. Of course, this is the same as minimizing the average of the variances of the estimated parameters and this criterion is therefore sometimes called A optimality with the 'A' standing for average. For simultaneous estimation of the two parameters in the simple linear model on the design space $[-1, 1]$, it is easy to verify that the design equally supported at the extreme points $x = -1$ and $x = 1$ is the A-optimal design. For the quadratic model on $[-1, 1]$, the A-optimal design is the symmetric design that assigns weight 0.5 at $x = 0$ and the rest equally at $x = -1$ and $x = 1$.

The matrix L can also be chosen to include only the sum of the variances of a subset of parameters β. For example, if we have a quadratic model and we only want to estimate the coefficients associated with the linear and quadratic terms, we set L to be the 3×3 zero matrix but with 1 in both the $(2, 2)$ and $(3, 3)$ entries. For any matrix L that can be written as $L = AA'$, the following property holds:

$$\text{Trace}[L\,M(\xi)^{-1}] = \text{Trace}[AA'M(\xi)^{-1}] = \text{Trace}[A'\,M(\xi)^{-1}A]. \qquad (9.5)$$

This means that when we can write the matrix $L = AA'$, we can specify the matrix A instead of L. For the simple linear regression model with $\beta = (\beta_0\ \beta_1)'$, we set $A' = (1\ 0)$ or $A' = (0\ 1)$ when we are interested to estimate only the parameter β_0 or β_1, respectively.

The L-optimality criterion is versatile and can also be used in other ways. For an interesting use of this criterion, consider the situation where it is known that the average of a continuous response can be adequately described by a quadratic model on a known design space $[x_{\min}, x_{\max}]$ and the goal is to estimate the turning point in the mean response curve. Here, $y = f'(x)\beta + \varepsilon_i$ with $f'(x) = (1, x, x^2)$ and $\beta = (\beta_0, \beta_1, \beta_2)'$. By differentiating the mean function and setting it equal to zero, it is easy to show that the turning point depends on the model parameters and occurs when the value of x is equal to

$$g(\beta) = -\frac{\beta_1}{2\beta_2}. \qquad (9.6)$$

Let $L = A(\beta)A'(\beta)$, where $A'(\beta)$ is the derivative vector of $g(\beta)$ with respect to the three parameters, that is $A'(\beta) = [0, -1/(2\beta_2), \beta_1/(2\beta_2^2)]$. Then, using a first-order Taylor's expansion, the asymptotic variance of the maximum likelihood estimator of $g(\beta)$ is approximately equal to

$$\text{Trace}[L\,M(\xi)^{-1}] = \text{Trace}[A'(\beta)\,M(\xi)^{-1}A(\beta)]. \qquad (9.7)$$

Chaloner (1989) showed that when the design space is $[-1, 1]$, the locally D-optimal design for estimating the turning point depends on the value of $g(\beta)$ and has the following complex structure:

1. if $g(\beta) > 1/2$, take one-half of the observations at $x = 0$, $1/4 + 1/(8g(\beta))$ of the observations at $x = 1$ and the rest at $x = -1$;

2. if $0 \le g(\beta) \le 1/2$, take one-half of the observations at $x = 1$ and the rest at $x = 2g(\beta) - 1$;

3. if $-1/2 \le g(\beta) \le 0$, take one-half of the observations at $x = 1$ and the rest at $x = 2g(\beta) + 1$.

We end this section by noting that the problem considered here also has an application to designing a study for the quadratic logistic regression model when interest is in maximizing the probability of a response. Fornius (2008) is perhaps the latest among several others who used this idea to use a sequential design to estimate the value of x that gives the optimum operating conditions when the response is a binary outcome.

9.6   Bayesian optimal designs

In practice, there is usually some information available at the beginning of a scientific study. Some may come from similar studies or from different experts in the field. The researcher may also carry out pilot studies when it is feasible to do so. We recall that pilot studies are small studies with the same subjects or experimental units to be used for the main study and carried out with the purpose to obtain preliminary estimates of key parameters in the proposed studies. For example, it is common to conduct a pilot study to determine estimates for parameters for the sample size calculation in the main study. A main research question here is how to use prior information at the onset to construct a better design.

Frequently, the source of the information is not unique and several similar studies may provide different information. This information may take on different forms, such as information on the model parameters or only as a certain function of the model parameters. A design strategy for dealing with this situation is to use a Bayesian framework. To fix ideas, .we suppose we have a simple linear model and earlier studies suggest that there are several plausible values for the slope parameter β_1. The first step in the Bayesian approach is to capture all prior information in the form of a probability distribution for the slope parameter. This distribution is often referred to as the *prior distribution*. The choice of the prior distribution is subjective and depends on the subject matter. In a simple situation say with three equally probable values for the slope parameter, the prior distribution $p(\beta_1)$ may be the discrete prior uniformly distributed at the three values. On the other hand, if these values came from three experts and the third expert is deemed to be twice as reliable as the other two equally reliable experts, the mass distribution for the discrete prior becomes $1/4$, $1/4$ and $1/2$. In general, information that is deemed to be more reliable should have a higher weight in the prior distribution. Of course, how to quantify how reliable the information is from one source versus another is a subjective matter.

The second step in the Bayesian paradigm is to combine information in the prior distribution and our model assumptions, expressed in the likelihood function. Bayesian inference is then based on the posterior distribution obtained by multiplying the prior distribution with the likelihood function. The reason for basing inference from the posterior distribution is that this distribution captures the prior information and information from the current data. There are other justifications for the Bayesian paradigm but we do not go into them here.

For design purposes, the Bayesian design criterion averages the proposed design criterion over the prior distribution. One can use a formal criterion and seek to optimize the criterion. Some possibilities for Bayesian design criteria are $E \ln \mathrm{Det}[M(\xi)]$, $E \, \mathrm{Det}[M(\xi)]$ or $\ln E \, \mathrm{Det}[M(\xi)]$, where in each case, the expectation E is over the prior distribution and ln represents the natural logarithm. The Bayesian optimal design is the one that maximizes the criterion over all designs on the design space. We do not discuss the pros and cons of these criteria, but note that the Bayesian optimal designs for the above criteria can be different.

Among the three, the most popular one is E ln Det$[M(\xi)]$ for estimating model parameters when prior information is available.

Let us follow up with an example of a Bayesian optimal design using the example in the previous section. The problem was to find a design to estimate the turning point for the quadratic linear model. Suppose that a prior distribution is now available for the x value of the turning point $g(\beta)$, and not for β. The matrix L and the vector A in the L-optimality criterion now depend on β and the Bayesian L-optimality criterion is

$$\begin{aligned}
\text{E Trace}[LM(\xi)^{-1}] &= \sigma_\varepsilon^2 \text{ E Trace}[A(\beta)A'(\beta)M(\xi)^{-1}], \\
&= \sigma_\varepsilon^2 \text{ Trace}[E A(\beta)A'(\beta)M(\xi)^{-1}].
\end{aligned} \quad (9.8)$$

Here $L = A(\beta)A'(\beta)$ with $A(\beta)' = [0, -1/(2\beta_2), \beta_1/(2\beta_2^2)]$ and E is the expectation with respect to the prior distribution for $g(\beta)$. If we further make a simplifying assumption that σ^2 and β_2 are independent of $g(\beta)$, then the optimization problem becomes minimize Trace$[LM(\xi)^{-1}]$ where the matrix L now takes on the form:

$$L = \begin{bmatrix} 0 & 0 & 0 \\ 0 & 1 & 2E[g(\beta)] \\ 0 & 2E[g(\beta)] & 4E[g(\beta)^2] \end{bmatrix}, \quad (9.9)$$

and the expectation is taken with respect to the prior distribution of $g(\beta)$. For instance, if we believe that the uniform distribution on the interval $[-1/2, 1/2]$ is a reasonable prior distribution for $g(\beta)$, we have $E[g(\beta)] = 0$ and $E[g(\beta)^2] = 1/12$ and L becomes

$$L = \begin{bmatrix} 0 & 0 & 0 \\ 0 & 1 & 0 \\ 0 & 0 & 1/3 \end{bmatrix}. \quad (9.10)$$

Computer algorithms for generating L-optimal designs are discussed in Chapter 11. Alternative approaches to designs for estimating the turning point in the quadratic model are given in Mandal and Heiligers (1992) and Fedorov and Mueller (1997). The former adopted a minimax approach and the latter re-parameterized the model to obtain more insights into the design problem using Bayesian, minimax and sequential techniques.

Bayesian optimal designs rarely have a formula capable of describing the optimal design points and the proportion of observations to be taken at each of the points. This is because the mathematical formulation of a Bayesian design problem is complicated and closed form solutions to the optimization problem rarely exist unless the model is very simple. As a consequence, properties of Bayesian optimal designs and their sensitivities to the prior distribution are difficult to study. In practice, Bayesian optimal designs are usually found from self-written algorithms using ideas from standard algorithms for finding optimal designs for linear models. Foundational work on algorithms for finding an optimal design is available in Wynn (1972), Atwood (1976) and Wu and Wynn

(1978a,b), among others. We describe computer algorithms for searching optimal designs in Chapter 11.

To implement a Bayesian approach for finding an optimal design, the following questions have to be addressed up front. As an illustration, consider the simple regression model and that we wish to find a Bayesian optimal design for estimating the slope parameter β_1. The first decision is to pick a suitable prior distribution for the slope parameter β. For example, should I use a uniform prior on $[\beta_L, \beta_U]$ for selected values of β_L and β_U? Or should I use a Gamma distribution on $[\beta_L, \beta_U]$ for the slope parameter, and if so, what are the parameters for the Gamma distribution? In the latter case with the Gamma distribution, how do I estimate the parameters for the Gamma distribution given the prior information? How do I know that I have chosen an appropriate prior? How do I know if the resulting design is very dependent on my choice of the prior distribution?

There are rarely definitive answers or strategies to the above questions; eliciting a prior distribution is part art and part science and unfortunately this important topic has been consistently given short shrift in the Bayesian literature. The bulk of the Bayesian work does not discuss or justify the choice of the prior distribution, even though it is well known that the choice of the prior distribution can be highly subjective and can affect the final results substantively. In the early development of Bayesian statistics, *conjugate prior distributions* were popular – these are prior distributions that ensure that the posterior distributions also belong to the same class of distributions. For instance, if one uses a normal prior distribution for the mean of a normally distributed variable, then the posterior is also a normal distribution. However, their popularity comes mainly from convenience and also from the fact that a closed form expression for the posterior distribution is possible. However, conjugate priors are less used now in part because they may not be realistic in practice and also as more powerful computing techniques become available to generate the posterior distribution.

Bayesian optimal design requires that the researcher comes up with a prior distribution that captures all existing information for the model parameters. As mentioned before, this task is invariably subjective and different researchers can arrive at quite different prior distributions. Part of the problem is that there is no standard way of quantifying the information available. The resulting Bayesian optimal design depends on the prior distribution and can vary substantially when different prior distributions are used. For instance, we see in the next chapter that if independent uniform priors are jointly used for the two parameters in the logistic model, the number of design points in the Bayesian optimal design depends on how spread out each of the prior distributions is. Chaloner and Larntz (1989) gave several examples to show how the prior distribution affects the design points in different types of optimal designs for the logistic model. The upshot is that it is important to check the sensitivities of the Bayesian optimal design to the prior distribution before the design is implemented. We prefer a Bayesian optimal design that is more robust to the prior distribution.

We conclude this section by emphasizing that the choice of the prior distribution must be deliberate and carefully selected. van Dongen (2006) used three real examples in the biological sciences and showed that great caution must be used in choosing the prior distribution. Unfortunately, very few papers discussed this important topic on prior elicitation. Chaloner and Duncan (1983, 1987) and Chaloner *et al.* (1993) are among the only couple of exceptions; the former considered prior elicitation for a hyper-parameter in a Binomial distribution while the latter dealt with prior elicitation for a toxoplasmosis prophylaxis trial conducted through the Community Programs for Clinical Research on AIDS. See also Gavasakar (1988) who compared two elicitation methods for a prior distribution for a Binomial parameter and Hahn (2005) who proposed a new method for prior elicitation using Markov Chain Monte Carlo methods. Practitioners can greatly benefit if Bayesians provide more illustrations and fully demonstrate how a prior distribution is constructed in practice using real examples and available data.

9.7   Minimax optimal design

An alternative to the Bayesian approach is to elicit information on only the plausible extreme values of the quantity or quantities of interest. Frequently, the quantity or quantities of interests are model parameters or functions of model parameters. To fix ideas, we discuss only the case where we wish to construct a minimax optimal design to estimate model parameters. This task is usually easier to do in practice than trying to elicit information from the researcher to construct a prior distribution for the Bayesian optimal design. In the minimax approach, we only need to specify a plausible region where we believe the true values of the model parameters lies. This region can be complicated where some parameters depend on values of other parameters in the model. For our purpose here, we focus on simpler plausible regions where we assume that the range of each parameter can be specified independent of other parameters. This means that when we wish to estimate, say $\beta' = (a, b, c)$, the plausible region takes on the form $[a_1 < a < a_2] \times [b_1 < b < b_2] \times [c_1 < c < c_2]$ for user-selected values a_i, b_i and $c_i, i = 1, 2$.

Our experience is that for many applications it is easier to elicit information on the extreme values of each parameter. The theory for the construction of minimax optimal design is the same regardless of the complexity of the plausible region. The only effect of a more complex plausible region on finding the minimax optimal design is the additional computational burden imposed on the optimization problem. Wong (1992) gave technical details for finding minimax optimal designs in general and noted that an analytical description for a minimax optimal design is notoriously difficult unless in very simple cases. Computer algorithms are required to generate minimax optimal designs for most problems, see, for example, Brown and Wong (2000). Unfortunately, proven general-purpose algorithms for finding minimax optimal design in practice are not available. The problem is somewhat simplified if one restricts the search to a smaller class of designs rather than the set of all designs on the given interval. For example, several types of minimax optimal designs are presented in Chapter 11 using tools from a web site, and these optimal designs are found within the class of all designs with number of design points equal to the number of parameters in the model for the mean response. Such optimal designs are called *minimally supported optimal designs*.

Here is a simple motivating example for using the minimax approach. Consider the design problem for predicting the response in a dose–response study using the simple linear regression model in the given protocol design interval properly scaled to $[-1, 1]$. If the only purpose is to predict the response at a particular dose level, say $x_0 = 0.5$, the optimal design is to take all observations at $x_0 = 0.5$. This optimal design is clearly unattractive for other purposes. In this case, it can be shown from the variance expression for the fitted response in Equation (2.23) of Chapter 2 that the optimal design is not unique and any design that allocates observations such that the average value of the x is equal to 0.5 is optimal for predicting the response at $x_0 = 0.5$. An example of such a design is to take 75% of the observations at $x = 5/6$ and the rest at $x = -0.5$. Another one is to take one-third of the observations at $x = 3/4$ and the rest at $x = 3/8$. It is easy to see that there are infinitely many optimal designs for predicting the response at $x = 0.5$.

The above example assumes that we know for sure that we wish to predict at $x_0 = 0.5$ before we construct the design. What happens if it is not known in advance that $x_0 = 0.5$ is the only dose level of interest to predict? The researcher may have a general idea where prediction is needed, but unable or unwilling to specify specific dose level or levels in advance of data collection. Under such a situation, one can design so that no matter which dose level in the plausible region is of ultimate interest, the prediction variance is not too large. The motivation for the minimax approach is based on the worst-case scenario. Within the plausible region where the researcher thinks prediction is needed later on, we consider the worst possible outcome case and attempt to minimize its effect. In other words, we look at all variances of the predicted values within the plausible region and find a design to minimize the largest of these variances. This way we have some global protection against the worst-case scenario where we really need to predict at the dose level that has the largest predicted variance. The reader may recall that we discussed such a design criterion in Section 2.5 of Chapter 2 under G optimality, with G standing for global.

So far, we have always assumed that the variance of the error term is constant. In practice, the variance of the response may vary as experimental conditions change. Wong and Cook (1993) applied a minimax approach to find an efficient design to estimate the overall mean response function when errors are heteroscedastic. This means that the variance of the error depends on where the observation is observed. In traditional design terminology, this dependence is usually represented by the inverse of a known positive function called the *efficiency function*, that is we assume that the var(y) is proportional to $1/\lambda(x)$. In this case, the information matrix is given by

$$M(\xi) = \sum_j \lambda(x) f(x_j) f'(x_j). \tag{9.11}$$

We recall that the G-optimality criterion for estimating the mean response $f'(x)\beta$ first focuses on the region where the largest variance of the fitted response occurs over the design space using design ξ, that is

$$\max_{x\in\Delta} \text{var}(f'(x)\hat{\beta}) = \max_{x\in\Delta} f'(x)M^{-1}(\xi)f(x), \qquad (9.12)$$

and the G-optimal design is the design that minimizes this quantity over all designs ξ on the design space Δ. Designs that minimize the above expression are called *heteroscedastic G-optimal designs*. The key difference here is that we are minimizing the variance of the fitted response given above, and not the weighted variance of fitted response itself, which is given by $\max_{x\in\Delta} \lambda(x)f'(x)M^{-1}(\xi)f(x)$. Designs that minimize this weighted variance are just the D-optimal designs (Fedorov, 1972, p. 71). Illustrative examples of differences in these two types of minimax optimal designs are given in Wong (1990). Here, we provide two examples when we have a simple linear model defined on the space $\Delta = [-1, 1]$ where (i) $\lambda(x) = c - x^2$ and (ii) $\lambda(x) = 1/(c + x)$. Evidently, we require $c > 1$ in both cases. We may use case (i) to model the variance of the response if we feel that the variance varies in the form of a symmetric parabola about 0 and reaches its maximum at the extreme ends of the design space. Case (ii) is more appropriate when we expect the variance of the response to vary linearly and reaches its minimum at $x = -1$. It can be shown that in case (i) the heteroscedastic G-optimal design is equally supported at $x = -1$ and $x = 1$ if $c \geq 3$; otherwise, it is supported equally at $x = \pm\sqrt{\sqrt{1 + c} - 1}$. In case (ii), the heteroscedastic G-optimal design is no longer symmetric and requires observations at $x = -1$ and $x = 1$. The proportion of observations allocated at $x = 1$ is $(c + 1)/(2c)$ for $c > 1$. In contrast, the D-optimal designs for both cases are always equally supported at two points. In case (i), the points are at $x = -1$ and $x = 1$ if $c \geq 3$; otherwise, the points are at $x = \pm\sqrt{c/3}$. In case (ii), the points are always at $x = -1$ and $x = 1$. The function $\lambda(x)$ can also contain parameters that may be distinct from the parameters in the mean function as well. Dette and Wong (1996) discussed these design issues and the web site in Chapter 11 generates optimal designs for a few forms of $\lambda(x)$.

In summary, the minimax design strategy is simpler to implement in practice because only the extreme possible values of each parameter in the model have to be elicited. It does not require the subjective construction of a prior distribution that can impact the final design in important ways. The main drawback is that minimax optimal designs do not have closed form descriptions and they have to be found numerically. Despite this limitation, the minimax design strategy is useful and popular in situations where little information on the model parameters is available at the design stage.

We end this section with the remark that maximin optimal designs operate in a similar way as minimax optimal designs. Instead of minimizing the worst possible outcome, which the minimax optimal design seeks to do, a maximin optimal design maximizes the minimal benefit. For example, in the motivating example for the minimax design strategy, the maximin optimal design is the design that maximizes the minimum efficiency for predicting the response at the point x_0 where the minimum is taken over all possible points x in the plausible region.

9.8   Multiple-objective optimal designs

This section discusses a formal and versatile design strategy for designing a study with multiple objectives. In pharmacokinetic studies, for example, estimation of different bioavailabilities of a drug is often of interest. Oftentimes, they include micro- and macro variables such as time to maximum concentration, the peak concentration and average time of the drug in the organ compartment. Some of these goals may be of different levels of interest to the researcher, and so the design should provide higher efficiencies for more important objectives in the study. Even when the overall objective is estimation, it may be desirable to use different criteria in the study to gauge the goodness of the design for that purpose. For example, Dette (1995) considered an optimality criterion that mixed D- and E optimality and Lee (1987, 1988) and Wong (1995b) combined D- and A criteria to find efficient designs for low-order polynomial models. The goal of a multiple-objective optimal design is to meet the several objectives of the study and provide efficiencies according to the importance of the objectives.

Multiple-objective optimal designs can also be used to address important concerns arising from model assumptions. Throughout, we assumed we have a fully specified statistical model and then proceeded to construct an optimal design for the assumed model and the given objective. In practice, model assumptions may be questionable and the researcher wants to design with this in mind. The consequence of constructing a design based on erroneous model assumptions can be devastating. For instance, if the simple linear model is correctly assumed, the design that places equal observations at both extremes on the design interval is an excellent design for many purposes; in particular, this design is *universally D optimal*, meaning that this design provides the smallest volume of the confidence area for the intercept and slope parameters among all designs on the given design interval. However, if there is doubt whether there is curvature in the mean response, this design is useless for checking whether the additional quadratic term is needed. This is because the optimal design for the linear model has only two points, and so does not provide enough degree of freedom to carry out a lack of fit test (Montgomery, 2000, p. 83). On a more intuitive level, to detect curvature, we need to have at least three distinct points to assess whether the values of the responses at these three points fall roughly in a straight line or not.

An obvious design strategy for dealing with concerns on adequacy of model assumptions is to find the optimal design for each of the different models and average them, with the hope that the resulting design is nearly optimal for each of the models at hand. The averaging can be done in a way that reflects the prior belief that each model has a certain chance of being the true model (Läuter, 1974, 1976). Another way is to find the optimal design for the 'largest' model and hope that the design is also relatively efficient for all other possible models. Here, 'largest' usually assumes a hierarchical class of possible nested models such as the class of polynomial models with degrees up to degree $(p - 1)$, where p is user selected. A Bayesian design strategy that incorporates the prior belief on the validity of each model may be used as well.

In this section, we describe a formal and effective way of designing a study with multiple objectives. We will use a simple example to demonstrate the design strategy to formally incorporate competing objectives and show how a multiple-objective optimal design is determined. Multiple-objective optimal designs are sometimes called *compromised designs* for obvious reasons. They are also related to *constrained optimal designs* and *compound optimal designs* as discussed below.

9.8.1   Constrained optimal design

Consider the situation where the researcher wants to use the simple linear model but now there is doubt whether the model is adequate. The researcher is concerned whether there is curvature in the mean response, and so he or she is also interested to use the quadratic model. The problem is that the optimal design for the simple linear model is equally supported at both ends of the design space, and so with only two points the researcher is unable to check whether there is curvature in the model. On the other hand, if we design for the quadratic model and the simple linear model proves to be the more appropriate model, the estimates for the simple linear model may be inefficient.

For this problem, our goals are to estimate the two parameters in the linear model and also be able to estimate whether a quadratic term is needed in the model. To fix ideas, assume that the more important objective is to estimate the coefficient for the quadratic term. This is the primary objective. The second objective is to estimate the slope and intercept coefficients in the simple linear model. The practitioner then quantifies the relative importance of the two objectives; this can be meaningfully accomplished by specifying the minimum efficiency required for the primary objective. The plan is to construct a design that will provide the specified minimum efficacy for the primary objective, and subject to this constraint does as best as possible for the secondary objective. Operationally, we proceed as follows.

We first formulate each of the objectives as a function of the information matrix. One can work with concave or convex functionals. Of course, when we work with a convex functional, we want to find a design that minimizes it, whereas we seek a design to maximize a concave functional. Our first task is to formulate the two goals in terms of convex or concave functionals. Technically, either formulation is the same. To fix ideas, let us work with concave functions and let $\phi_1(\xi)$ be the primary criterion for estimating the coefficient in the quadratic term and let $\phi_2(\xi)$ be the secondary criterion for estimating the two parameters in the linear model. Assume that the design space is given and ξ is an arbitrary design on the design space. Define

$$f_1'(x_j) = (1, x_j, x_j^2) \text{ and } f_2'(x_j) = (1, x_j)$$

$$M_1(\xi) = \sum_j w_j f_1(x_j) f_1'(x_j) \text{ and } M_2(\xi) = \sum_j w_j f_2(x_j) f_2'(x_j). \quad (9.13)$$

The two information matrices $M_1(\xi)$ and $M_2(\xi)$ are inversely proportional to the asymptotic variance–covariance matrices of the parameter estimates in the quadratic and linear models, respectively. In particular, the $(3, 3)$ element in $M_1(\xi)^{-1}$ is proportional to the variance of the estimate for coefficient of the quadratic term in the quadratic regression model. This element is $c'M_1(\xi)^{-1}c$ with $c' = (0, 0, 1)$, and we want this positive quantity to be small or equivalently want $[-c'M_1(\xi)^{-1}c]$ to be large.

The design sought is the one that maximizes $\text{Det}[M_2(\xi)]$ over all designs on the given design space, subject to the constraint that $[c'M_1(\xi)^{-1}c]$ is sufficiently small. Equivalently, because the logarithmic function (log) is a concave increasing function, we want to maximize $\log \text{Det}[M_2(\xi)]$ subject to the constraint that $[-c'M_1(\xi)^{-1}c]$ is sufficiently large. Accordingly, we formulate our primary criterion as $\phi_1(\xi) = -c'M_1(\xi)^{-1}c$ and our secondary criterion as $\phi_2(\xi) = \log \text{Det}[M_2(\xi)]$. Both are concave functions over the space of information matrices. The optimization problem can now be succinctly formulated as

$$\text{Maximize } \phi_2(\xi) = \log \text{Det}[M_2(\xi)]$$

$$\text{subject to } \phi_1(\xi) = [-c'M_1(\xi)^{-1}c] \geq \text{Constant.} \quad (9.14)$$

An immediate question is what is the constant in the above formulation. Should it be 12? Should it be 0.4? This constant is difficult to specify as it is. To overcome this problem, we rewrite the objective function and constraint in terms of efficiency, which is easier to interpret and quantify. Let the optimal designs under the primary and secondary criteria be ξ_1^* and ξ_2^*, respectively, and assume that the design interval is $[-1, 1]$ for illustration. The optimal design ξ_1^* for estimating just the coefficient in the quadratic term is a D_s-optimal design and as pointed out in Section 9.3, this design has support at $\{-1, 0, 1\}$ with weight at $x = 0$ equal to $1/2$ and the rest equally supported at $x = -1$ and $x = 1$. The optimal design under the secondary criterion, ξ_2^*, for estimating the parameters in the simple linear model is equally weighted at $\{-1, 1\}$. In terms of efficiencies of ξ under the two criteria, we have $E_1(\xi) = c'M_1(\xi^*)^{-1}c/[c'M_1(\xi)^{-1}c]$ and $E_2(\xi) = \{\text{Det}[M_2(\xi)]/\text{Det}[M_2(\xi^*)]\}^{1/2}$. The value of the constant in the above optimization problem may now be chosen as ϕ_1^*/e_1, where ϕ_1^* is the optimal value of ϕ_1 and e_1 is the user-specified minimum efficiency required of the design under the primary objective. Clearly, ϕ_1^* is computable because $\phi_1^* = \phi_1(\xi_1^*)$. For example, if we want a design to ensure that we can estimate the coefficient of the quadratic term with 95% efficiency, we set $e_1 = 0.95$. Typically, we would want to choose e_1 close to unity to obtain an optimal design with high efficiency for the primary criterion $\phi_1(\xi)$. This formulation of the constrained optimization problem is appealing to practitioners, but unfortunately does not yet offer a clue on how to find such an optimal design. The next section on compound optimal design describes an indirect way of finding constrained optimal designs.

9.8.2   Compound optimal design

Cook and Wong (1994) considered a compound optimality criterion that is a convex combination of the two concave criteria $\phi_1(\xi)$ and $\phi_2(\xi)$. For each $\lambda \in [0, 1]$, define

$$\phi_\lambda(\xi) = \lambda\phi_1(\xi) + (1 - \lambda)\phi_2(\xi). \qquad (9.15)$$

Then for each fixed λ value, $\phi_\lambda(\xi)$ is still concave and so we can find the optimal design directly as if this is a single objective optimal design problem. Denote this design that maximizes $\phi_\lambda(\xi)$ by ξ_λ and call it a *compound optimal design*.

How does one choose λ and what is the practical interpretation of λ? Does choosing $\lambda = 0.5$ mean we are equally interested in the two objectives? It turns out that the choice of λ depends on the value of e_1 specified in the constrained optimization problem. Their analytical relationship is often complicated and not necessarily insightful. In practice, we construct an efficiency plot to solve the constrained optimization problem by solving the compound optimal design problem indirectly. Procedurally, we first determine compound optimal designs ξ_λ for values of λ starting from 0 to 1 in small increments of, say 0.1 or 0.05. Of course, at the extreme values when λ takes on the values 0 and 1, we obtain ξ_2^* and ξ_1^* and they correspond to the optimal designs for ϕ_2 and ϕ_1, respectively.

The next step to find the desired constrained optimal design is to construct an efficiency plot using the compound optimal designs. The efficiency plot graphs $E_1(\xi_\lambda)$ and $E_2(\xi_\lambda)$ versus values of λ in the interval $\lambda \in [0, 1]$. The graph of $E_1(\xi_\lambda)$ is always an increasing function of λ and the graph of $E_2(\xi_\lambda)$ is always a decreasing function as λ increases. They usually cross at some point and the value of λ for which it does is the value λ^* that should be used in the compound optimality criterion if we want to have a constrained optimal design that provides equal efficiencies under both objectives. This situation corresponds to the case when both objectives are equally important. Otherwise, corresponding to the user-specified value of e_1, a horizontal line is drawn across the efficiency plot when $E_1(\xi_\lambda) = e_1$ and the value of λ that corresponds to where the horizontal line meets $E_1(\xi_\lambda)$ is the value that should be used to generate the desired constrained optimal design. The technical details are given in Cook and Wong (1994).

The efficiency plot for this example is displayed in Figure 9.1. From the figure, one concludes that setting $\lambda = 0.25$ ensures an equal-interest dual-objective optimal design with an efficiency of 84% under both criteria. If the user wants to have a design that will guarantee an efficiency of 96% or higher for estimating the quadratic term coefficient, we set $e_1 = 0.96$, and from the plot the constrained optimal design sought is the compound optimal design with $\lambda = 0.5$.

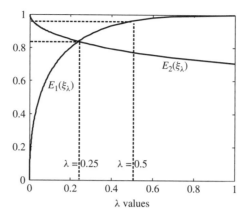

Figure 9.1 Plot of efficiencies $E_1(\xi_\lambda)$ and $E_2(\xi_\lambda)$ versus λ.

As expected, $E_1(\xi_{0.5}) = 0.96$, the minimum efficiency required. The efficiency under the secondary criterion of this compound optimal design is given by where the vertical dotted line at $\lambda = 0.5$ meets $E_2(\xi_\lambda)$; this value is 0.78. This implies that under the user-specified constraint, the best we can do is to estimate the two parameters in the simple linear model with about 78% efficiency.

The above approach for finding a multiple-objective optimal design is attractive in several ways. Without going into details, we note that the advantages of this approach include the following:

1. if the primary and secondary roles of the criteria are reversed, one can deduce corresponding results from the efficiency plot by replacing λ by $(1 - \lambda)$ on the λ axis;

2. the approach relies on existing methods for finding a single-objective design;

3. the efficiency plot is informative as it illustrates the trade-off between the two goals visually; if the slopes in the efficiency plots are steep, this suggests the goals are competitive meaning that much of one has to be given up for a gain in the other criterion;

4. this approach generalizes directly to situations when we have more than two objectives because a convex combination of concave functionals is still concave and no new theoretical issues arise; however, the multi-dimensional efficiency plot makes it harder to use and interpret.

We end this chapter with some very brief remarks on optimal designs for model discrimination when the researcher wants to find the best fitting model among a few possible models. Early work on finding an optimal design to discriminate among competing models includes Hill, Hunter and Wichern (1968), Atkinson and Fedorov (1975a,b) and Hill (1978a). Sometimes, in rare instances, different models have the same optimal design so that discrimination among the models is no longer a problem. Hill (1978b) provided an example where there are three very different rival linear models and they all have the same optimal design.

We mention this topic in passing only because we feel model discrimination by itself is often not the end goal in the study. For example, model discrimination is typically followed by estimating parameters or functions of parameters in the selected model. Hence, the problem can be subsumed more generally as a multiple-objective optimal design problem.

9.9   Summary

In this chapter, alternative optimal designs commonly used in practice are described. They include optimal designs for estimating only selected parameters in the model, extrapolation optimal designs for making inference outside the design space and L-optimal designs useful for estimating functions of model parameters. The Bayesian framework for designing a study, including its advantages and potential difficulties, is also discussed. Minimax optimal designs provide yet another design strategy when prior information is minimal and we are concerned about the 'worst-case' scenario. Finally, a formal design approach for handling multiple objectives in the study is provided. Naturally, the objectives are competitive, meaning that one has to give up efficiency under one criterion to gain efficiency for another criterion. Therefore, these multiple-objective optimal designs are sometimes referred to as *compromised designs* or *multipurpose designs* as well. We recall that the efficiency plot is constructed from evaluating efficiencies of compound optimal designs under different criteria and the plot provides an effective visual assessment of the trade-off between the competing objectives. The sought constrained optimal design is then found indirectly from the efficiency plot.

10

Optimal designs for nonlinear models

10.1 Introduction

Most of the previous chapters were concerned with linear models. We now turn our attention to designing for nonlinear models. Nonlinear models are widely used in the biological sciences and health sciences. The interested reader may want to refer to several monographs where there are plenty of examples of use of nonlinear models and their analysis in various fields, see Lindsey (2001) and Motulsky and Christopoulos (2004), for example. However, design issues for such models were not much discussed historically largely because of their difficulty to solve. An early paper on designs for nonlinear models is Box and Lucas (1959).

Most of the current approaches to design a study for a nonlinear model simply linearize the nonlinear mean function via a first-order Taylor's approximation and work with the simplified model. Pázman (1993) is one among a handful who did not linearize the mean function and chose to work with the curvature of the nonlinear function. This approach is theoretically more appealing than the linearization approach, but results in a very difficult optimization problem from the mathematical point of view. To date, only very limited useful results are available from this approach, which also appears to be workable only for simple models. For many realistic models involving several parameters, this approach has technical difficulties. Accordingly, in the following sections, we adopt the linearization approach and illustrate how this method works for a very simple nonlinear model.

There are a few articles that review design issues and optimal designs for nonlinear models. Ford, Kitsos and Titterington (1989) is an early one. Wong (1999)

An Introduction to Optimal Designs for Social and Biomedical Research   M. P. F. Berger & W. K. Wong
© 2009, John Wiley & Sons, Ltd

reviewed optimal designs for multiple-objective design problems that included models with continuous or binary outcomes, and Khuri *et al.* (2006) reviewed design issues for generalized linear models.

Throughout this chapter, we assume all errors are independently and normally distributed with mean zero and constant variance. Unlike design books currently available, we choose not to deal with technical details; instead, we present design issues and concepts for nonlinear models and refer the interested reader to references for mathematical details. The next chapter introduces a web site that allows visitors free access to generate optimal designs for a variety of nonlinear models.

10.2   Linear models versus nonlinear models

Linear models are appropriate if the mean response can be expressed as a linear combination of the covariates. One way to determine whether we have a linear model is to differentiate the mean function with respect to the vector of model parameters and see if the derivative is free of the parameters. If it is, we have a linear model; otherwise, we have a nonlinear model.

For example, consider the model $y_i = \beta_0 + \beta_1 x_{1i} + \beta_2 x_{2i} + \varepsilon_i$ (Model 1.1, Chapter 1) where we have three parameters $\beta = (\beta_0, \beta_1, \beta_2)'$. The derivative of the mean function with respect to each of these parameters is $\mathrm{d}E(y_i)/\mathrm{d}\beta = (1, x_{1i}, x_{2i})'$, which is free of model parameters. So this model is linear, as expected. On the other hand, suppose we have a study where the outcome y_i is modelled as

$$y_i = \exp(ax_i) + e_i. \tag{10.1}$$

Clearly, the parameter a controls the shape of the response curve. If the parameter a is positive, the mean response increases when the control variable x_i increases. On the other hand if the parameter a is negative, the mean response decreases as the control variable decreases in value. When $a = 0$, the response is independent of the control variable and is expected to take on the value of unity on average. In this model, the mean response function is not a linear combination of x_i and the model parameter. One can verify this by differentiating the mean response $E(y_i) = \exp(ax_i)$ with respect to the parameter a. The derivative $\mathrm{d}E(y_i)/\mathrm{d}a = x_i \exp(ax_i)$ is dependent on the parameter a. Model (10.1) is therefore a nonlinear model.

There are many popular nonlinear models in the biological and health sciences. We briefly mention four nonlinear models here and describe their use in practice. The first one is commonly called the *Arrhenius equation*.

10.2.1   The Arrhenius equation

A Swedish chemist Svante Arrhenius discovered the following relation between temperature and reaction rate:

$$y_i = a \exp(-b/t_i) + \varepsilon_i. \tag{10.2}$$

The reaction rate y_i is the product of a factor a, representing the collision frequency between two molecules and an exponential term $\exp(-b/t_i)$, which describes the fraction of molecules with a minimum energy required to react. The parameter b is usually expressed implicitly as $b = (E_a/R)$, where E_a is the activation energy and R is a gas constant. The variable t_i is the temperature reading in the Kelvin scale. The Arrhenius equation is plotted in Figure 10.1a for degrees ranging from 273 to 373 (0–100 °C). The parameters are $a = 10$ and $R = 8.314$ J mol^{-1}, and four different values for the activation energy are plotted, namely $E_a = 65$, 75, 85 and 95 J mol^{-1}. Applications of the Arrhenius equation to pharmaceutical stability testing are quite common and straightforward; an example is in measuring shelf-life of vitamins or drugs on display in a pharmacy store.

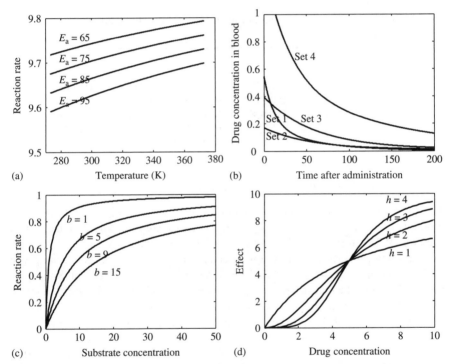

Figure 10.1 Four typical nonlinear models: (a) Arrhenius equation, (b) compartment model, (c) Michaelis–Menten model and (d) Emax model.

10.2.2   The compartmental model

Another example of a nonlinear model is the compartmental model. Compartmental models in various forms have been widely used in pharmacokinetic research (see, e.g. Wise, 1985 and Faddy, 1993). For instance, if y_i is the amount of drug in a compartment (or organ) at time t_i after administration into the body, then a common form of the compartmental model to study the flow of drug through the body is

$$y_i = a \ \exp(b \ t_i) + c \ \exp(d \ t_i) + e_i. \qquad (10.3)$$

The parameters a and c are 'linear' parameters and the main parameters of interest are b and d. These are all macro parameters meaning that they by themselves may not have any physical interpretation but are functions of micro parameters that have biological meanings. Because each macro parameter is usually a very complicated function of the micro parameters, sometimes macro parameters are estimated first. Landaw (1980) gave details and discussed various types of related models. Figure 10.1b shows mean response from the compartmental model for four sets of parameters displayed in Table 10.1.

Table 10.1   Four sets of parameters for the compartmental model.

	a	b	c	d
Parameter set 1	0.2020	0.0180	0.3450	0.0881
Parameter set 2	0.0093	0.0003	0.1620	0.0181
Parameter set 3	0.0108	0.0002	0.3870	0.0168
Parameter set 4	0.5220	0.0071	0.8370	0.0339

10.2.3   The Michaelis–Menten model

The third example of a nonlinear model we give is a popular one used in the biological sciences and concerns enzyme-kinetics studies. The Michaelis and Menton (1913) model is widely used because of its simplicity and effectiveness in describing properties of saturation effects. Some examples of its use are given in Houston and Kenworthy (2000), Holmberg (1982) and Yu and Rappaport (1996), Dunn (1988) and Lopez-Fidalgo, Tommasi and Trandafir (2008), among others. The statistical analysis of the model has also been frequently discussed, see Raaijmakers (1987) and the references therein. The model is given by

$$y_i = \frac{a \ x_i}{b + x_i} + e_i. \qquad (10.4)$$

Here, y_i is the velocity of a response and x_i is the substrate concentration. The parameters a and b are assumed to be positive, and while theoretically x_i is positive and unbounded, there is usually an upper bound placed in the substrate concentration in practice. The parameter a is the maximum value that the velocity (reaction rate) can reach and the parameter b is the so-called Michaelis–Menten constant.

Figure 10.1c shows a saturation curve for an enzyme. The velocity (reaction rate) is plotted against the substrate concentration using different values of the

Michaelis–Menten constant b: 1, 5, 9 and 15. The parameter a for this plot is equal to 1.0.

10.2.4 The Emax model

A very often used model in pharmacodynamics to relate the drug concentration to the observed effect is the sigmoid Emax model. Derendorf and Meibohm (1999) gave a review of this model and it is given by

$$y_i = \frac{a\, x_i^h}{b + x_i^h} + e_i. \tag{10.5}$$

The effect y_i is a function of the concentration of the drug x_i and depends on the maximum effect a and the parameter $b = C_{0.5}^h$, where $C_{0.5}$ is the concentration of the drug that produces half the maximal effect. The parameter h is called the *shape factor*. Figure 10.1d shows the mean response from the Emax model when $a = 10$ and $C_{0.5} = 5$ for different values of the shape parameter $h = 1, 2, 3$ and 4. From the figure, we observe that as the h value increases, the mean function becomes more S shaped.

10.3 Design issues for nonlinear models

We now use the simple nonlinear model in Equation (10.1) to illustrate design issues that arise for a nonlinear model: $y_i = \exp(ax_i) + e_i$. The response y_i may represent the weight of the substance left after it is exposed to a reactant and x_i is the time since exposure. A common application of such a model is in modelling decay of a chemical substance over time. In this case, the shape parameter a is negative and the design space is $[0, \infty]$.

One important consideration in optimal design theory is to focus on the normalized information matrix. For simplicity, let us consider Model (10.1) and show how the information is arrived at for an arbitrary design ξ for the parameter a. As explained in Section 10.2, the derivative of $f(x_i) = \exp(ax_i)$ with respect to the parameter a is $\nabla f(x_i) = x_i \exp(ax_i)$. The formula for the (normalized) information matrix for a design ξ supported with weight w_i at x_i on the design space is

$$M(\xi) = \sum_i w_i \nabla f(x_i) \nabla f'(x_i) = \sum_i w_i [x_i^2 \exp(2ax_i)]. \tag{10.6}$$

The information matrix $M(\xi)$ is a scalar in this example because there is only one parameter a in the mean function. If the model has two or more parameters, then $M(\xi)$ is a square matrix. As a numerical example, suppose that ξ is a two-point design with half of its observations at $x_i = 0.75$ and the rest at $x_i = 2.0$. For $a = 0.3$, a direct calculation shows that the numerical value of the information is $M(\xi) = \frac{1}{2}[0.75^2 \exp(2 \times 0.3 \times 0.75)] + \frac{1}{2}[2^2 \exp(2 \times 0.3 \times 2)] = 7.0813$.

It is clear from this computation that the amount of information depends on the x_i- and w_i values.

A noticeable feature, however, is that the information matrix depends on the unknown parameter a, which we are trying to estimate! Therefore, strictly speaking, the information matrix $M(\xi)$ should be written as $M(\xi, a)$. Here, we retain the former notation for simplicity, but in general, we write the information matrix as $M(\xi, \beta)$ with β being the vector of model parameters in the mean response function.

10.3.1 Local optimality

A main distinguishing feature in designing for linear or nonlinear models is that the information matrix depends on unknown parameters. As such, an optimal design for a nonlinear model is only optimal *locally* for given values of the parameters (Chernoff, 1953). This is in contrast to the case for linear models where the optimal design does not depend on the model parameters. However, we caution the reader that if the variance of the response in the linear model depends on unknown parameters in a known way, then the information matrix of the linear model for an arbitrary design will also depend on the unknown parameters. Some examples for this situation are given in the next chapter when we introduce a web site for generating a variety of optimal designs for nonlinear models.

It is instructive to consider Model (10.1) again, with only a single unknown parameter a. Because there is only one parameter, one may just want to take observations only at a single time point to observe the response. This strategy is attractive especially if it is expensive to take observations at different sites. For our purpose here, we assume the design ξ is a one-point design, and, say, it takes all observations at the point x_0. The information matrix for ξ becomes $M(\xi) = x_0^2 \exp(2ax_0)$.

If a is positive, the information is a strictly monotonic increasing function of x_0, and so the best one-point design is to take all observations at a time point as far from zero as possible. In practice, the design space is truncated at some time point representing when the study ends. This is the time point where one should allocate all resources to observe the outcome y_i. The interesting case is when a is negative, representing that the mean response decreases over time. Differentiating $M(\xi)$ with respect to x_0 and setting it equal to 0 yields

$$2x_0 \exp(2ax_0) + 2ax_0^2 \exp(2ax_0) = 0$$

or   (10.7)

$$2 \exp(2ax_0)[x_0 + ax_0^2] = 0.$$

The solution to this equation is either $x_0 = 0$ or $x_0 = -1/a$, suggesting that the optimal one-point design to estimate the negative parameter a is to take all observations at $x_0 = 0$ or $x_0 = -1/a$. To determine which of these two points is optimal, we verify whether the second derivative of $M(\xi)$ is negative at $x_0 = 0$ and $x_0 = -1/a$. A simple calculation shows that the second derivative is negative only when $x_0 = -1/a$. This means that the one-point optimal design is to observe the response only at $x_0 = -1/a$.

The interesting feature here is that the optimal design depends on a, the parameter which is what we want to estimate. The design that takes observations only at $x_0 = -1/a$ for Model (10.1) cannot be implemented because we do not know the value of a, and if we know the value of a, we would not have a design problem in the first place.

Design problems for nonlinear models are much harder than that for linear models in large part because, as just noted above, the optimal design for nonlinear models depends on model parameters, which we are trying to estimate. Much of the recent research work in optimal design for nonlinear models involves different ways of making the optimal design less dependent on the values of the model parameters. However, not all parameters in the nonlinear model influence our choice of the optimal designs. In fact, some parameters have no effect on the optimal design at all. These parameters are sometimes called *partially linear parameters* (Hill, 1980; Khuri, 1984), and they can be ignored from the design perspective. A simple example is the Arrhenius model in Equation (10.2) with two parameters, and the parameter a enters the model linearly. This is verified by noting that the information matrix for this model for any design depends on a in an unimportant way. Specifically, this means that the optimization problem is independent of the value of a. For instance, if one is interested in the D-optimality criterion, the determinant of the information matrix for Model (10.2) for any design is proportional to a^2, and so maximizing the determinant involves picking design points only for a given value of the parameter b.

Up to now we have explained information for a model with only one parameter. If the nonlinear model contains p parameters in β and the mean response function is $f(x_i, \beta)$, then the gradient of $f(x_i, \beta)$ is the $p \times 1$ vector $\nabla f(x_i, \beta)$ whose components are equal to the derivative of $f(x_i, \beta)$ with respect to each of the p parameters in β, that is:

$$\nabla f(x_i, \beta) = \begin{pmatrix} \dfrac{\mathrm{d}f(x, \beta)}{\mathrm{d}\beta_1} \\ \dfrac{\mathrm{d}f(x, \beta)}{\mathrm{d}\beta_2} \\ \vdots \\ \dfrac{\mathrm{d}f(x, \beta)}{\mathrm{d}\beta_p} \end{pmatrix}. \tag{10.8}$$

The information matrix of a design supported with weight w_i at x_i is then defined by $M(\xi) = \sum_i w_i \nabla f(x_i) \nabla f'(x_i)$. Unlike linear models, this matrix now depends on some or all the parameters in β. We recall from the previous chapter that when we have a large sample in the design, this information matrix is inversely proportional to the covariance matrix of the estimator of β.

Here is an illustrative calculation for the information matrix of an arbitrary design for the Michaelis–Menten model. Let $f(x_i, \beta) = ax_i/(b + x_i)$, where $\beta = (a, b)'$. The derivatives of the mean response in the Michaelis–Menten model with respect to a and b are elements in the 2×1 vector:

$$\nabla f(x_i, \beta) = \begin{pmatrix} \dfrac{\mathrm{d} f(x, \beta)}{\mathrm{d} a} \\ \dfrac{\mathrm{d} f(x, \beta)}{\mathrm{d} b} \end{pmatrix} = \begin{pmatrix} \dfrac{x_i}{b + x_i} \\ \dfrac{-ax_i}{(b + x_i)^2} \end{pmatrix}. \tag{10.9}$$

The information matrix $M(\xi)$ is now

$$M(\xi) = \sum_i w_i \begin{pmatrix} x_i/(b + x_i) \\ -ax_i/(b + x_i)^2 \end{pmatrix} \begin{pmatrix} x_i/(b + x_i) & -ax_i/(b + x_i)^2 \end{pmatrix}$$

and

$$M(\xi) = \sum_i w_i \begin{pmatrix} x_i^2/(b + x_i)^2 & -ax_i^2/(b + x_i)^3 \\ -ax_i^2/(b + x_i)^3 & a^2 x_i^2/(b + x_i)^4 \end{pmatrix} \tag{10.10}$$

Upon inversion, the resulting matrix is proportional to the asymptotic variance–covariance matrix of the parameter estimators of a and b. Ignoring an unimportant multiplicative constant, the first and second diagonal elements in $M(\xi)^{-1}$ are the variances of the estimators of a and b, respectively. The off-diagonal element in $M(\xi)^{-1}$ is proportional to the covariance of the estimators of a and b. There are two remarks about the Michaelis–Menten model, which are worth mentioning here:

1. It is clear that if the design space is $[0, x_{\max}]$ for some user-specified $x_{\max}$, we should not take observations at $x_i = 0$ because all elements in the information matrix are 0 and no new information is obtained. For this design space, we note that taking observations at two points including 0 still does not help in terms of estimating the two parameters. This is because the information matrix provides no information at $x_i = 0$ and with only one non-zero value for x_i, the resulting information matrix is not invertible. As a result, a two-point design including a design point at 0 is not able to estimate the two parameters in β. For the information matrix to be invertible, we need to have support at least at two non-zero design points.

2. The information matrix is usually complicated enough so that software is required to do the calculation. We see here that even when we use a design supported at two non-zero points for the simple model with two parameters, each element in the information matrix is a sum of two terms that do not simplify easily. So, in general, information matrix computed

under a nonlinear model is complicated and this partially explains why design problems for nonlinear models are hard and closed form solutions rarely exist.

In practice, designing for nonlinear models is an interactive process. In the first place, one should have some rough idea about what the model parameters are. These supposed values are called *nominal values* or *best guesses* in the literature and they are typically available from pilot studies or experiments with similar units. One then designs sequentially; an optimal design is found based on the nominal values and the next study will use nominal values found from the estimates from the previously designed study. One then hopes that after a couple of iterations, the estimated model parameters become stable.

10.4   Alternative optimal designs with examples

We now describe optimal designs for nonlinear models other than locally D-optimal designs for estimating the full set of parameters in the model. Locally D-optimal designs are widely used and arguably overused because of their simplicity and ease of construction. We must, however, keep in mind that we design the study to attain a given set of objectives in an efficient manner. In particular, when the goal of the study is not to estimate all parameters in the model, our design criterion has to be modified appropriately to reflect reality.

10.4.1 D_A or D_s-optimal design

A compelling case for use of a D_A- or D_s-optimal design is in the biological sciences where research interest is naturally targeted on a selected parameter in some nonlinear models. For example, consider the Michaelis–Menten model $y_i = ax_i/(b + x_i) + \varepsilon_i$ discussed earlier. The observation errors ε_i are assumed to be independent and have mean zero and constant variance. The parameter a is the maximum velocity theoretically attainable, and as noted earlier the parameter a enters the model linearly so that design issues are not affected by the nominal value of a. The Michaelis–Menten parameter b is the value of the substrate at which the velocity is one-half the maximum velocity and this parameter is of primary biological interest. This constant is frequently used as a defining characteristic of the substance under study or for comparison purposes. In practice, we may not always want to find the optimal design for estimating both a and b. We may just be interested in finding an efficient design for estimating only b.

The locally D-optimal for estimating both a and b can be found directly using calculus. The optimal design for estimating a alone or b alone can be found using Elfving's theorem, which is widely discussed in design monographs, see Atkinson and Donev (1992), Silvey (1980) and Pukelsheim (1993), for example. The theorem uses a geometrical but old argument that works very well for estimating a linear combination of parameters in a two-parameter model. Indeed, mathematical formulae for all these optimal designs are available. We do not list the D_A-optimal designs here, but refer the reader to the next chapter where we provide a web site that generates a variety of optimal designs for the Michaelis–Menten model after the design parameters are supplied. In this case, the design parameters are the design space, the nominal value for b and also the form of the parameterization of the Michaelis–Menten model itself. The nominal value of a is optional and is only used to generate the plot of the mean response function versus values of the substrate concentration. As we just mentioned, the locally D-optimal designs for estimating each parameter or both the parameters in the Michaelis–Menten model do not depend on the nominal value of a.

In practice, design problems are not always concerned with parameter estimation. There are many studies whose aims are to estimate not simply model parameters, but functions of model parameters. We describe a few examples in the next few sections.

10.4.2   Extrapolation optimal design

The same rationale for using extrapolation optimal design discussed in Chapter 9 applies for nonlinear models as well. The theory is similar once the nonlinear model function is replaced by its first-order linear approximation. In particular, we first formulate the extrapolation design problem as one for seeking an L-optimal design. The extrapolation optimal design is usually found using complicated mathematics, and in many cases formulae for the optimal designs are possible. Examples of extrapolation optimal designs for the Michaelis–Menten model can be generated directly from the web site described in Chapter 11. For instance, if the two parameters are $a = b = 1$ and the design space is $[0, 10]$, then the optimal extrapolation design for $x_0 = 12$ is supported at two points: $x_1 = 0.604$ and $x_2 = 10$ with weight at $x_2 = 10$ equal to $w_2 = 0.942$ and the rest at $x_1 = 0.604$. The accompanying plot in the output on the webpage confirms the optimality of two-point extrapolation optimal design. Dette, Kiss and Wong (2008) discussed extrapolation optimal designs for the Michaelis–Menten model and investigated its robustness under a variation of optimality criteria. The web site in Chapter 11 also generates extrapolation optimal designs for other models.

10.4.3   Optimal design for estimating percentiles

In toxicology experiments with mice, it is of frequent interest to estimate the dose that produces a certain percentage of occurrence of an event. The interest is on observing a binary outcome, such as whether fetal death has occurred or whether fetal malformation of limbs has occurred and what dose will result in say 5% of the animals experiencing the event. For example, there has been ongoing interest in estimating LD5, which is the dose that results in 5% animal death; sometimes LD2 or lower doses may also be of interest. On the other extreme, for efficacy studies, interest is in estimating high percentiles, such as ED90; this dose is expected to produce 90% success rate. Here, ED represents effective dose while LD represents lethal dose (Kalish, 1990). These percentiles have very different motivations, but the method for finding an optimal design for estimating either of them is the same.

We assume a statistical model with a binary outcome and find an expression for the percentile in terms of the dose level. We focus on the popular logistic model, for simplicity, and recall that the probability of a response (occurrence of an event) at dose level x_i is given by

$$p(x_i) = \frac{1}{1 + \exp[-\beta_1(x_i - \beta_0)]}. \qquad (10.11)$$

Here, β_0 and β_1 are the model parameters, and we assume that their nominal values are available. The independent variable, dose level, is assumed to be defined on a user-selected region, which is sometimes a log scale. A discussion on whether to work with dose or log dose is given in Motulsky and Christopoulos (2004). If this model holds, the median dose is the value of the independent variable that results in 50% response rate. Setting $p(x_i) = 1/2$ implies immediately that the median dose is given by $x_i = \beta_0$, independent of the value of β_1. The same procedure can be used to find the dose that results in a user-specified percent of response $p = p(x_i)$. The dose that results in $100p\%$ of response is

$$x_p = \beta_0 + \frac{\text{logit}(p)}{\beta_1}, \qquad (10.12)$$

where $\text{logit}(p) = p/(1 - p)$. This dose x_p is also called the $100p$th percentile for the logistic distribution, and we note that it is a function of the model parameters. Consequently, to find the optimal design for estimating the $100p$th percentile, we first apply the delta method to obtain the asymptotic variance of the estimated x_p and then show that the design problem is similar to the one for finding an L-optimal design discussed in the previous chapter. Wu (1988) gave details and many examples of locally optimal designs for estimating percentiles in this and related models.

10.5   Bayesian optimal designs

Bayesian optimal design requires that the researcher comes up with a prior distribution that captures all existing information for the model parameters. This task is invariably subjective and different researchers can arrive at quite different prior distributions. Part of the problem is that there is no standard way of quantifying the information available. The resulting Bayesian optimal design depends on the prior distribution and can vary substantially when different prior distributions are used. For instance, if independent uniform priors are jointly used for the two parameters in the logistic model, the number of design points in the Bayesian optimal design depends on how spread out each of the prior distributions is. Chaloner and Larntz (1989) gave several examples using different types of optimal designs for the logistic model. The upshot is that different prior distributions can affect the optimal designs substantially, and so it is important to check the sensitivities of the Bayesian optimal design to the prior distribution before the design is implemented. For the reason given in the next paragraph, we prefer a Bayesian optimal design that is more robust to the prior distribution. See Chaloner and Verdinelli (1995) for a review on Bayesian optimal design problems.

Here are some examples of Bayesian optimal designs for estimating percentiles. In toxicology, there is often interest in estimating low percentiles. Recall that these are percentiles or dose levels that result in a certain percent of animals experiencing an event of interest after administration of an agent. Typical events could be death or malformations at birth. The optimal designs for estimating these threshold values are usually found from an iterative algorithm. For example, Zhu, Ahn and Wong (1998) found Bayesian optimal designs for estimating the LD1, LD5 and LD10 for the logistic model in Equation (10.11) using two independent uniform priors for the two parameters. The design space was standardized to $[-1, 1]$. The prior distribution for β_0 was uniform on $[-0.1, 0.1]$ and the prior distribution for β_1 was uniform on $[6.9, 7.1]$. Table 10.2 shows the optimal designs and their relative efficiencies for estimating the other two percentiles. All optimal designs are two points requiring a lot more observations near the dose level $x_i = -0.3$ than the other design point. The reason for the two-point designs is in part due to the fact that both uniform priors have a relatively small range and so information is deemed to be reliable. Consequently, the optimal design does not require additional points to sample information on other possible parameter values. We also note that the relative efficiency of the LD10 optimal design for estimating LD1 is about 66%, suggesting that these optimal designs are not robust for estimating other percentiles in the study.

Table 10.2   Bayesian optimal designs for estimating low percentiles in a two-parameter logistic regression model and their relative efficiencies for estimating LD1, LD5 and LD10. The independent prior distributions for the two parameters β_0 and β_1 are U$[-0.1, 0.1]$ and U$[6.9, 7.1]$.

Criterion	Design points	w_1	RE$_{LD1}$	RE$_{LD5}$	RE$_{LD10}$
LD1	$-0.346, 0.319$	0.756	1	0.921	0.794
LD5	$-0.341, 0.265$	0.881	0.883	1	0.923
LD10	$-0.311, 0.145$	0.916	0.662	0.904	1

Note. Data from Table 1 in Zhu, Ahn and Wong (1998).

Table 10.3 is similar to Table 10.2 except that now a more diffuse set of prior distributions is used for both parameters. The prior distribution for β_0 was uniform on $[-1, 1]$ and the prior distribution for β_1 was uniform on $[6, 8]$. The range of the prior distribution of β_1 is much wider than before, suggesting that prior information is not precise. In fact, the variance for the prior for β_1 is $(8 - 6)^2/12 = 4/12$, which is 100 times larger than the previous case when the variance for β_1's prior is $(7.1 - 6.9)^2/12 = 0.04/12$. The optimal design points for the set of more diffuse priors are now six versus two for the less diffuse set of

Table 10.3   Bayesian optimal designs for estimating low percentiles in a two-parameter logistic regression model and their relative efficiencies for estimating LD1, LD5 and LD10. The independent prior distributions for the two parameters β_0 and β_1 are U[−1, 1] and U[6, 8].

Criterion	Designs (weights)	RE_{LD1}	RE_{LD5}	RE_{LD10}
LD1	−1.05, −0.567, −0.165, 0.214, 0.575, 1.09 (0.207, 0.181, 0.211, 0.223, 0.126, 0.051)	1	0.977	0.961
LD5	−1.04, −0.575, −0.196, 0.150, 0.451, 0.751 (0.231, 0.184, 0.192, 0.204, 0.146, 0.043)	0.943	1	0.975
LD10	−1.03, −0.573, −0.194, 0.154, 0.465, 0.77 (0.243, 0.189, 0.184, 0.176, 0.141, 0.067)	0.938	0.990	1

Note. Data from Table 2 in Zhu, Ahn and Wong (1998). Weights are in brackets.

priors as shown in Table 10.2. In other words, the number of support points in an optimal design may change as the uncertainty in the postulated model changes. In this case, the more diffuse the prior distribution is, the more points the optimal design will need to have. Table 10.3 also shows that, unlike the optimal designs in Table 10.2, these three optimal designs constructed under a set of more diffuse priors are also relatively robust for estimating the other two percentiles. In all cases, the minimal relative efficiency of any one of these optimal designs for estimating the two other percentiles is at least 93%.

10.6   Minimax optimal design

The minimax optimality concept described for linear models in Chapter 9 can be straightforwardly applied for nonlinear models. There will be additional computational burden simply because the optimization problem is more complex. This is particularly true when the plausible region for the parameters is quite involved. We should first, however, emphasize that maximin optimal designs operate in a similar way as minimax optimal designs do. Instead of minimizing the worst possible outcome, which is what the minimax optimal design seeks to do, a maximin optimal design maximizes the minimal benefit.

King and Wong (2000) employed a minimax approach to design an optimal design for estimating the two parameters in the logistic model in Equation (10.11): $p(x_i) = \{1 + \exp[-\beta_1(x_i - \beta_0)]\}^{-1}$. Note that this model is equal to the model in Equation (1.6) of Chapter 1 and is also described in Chapter 5 with a different notation. King and Wong (2000) assumed that the plausible region for the two parameters can be split independently of one another into $[\beta_{0,\min} \le \beta_0 \le \beta_{0,\max}] \times [\beta_{1,\min} \le \beta_1 \le \beta_{1,\max}]$, for user-selected constants $\beta_{0,\min}$, $\beta_{0,\max}$, $\beta_{1,\min}$ and $\beta_{1,\max}$. Table 10.4 compares minimax D-optimal design with Bayesian D-optimal design in two cases: (i) We used independent uniform priors with $\beta_0 \sim U[-0.3, 0.3]$ and $\beta_1 \sim U[6, 8]$ versus a small plausible region given by $[-0.3, 0.3] \times [6, 8]$ for the minimax criterion. (ii) We used independent uniform priors with $\beta_0 \sim U[-1, 1]$ and $\beta_1 \sim U[6, 8]$ versus a large plausible region $[-1, 1] \times [6, 8]$ for the minimax criterion.

Table 10.4 shows both the Bayesian and minimax optimal designs for the two cases. The Bayesian optimal designs were obtained from Chaloner and Lartnz (1989) and the minimax optimal designs were obtained from King and Wong (2000). Both types of optimal designs have more points in case (ii) than in case (i). This is as expected because a larger plausible region or a more diffuse prior distribution reflects greater uncertainty in the possible values for the two parameters. This observation has been reported in the literature for other settings as well, see, for example, Chaloner and Larntz, (1989). The proximity of the two types of designs can be measured using the efficiency. If one chooses the Bayesian D-optimal design as the base design to compare with, one can verify the efficiencies of the minimax D-optimal designs relative to the Bayesian optimal designs for the two situations presented below are 0.936 and 0.887.

Table 10.4   Bayesian optimal designs and minimax optimal designs for estimating the two parameters in a logistic model with two sets of priors and two sets of plausible regions. The numbers in the brackets are the weights associated with the design points.

Plausible region	Bayesian D-optimal design	Minimax D-optimal design
Case (i) $-0.3 \leq \beta_0 \leq 0.3$ $6 \leq \beta_1 \leq 8$	−0.308 (0.368) 0.000 (0.264) 0.308 (0.368)	−0.436 (0.269) 0.000 (0.462) 0.436 (0.269)
Case (ii) $-1 \leq \beta_0 \leq 1$ $6 \leq \beta_1 \leq 8$	−0.953 (0.118) −0.566 (0.156) −0.267 (0.153) 0.000 (0.146) 0.267 (0.153) 0.566 (0.156) 0.953 (0.118)	−1.151 (0.152) −0.738 (0.179) −0.313 (0.118) 0.000 (0.102) 0.313 (0.118) 0.738 (0.179) 1.151 (0.152)

Note. Data from Table 2 in King and Wong (2000).

In summary, the minimax design approach offers another practical way to design a study that takes into account available information on the model parameters. This design strategy is conceptually appealing because only the extreme possible values of each parameter in the model have to be elicited. It does not require the subjective construction of a prior distribution that can impact the final design in important ways. The main drawback is that minimax optimal designs do not have closed form and have to be found numerically using algorithms that are not widely available. Despite this limitation, the minimax design strategy is useful and popular in situations where little information on the model parameters is available at the design stage.

We end this section by noting that logistic models are very popular for modelling binary outcomes, and Chapter 5 is devoted solely to discussing design issues for logistic regression models. Various design strategies for the logistic regression models have been proposed by Begg and Kalish (1984), Chaloner and Larntz (1989), King and Wong (2000), Minkin (1987), Ouwens, Tan and Berger (2006), Sitter (1992), Tsutakawa (1972), Tekle, Tan and Berger (2008) and Zhu and Wong (2000a,b), among many others. See Chapter 5 for further references.

10.7   Multiple-objective optimal designs

The concept and construction of a multiple-objective design discussed in Chapter 9 for linear models apply to nonlinear models as well. As such, we will provide only examples of these designs for nonlinear models.

Huang and Wong (1998) used a simple model suggested by Neter, Kutner and Wasserman (1985) to study the relationship between a prognosis index for long-term recovery y_i and the number of days of hospitalization x_i. The range of hospitalization days under investigation was from 0 to 90 days, and so we set the design space as [0, 90]. The proposed two-parameter model is an extension of the model in Equation (10.1) and is given by

$$y_i = b \ \exp(-ax_i) + e_i \text{ for } 0 \le x_i \le 90. \tag{10.13}$$

The nominal parameter values are $b = 56.66$ and $a = 0.038$. To estimate the parameters, Neter, Kutner and Wasserman (1985) sampled records randomly across 15 different number of days of hospitalization: 2, 5, 7, 10, 14, 19, 26, 31, 33, 38, 45, 52, 53, 60 and 66. This means that if the total number of observations to be sampled is $N = 150$ observations, then they are sampled so that we have 10 patients' record for each of the 15 x values. If other values of N were used and N is at least 15, then approximately $N/15$ records are sampled at each of the x values.

This design could be optimized as shown in Section 10.4. But multiple-objective optimal design provides more flexibility. Suppose that in addition to estimating the parameters, we also want to estimate the mean recovery index value of patients who are hospitalized between 2 and 10 days. In this case, one may choose D optimality as one criterion and L optimality as the other. Following Huang and Wong (1998), we call the latter optimal design an integrated optimal design, and we want to design a study to balance the trade-off between the two objectives by compromising on the D-optimal design and the integrated optimal design. For this purpose, we set $f(x_i) = b \exp(-ax_i)$ and differentiate $f(x_i)$ with respect to a and b to obtain its gradient $\nabla f(x_i)$. It is straightforward to verify that

$$\nabla f(x_i) = \begin{pmatrix} \exp(-ax_i) \\ -x_i \exp(-ax_i) \end{pmatrix}. \tag{10.14}$$

The normalized information matrix for a design ξ on the design space is

$$M(\xi) = \sum_i [w_i \nabla f(x_i) \nabla f'(x_i)]. \tag{10.15}$$

The L optimality is $\phi_1(\xi) = \sum \nabla f(x_i) M(\xi)^{-1} \nabla f'(x_i) = \text{Trace}[AM(\xi)^{-1}]$ and the user-selected matrix is $A = \sum \nabla f(x_i) \nabla f'(x_i)$. L-optimal designs may be found analytically; otherwise, they can be found using computer algorithms discussed in Fedorov (1972). Table 1 in Huang and Wong (1998) is partly reproduced in Table 10.5, in which the different efficiencies of the single-objective optimal design under the two objectives are shown.

Table 10.5 Relative efficiencies of the original design ξ_0, the integrated optimal design ξ_I and the compound optimal design ξ_λ with $\lambda = 0.65$.

Design	Design points and weights	D-efficiency	Integrated efficiency
ξ_0	2,5,7,10,14,19,26,31, 34,38,45,52,53,60,65 (mass 1/15)	0.580	0.355
ξ_I	1 15.66 (0.535 0.465)	0.865	1
ξ_λ	1 24.36 (0.554 0.446)	0.987	0.95

Note. Data from Table 1 in Huang and Wong (1998).

For this particular problem, we wanted a compromised design that will provide an efficiency of at least 95% for estimating the mean recovery index for patients hospitalized between 2 and 10 days, and subject to this requirement, the design estimates the two parameters in the model as accurately as possible. Proceeding as in Chapter 9, we first construct the efficiency plot and then use it to deduce the desired design. We omit this plot for space consideration, but note that the desired design is given by the compound optimal design with $\lambda = 0.65$. This design tells us to sample records from patients hospitalized for 1 and 24.36 days only and about 55.4% of the records should come for patients hospitalized for 1 day only. This multiple-objective design has a D-efficiency of 98.7% and an L-efficiency of 95%, as expected. In contrast, the corresponding efficiencies for the implemented design are only 58% and 35.5%, showing once again that careful design of the study can increase the overall efficiency by providing more accurate estimates at the same cost.

Multiple-objective designs were also used in Rosenberger and Grill (1997) who conducted a psychophysical experiment where subjects sequentially received different levels of a stimulus, and data were recorded on response or non-response to the stimulus. The goal of the study was to elicit information efficiently about the relationship between stimulus levels and responses by simultaneously estimating the 25th, 50th and 75th quantiles of the stimulus–response curve. Using a logistic model, Zhu and Wong (2000b) applied a Bayesian approach and designed a psychophysical study to estimate the three percentiles simultaneously with possibly unequal interest.

10.8 Optimal design for model discrimination

A common design strategy for discriminating among competing models is to assume that we have a class of nested models, which contains the 'true' model. The hope is that one of the models in the class will provide an adequate fit to the data. These models are usually nested, meaning that the models get increasingly large in terms of the number of parameters in the mean function, and when other models in the class can be obtained by setting some of these parameters equal to specific values, zero being the most common. A simple example of such a class is polynomial models of degrees up to $(p - 1)$. There are p such models in the class beginning with the 'smallest' having degree equal to 0 to the 'largest' of degree $(p - 1)$. The intermediate models are obtained by setting more and more coefficients associated with the highest terms in the polynomial terms equal to 0.

Here are a few simple illustrations of nested nonlinear models. The first is concerned with the logistic model described in Equation (10.11). An immediate generalization or 'larger' model is the power logistic model defined by

$$p(x_i) = \frac{1}{\{1 + \exp[-\beta_1(x_i - \beta_0)]\}^s}. \tag{10.16}$$

This model was introduced by Prentice (1976) and described by Gaudard *et al.* (1993) to model skewed binary responses. We recall the simple logistic model is symmetric in the sense that the probability of a response is the same at two points equally spaced from the median. When this symmetry assumption is questionable, skewed binary responses can be modelled using values of s other than unity. If $s > 1$, responses are skewed to the right and if $s < 1$, the responses are skewed to the left. Of course, in practice, the true value of s is usually unknown and will have to be estimated from the data.

In enzyme-kinetics studies, another example of such a nested class may consist of just two models: the well-known Michaelis–Menten model and the Emax model, both described in Section 10.2. The larger model is the Emax model defined in Equation (10.5). The extra parameter h in the Emax model permits the shape of the response curve to be skewed and takes on different steepness as the concentration of the substrate varies. Clearly, when $h = 1$, the Emax model reduces to the Michaelis–Menten model. For design purposes, one assumes the Emax model and one seeks to design to estimate h as accurately as possible and at the same time also have efficient estimates for the parameters in the selected model. Dette, Melas and Wong (2005) discussed this design problem and provided an algorithm for generating a variety of optimal designs for this model.

Another nesting of models involving the Michaelis–Menten model is considered by Lopez-Fidalgo, Tommasi and Trandafir (2008). They discriminated between the Michaelis–Menten model and what they called the 'modified' Michaelis-Menten (MMM) model defined by

$$y_i = \frac{a \, x_i}{b + x_i} + F x_i + e_i. \tag{10.17}$$

Clearly, when $F = 0$ in the MMM model, we have the Michaelis–Menten model. The problem for us is to find a design to estimate the parameter F in the MMM model accurately and at the same time provide good estimates for the parameters (a, b). In such design problems, one may also wish to weigh in on the competitive goals; clearly, having a more precise estimate of F requires more resources. With a fixed amount of resources N, the total number of observations available, these additional resources have to come at the expense of resources required for estimating (a, b). A similar situation arises if the researcher feels estimating (a, b) is more important than F. How does one design to account for these competitive needs when they exist and are of unequal interest to the researcher? One answer is to use a multiple-objective approach discussed in the previous section and come up with a compromised optimal design.

More complicated examples involving more than two models in the nested class abound in the literature. The strategy is to embed the postulated model into 'larger' models with more parameters with the caveat that these larger models reduce to successively simpler models when the parameters take on specified values. Sigmoidal models provide an excellent illustration. These models are widely used in studies that involve a 'growth equation' such as in botany, animal science, medical studies, immunoassay and bioassays. For example, consider the exponential regression model and the Weibull regression model defined by

$$y_i = a - b \, \exp(-c t_i) + e_i$$

and

$$y_i = a - b \, \exp(-c t_i^h) + e_i, \tag{10.18}$$

respectively. Baba (2002) aptly applied the exponential regression model with $a = 0$ to study the relationship between minute ventilation and oxygen uptake during incremental exercise. Manifestly, when $h = 1$ in the Weibull regression model, we obtain the exponential regression model. So estimating and subsequently testing whether $h = 1$ is of interest here to discriminate between the two models. An optimal design should minimize the variance of the estimated parameter h. Design issues are further discussed in Dette and Pepelyshev (2008) and their methods for finding these and other types of optimal designs are implemented in our web site discussed in the next chapter.

Models that extend the logistic model are called *3PL* (*three-parameter logistic*), *4PL* (*four-parameter logistic*) or *5PL* (*five-parameter logistic*) *model*, and sometimes more generally they are referred to as *generalized logistic model*. We will not display these models because of their mathematical complexity, but note that there is specialized software for working with these models. Some examples are the (i) NLREG Software at http://www.nlreg.com, (ii) GraphPad Prism at http://www.graphpad.com and (iii) StatLIA Software at http://www.brendan.com. These sites provide background, estimation issues and applications of the models to real problems. For example, NLREG that mainly performs nonlinear regression and curve fitting has an illustrative 4PL model for studying heart rate and its relation to blood pressure. Some exemplary papers in this area are Volund (1978) and Gottschalk and Dunn (2005).

10.9   Summary

In this chapter, several types of optimal designs for different nonlinear models were discussed. The key message here is that after the nonlinear mean response function in the nonlinear model is approximated by a linear function, design techniques that we used for linear models can be applied. As emphasized in this chapter, the search for optimal designs for nonlinear models is more difficult than for linear models in large part because optimal designs for nonlinear models depend on the very values of the parameters that we want to estimate. Bayesian methods and minimax or maximin design methods are possible alternatives but they may be complicated to apply and do not usually result in closed form solutions. Numerical computations may be difficult and sometimes algorithms for generating minimax optimal designs may not converge. In the next chapter, we review some algorithms for computing optimal designs and present a web site that freely generates optimal designs for some linear and nonlinear models discussed in Chapters 9 and 10.

11

Resources for the construction of optimal designs

11.1 Introduction

One of the reasons why optimal design techniques have not been frequently used in practice is that design issues tended to be treated lightly in the curriculum of many statistics programmes and emphasis has always been on data analysis. As such, statisticians may not be very familiar with the use of optimal design techniques. Another reason is that computer programs for general purpose are not available for handling the often complicated and time-consuming computation of optimal designs. This is especially so in social and biomedical research.

Computer algorithms are increasingly used to generate optimal designs. Some are more successful than others. In practice, however, whether the algorithm converges to a theoretical optimal design or not is less of an issue; what is useful is that the algorithm finds a design close to the optimum. There are theoretical methods of determining the efficiency of the generated design without knowing the optimum. If the efficiency is close to 100%, we use the generated design as the optimal design for all practical purposes. The generated optimal design also provides very helpful hints how the theoretical optimal design looks like – how many design points and where they are roughly distributed, along with the weights at these points. This information often allows us to then use Mathematica, Scientific Workplace or Maple software to find the optimal design analytically. Without information from the generated design, we can only guess on the structure of the optimal design, and if we are wrong, then it is difficult to

An Introduction to Optimal Designs for Social and Biomedical Research   M. P. F. Berger & W. K. Wong
© 2009, John Wiley & Sons, Ltd

find an analytical description for the optimal design, especially for complicated models.

The origin of the computation of optimal designs lies in the exchange algorithm (Fedorov, 1972). The theory behind this algorithm is more or less based on common sense and actually starts with the sequential selection of optimal design points.

In the following section, we first describe the sequential construction of a design using an example and then we follow up with a short description of the exchange algorithm. We also discuss different algorithms with some technical details, along with optimal design software. In Section 11.6 we introduce a web site for generating several types of tailor-made optimal designs for a variety of models.

11.2 Sequential construction of optimal designs

The method of finding an optimal design sequentially is a fast and easy to understand procedure. One could say that the method is actually a greedy procedure, which tries to obtain as much information as possible in the shortest possible time. We will first describe and illustrate the sequential procedure in detail for a linear regression model. This involves matrix algebra, but we will try to keep notation as simple as possible.

Suppose that we have a linear model $y = X_N \beta + \varepsilon$, where an $N \times 1$ response vector y is modelled as the product of an $N \times p$ design matrix X_N and a $p \times 1$ parameter matrix β. The N rows of X_N represent the N experimental conditions under which the responses were observed and the p columns of X_N represent the values of the p variables for each experimental run. There are p unknown parameters in β. For example the quadratic polynomial regression model $y_i = \beta_0 + \beta_1 x_{1i} + \beta_2 x_{1i}^2 + \varepsilon_i$ in Equation (3.17) of Chapter 3 can be written in matrix form as

$$
\begin{bmatrix} y_1 \\ y_2 \\ \vdots \\ y_N \end{bmatrix} = \begin{bmatrix} 1 & x_1 & x_1^2 \\ 1 & x_2 & x_2^2 \\ \vdots & \vdots & \vdots \\ 1 & x_N & x_N^2 \end{bmatrix} \begin{bmatrix} \beta_0 \\ \beta_1 \\ \beta_2 \end{bmatrix} + \begin{bmatrix} \varepsilon_1 \\ \varepsilon_2 \\ \vdots \\ \varepsilon_N \end{bmatrix}. \tag{11.1}
$$

As has been noted before, apart from an unimportant multiplicative constant, the asymptotic variance–covariance matrix of the parameter estimators $\text{Cov}(\hat{\beta})$ is equal to the inverse of the information matrix for the parameters β:

$$
M_N = \sum_i^N f(x_i) f'(x_i) = X_N' X_N, \tag{11.2}
$$

where $f'(x_i) = [1 \; x_i \; x_i^2]$ is the ith row vector of X_N. For the quadratic polynomial regression model, the information matrix reduces to

$$M_N = \begin{bmatrix} N & \sum_i x_i & \sum_i x_i^2 \\ \sum_i x_i & \sum_i x_i^2 & \sum_i x_i^3 \\ \sum_i x_i^2 & \sum_i x_i^3 & \sum_i x_i^4 \end{bmatrix}. \tag{11.3}$$

In previous chapters, we explained that in order to find a D-optimal design we have to minimize the determinant of the asymptotic variance–covariance matrix $\mathrm{Cov}(\hat{\beta})$. Because the determinant of a matrix has the property $\mathrm{Det}[A] = 1/\mathrm{Det}[A^{-1}]$, where A is any non-singular square matrix, we can also obtain a D-optimal design by maximizing the determinant of the information matrix, since asymptotically the variance–covariance matrix is $\mathrm{Cov}(\hat{\beta}) = \sigma_\varepsilon^2 M_N^{-1}$. The determinant of the information matrix can be expressed as

$$\mathrm{Det}[M_N] = \mathrm{Det}[X_N'X_N]. \tag{11.4}$$

The sequential procedure of finding an optimal design consists of the sequential addition of design points to the current design. After a single design point x_a is added, the design matrix X_N will have an extra row $f'(x_a)$ for that design point x_a. For the quadratic model in Equation (11.1), this row is $f'(x_a) = [1 \; x_a \; x_a^2]$ and the order of the new design matrix X_{N+1} becomes $(N + 1) \times p$. The information matrix for an $(N + 1)$-point design can be written as the sum of the information matrix of the N-point design and the information matrix of the additional point x_a:

$$M_{N+1} = X_N'X_N + f(x_a)f'(x_a). \tag{11.5}$$

It can be shown that the determinant of the information matrix after addition of the row $f'(x_a)$ can now be written as

$$\mathrm{Det}[M_{N+1}] = \mathrm{Det}[M_N]\{1 + f'(x_a)[X_N'X_N]^{-1}f(x_a)\}, \tag{11.6}$$

see Rao (1973, p. 32, complement 2.5), for example. This equation shows that as design points are added sequentially, the D-criterion value can be expressed as a function of the D criterion of the preceding design and the standardized variance of the predicted new observation, that is $s(x_a, \xi) = Nf'(x_a)[X_N'X_N]^{-1}f(x_a)$.

The sequential procedure searches for a design point x_a in the region $x_{\min} \leq x_i \leq x_{\max}$ that maximizes $s(x_a, \xi)$ in every step and expands the design matrix by adding the corresponding row $f'(x_a)$ to the current design matrix. The new design matrix with the added row $f'(x_a)$ is then used in the next step to compute $s(x_i, \xi)$ again. This sequential procedure is basically the same as the algorithm proposed by Dykstra (1971), Wynn (1970, 1972) and Wu and Wynn (1978a,b), who showed that the resulting design converges to the optimal design as $N \to \infty$.

Let us return to the quadratic regression model. Suppose that we start with a three-point design with x values $x_1 = -1, x_2 = -0.5$ and $x_3 = 0.75$. The corresponding 3×3 design matrix for a quadratic polynomial model is

$$X_3 = \begin{bmatrix} 1 & -1 & 1 \\ 1 & -0.5 & 0.25 \\ 1 & 0.75 & 0.5625 \end{bmatrix}. \tag{11.7}$$

It should be noted that the procedure requires a sufficient number of observations to start with, because we have to ensure that the inverse $[X_N'X_N]^{-1}$ exists. A simple condition that will guarantee the inverse exists is to have the number of different design points greater than or equal to the number of parameters, that is $N \geq p$.

In Figure 11.1, the sequential procedure is visually displayed in six steps (plots). The design points are selected within the region $-1 \leq x_i \leq 1$. The standardized variances are plotted for numbers of observations ranging from $N = 3$ to $N = 8$. In each step (plots of Figure 11.1), a design point x_a is added that has the largest $s(x_a, \xi)$ value, that is the design point that ensures the largest increase of the Det$[M_N]$ value in Equation (11.6). For example, after the $N = 3$ starting points, the largest $s(x_a, \xi)$ value is found for $x_a = 1$. So, the design point $x_a = 1$ is added to the design matrix X_3 and the design matrix is now

$$X_4 = \begin{bmatrix} 1 & -1 & 1 \\ 1 & -0.5 & 0.25 \\ 1 & 0.75 & 0.5625 \\ 1 & 1 & 1 \end{bmatrix}. \tag{11.8}$$

In the next plot for $N = 4$, the largest $s(x_a, \xi)$ value is found for $x_a = 0.02$, and so on.

The plots in Figure 11.1 show that the maximum $s(x_a, \xi)$ value decreases as N increases. These maximum $s(x_a, \xi)$ values are plotted in Figure 11.2 as a function of the first 100 steps ($N = 100$). According to the general equivalence theorem (Kiefer and Wolfowitz, 1960), a D-optimal design satisfies the condition that $s(x_a, \xi) \leq p$. Figure 11.2 shows that the maximum $s(x_a, \xi)$ values decrease rapidly to $p = 3$ (dotted horizontal line) indicating that we are very close to the D-optimal design after $N = 100$ steps. The corresponding $s(x_a, \xi)$ function is also plotted in Figure 11.2, and shows that a D-optimal design for the quadratic model has three distinct design points $[-1\ 0\ 1]$. The frequencies of design points are shown as histograms in Figure 11.3. It can be seen that after $N = 100$ runs, the generated design approaches the equally weighted D-optimal design very well. We also display the starting points $x_1 = -1, x_2 = -0.5$ and $x_3 = 0.75$, together with the design point $x_i = 0.02$ that was selected as the fifth point in Figure 11.3.

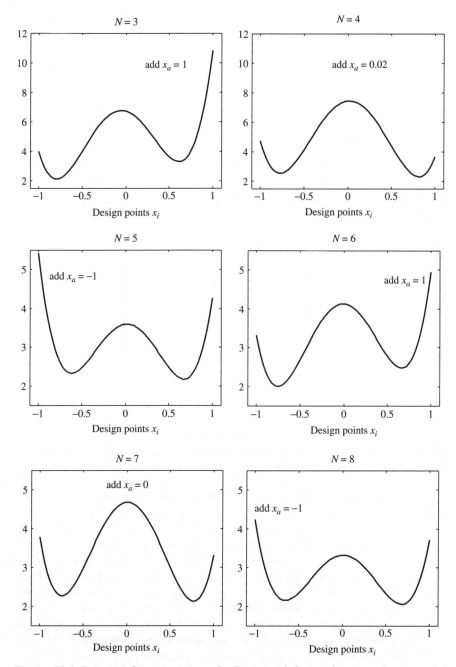

Figure 11.1 Sequential construction of a D-optimal design for a quadratic model
$y_i = \beta_0 + \beta_1 x_i + \beta_2 x_i^2 + \varepsilon_i.$ *(Plots of $s(x_i, \xi)$ in the region $-1 \le x_i \le 1$).*

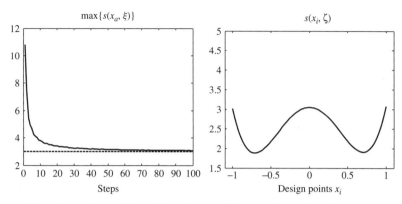

Figure 11.2 *Plots of* max$\{s(x_a, \xi)\}$ *values and the final* $s(x_i, \xi)$ *plot after* $N = 100$ *iterations.*

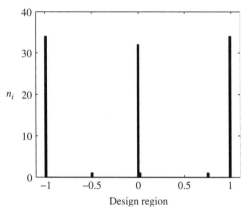

Figure 11.3 *Histogram of sequentially obtained design after* $N = 100$ *runs for the second degree polynomial model.*

Of course, this sequential procedure may not converge to the optimal design as fast as the one shown here; in many cases, convergence may take a longer time to achieve because the drop of the max$\{s(x_a, \xi)\}$ values may take many more steps (iterations). This is especially so for more complicated models such as linear mixed models. The success of the procedure also depends on the grid of the design points in the region $x_{\min} \le x_i \le x_{\max}$, which is used to compute the $s(x_i, \xi)$ function. Too large a grid may lead to inaccurate selection of the design points, and a grid that is too small will require more computing time to converge. Moreover, because design points are added sequentially, the final design may be quite different from the optimal design. In particular, the generated design tends to have many more points than the theoretical optimal designs because the values of the sequentially added points tend to cluster around the optimal design points and we will need a rule to collapse these surrounding points to a few points. This drawback of the procedure led to the notion that improvements may not only be obtained by addition of most informative designs points but also by substitution of less informative design points by more informative ones.

11.3   Exchange of design points

The sequential procedure in Figure 11.1 started with three arbitrarily chosen design points $x_1 = -1, x_2 = -0.5$ and $x_3 = 0.75$. However, two of these points are rather poor and not very informative. It may now be worthwhile to select better starting design points or to delete these starting points from the design after the first few runs. Equation (11.6) can be expanded (Harville, 1997, Theorem 18.1.1) as

$$\mathrm{Det}[M_{N\pm1}] = \mathrm{Det}[M_N]\{1 \pm f'(x_a)[X_N{}'X_N]^{-1}f(x_a)\}. \qquad (11.9)$$

Improvements of the $\mathrm{Det}[M_N]$ value are established not only by adding rows $f'(x_a)$ with the largest values of $s(x_a, \xi)$ to the design matrix X but also by deleting rows $f'(x_a)$ with the smallest $s(x_a, \xi)$ values from X. This suggests an exchange algorithm. Suppose that a row $f'(x_a)$ of X is substituted by another row $f'(x_b)$. Then it can be shown that the determinant of the information matrix will change by a factor:

$$\{f'(x_a)[X_N{}'X_N]^{-1}f(x_b)\}^2 + \{1 + f'(x_a)[X_N{}'X_N]^{-1}f(x_a)\}$$
$$\{1 - f'(x_b)[X_N{}'X_N]^{-1}f(x_b)\}. \qquad (11.10)$$

If after $N = 100$ we would want to replace the two starting points $x_2 = -0.5$ and $x_3 = 0.75$ together with the fifth design point $x_5 = 0.02$ by one of the D-optimal design points $[-1\ 0\ 1]$ in such a way that all three D-optimal design points are equally weighted, then the determinant of the information matrix will increase by a factor 1.0197. It is clear that exchanging these three design points by three D-optimal design points will not improve the $\mathrm{Det}[M_N]$ value very much. But, if relatively more design points are exchanged, this factor will increase in value. The factor in Equation (11.10) forms the basis for the so-called exchange algorithms.

11.3.1   Exchange algorithms

The first exchange algorithm was proposed by Fedorov (1972). Deletion and addition of design points is combined by evaluating all possible exchanges of pairs of design points. Each pair of the already selected design points and the candidate design points in the design space are evaluated in each step or iteration and those that result in the largest increase in criterion value $\mathrm{Det}[M_N]$ are selected. Improvements to speed up the algorithm have been suggested by Cook and Nachtsheim (1980) and Johnson and Nachtsheim (1983). Nguyen and Miller (1992), among others, concluded that the speed of convergence of these algorithms depends on the starting designs. Galil and Kiefer (1980) gave ideas to select a good starting design and Miller and Nguyen (1994) provided a computer program for the Fedorov exchange algorithm. Recently, Pronzato (2003) and Harman and Pronzato (2007) suggested refinements of the method of deleting non-informative points to accelerate the search for D-optimal designs. Modifications of the exchange algorithm for designs with correlated errors have been suggested by Ucinski and Atkinson (2004) and Stehlik (2006).

A well-known search algorithm that also adds and deletes design points in the DETMAX algorithm was proposed by Mitchell (1974). This algorithm separates the search procedure that adds the most informative design points from the search procedure that deletes the least informative design points. The DETMAX algorithm uses separate clusters of steps in which more than one design points are added and deleted, and the size of the clusters increase as the algorithm proceeds.

Another often used algorithm is the BLKL exchange algorithm proposed by Atkinson and Donev (1992). It is now implemented in the computer package GENSTAT (http://www.vsn-intl.com/genstat/). The search is made faster by locating the most likely to be exchanged design points and the most likely candidate design points in advance based on the variances of the predicted responses. The BLKL algorithm can also handle problems with both qualitative and quantitative variables. Further details can be found in Atkinson and Donev (1992).

Standard designs can be constructed using SAS (1999), where the FACTEX® procedure can be used to obtain fractional factorial designs. An algorithm to generate large D-optimal factorial designs is presented by Kuhfeld and Tobias (2005). The SAS (1999) OPTEX® procedure offers a variety of search algorithms to evaluate D- and A-optimal designs, ranging from a simple sequential search to the more computer intensive exchange algorithms. For linear models with independent errors, its use is straightforward. Adjustments can be made on the initialization method and the number of iterations. See Atkinson, Donev and Tobias (2007) for more details on the OPTEX coding for various optimal design problems.

Finally, we note that exchange algorithms cannot guarantee that a global optimum is found. Although such optimal designs can be found by exhaustive combinatorial optimization methods, these combinatorial methods are useful in practice only for a few types of small design problems. For large design problems, combinatorial optimization can be very computer intensive. In most cases, however, exchange algorithms will produce a design that is close to the optimal one.

11.4   Other algorithms

A variety of optimization algorithms have been proposed in different fields. Simulated annealing (Kirkpatrick, Gellat and Vecchi, 1983; Meyer and Nachtsheim, 1988), for example, is a generic probabilistic algorithm that attempts to locate a good approximation to the global optimum of a given function. It is an adaptation of the Metropolis–Hastings algorithm and has been used by Haines (1987) and Lejeune (2003) to construct optimal experimental designs. Because simulated annealing may converge slowly, Atkinson (1992) proposed a segmented search to speed up convergence.

Another class of algorithms to construct optimal designs originates from genetics. Recently, Jin, Chen and Sudjianto (2005) proposed a stochastic evolutionary algorithm, which seems very fast in terms of computer time. Heredia-Langner *et al.* (2003) proposed a generic algorithm from computer science for the construction of D-optimal designs. Sexton *et al.* (2006) compared a genetic algorithm with an exchange algorithm for the D-optimality criterion and concluded that the genetic algorithm at first gives a rapid drop of the D-criterion value, but subsequently converges slower than the exchange algorithm. Poland *et al.* (2001) showed that although genetic algorithms may be slower, they can improve the final design. More research, however, is needed to decide on the most efficient algorithm for optimal designs.

The statistical toolbox in Matlab and the Matlab unconstrained optimization algorithms (Matlab, 2004) can also be used to find optimal designs. The fminsearch algorithm, for example, uses the Nelder–Mead simplex search described by Lagarias *et al.* (1998). This algorithm is a direct search method that does not use numerical or analytic gradients. A simplex is an m-dimensional regularly sided figure characterized by $m + 1$ distinct vertices. For example, when $m = 2$ the simplex is a triangle and when $m = 3$, the simplex is a pyramid. In each step of the search, a new point in or near the current simplex is generated. The objective function value at the new point is compared with the value of the objective function at the vertices of the simplex. Usually, the new point replaces one of the vertices and leads to a new simplex. Steps are repeated until the diameter of the simplex is less than a priori specified tolerance level. Finally, the algorithms mentioned in this section are not designed to handle nonlinear models or mixed-effects models. However, with adjustments most of these algorithms can be used for nonlinear models (Dror and Steinberg, 2006).

We remind readers that the optimality criteria are generally complex functions with several local optima and most multidimensional optimization algorithms may stop at a local optimum. The standard treatment of this problem is to repeat the process with different starting values and see whether we obtain the same result. If we do, we may conclude with some confidence that a global optimum is found. Otherwise, the optimum we find may very well be only locally optimum.

11.5 Optimal design software

GOSSET. This is a general purpose program for designing experiments. Sloane and Hardin (2003) developed the program Gosset which is a powerful and flexible computer program for the construction of optimal designs. It can be used to find optimal designs, such as D-, A- and E-optimal designs. It does not, however, run on all operating platforms. It can only be run on Mac, Unix or Linux operating systems.

AlgDesign Package. This package calculates exact and approximate theory experimental designs for D-, A- and I-optimality criteria. It consists of a number of routines written by Wheeler (2004). Among others, it contains routines for Fedorov's exchange algorithm, full factorial optimal designs and blocked experimental designs. Further information can be found on http://www. r-project.org/ and from the author R.E. Wheeler at bwheeler@echip.com.

Design-Expert® 7.1. The package contains a variety of design creation tools to compose factorial designs with a large number of factors, response survey designs and mixture designs. Fractional factorial designs can also be specified with multiple constraints to obtain a restricted design region. See www.statease.com.

MINITAB® version 15. The recent version of MINITAB (2008) offers D-optimal response survey designs and mixture designs. The corresponding command is OPTDESIGN. Minitab gives various tools to create experimental designs and analyse and plot the results.

OPTIMAL DESIGN. Software for longitudinal and multilevel research. This software is developed with support from the National Institute of Mental Health and the William T. Grant Foundation. It can assist researchers in their computation of optimal sample sizes for linear and logistic multilevel models with two and three levels of nesting and repeated measurements. Power functions are also supplied. The software is regarded as a 'beta version', meaning that it is distributed for use under the condition that those who use it are asked to promptly report difficulties or errors to Andres Martinez (amzzz@umich. edu), Stephen W. Raudenbush (sraudenb@uchicago.edu) and/or Jessaca Spybrook (jessacah@umich.edu). The program is available on http://sitemaker. umich.edu/group-based/optimal_design_software (Raudenbusch *et al*, 2004).

POLS. This is an interactive computer program for optimal designs of longitudinal cohort studies. Large-scale longitudinal studies are carried out in many different fields of science. This computer program helps planners to design their longitudinal study by identifying the optimal cohort design, the optimal number of repeated measurements per subject and the optimal allocation of time points within a given study period (design space). Further, users can specify costs of sampling and measuring subjects, and compute loss of efficiency of any alternative design (Tekle, Tan and Berger, 2009).

ODMixed. A tool for heterogeneous longitudinal studies with dropout. Apart from the computation of optimal designs for longitudinal studies with the linear mixed-effects model, this program computes D-optimal designs for studies with a heterogeneous error variance structure and with anticipated dropout. The loss of efficiency is computed to facilitate researchers in deciding which optimal design they should use for their study when the heterogeneous error structure and the percentage of random dropout are not known in the design stage (Ortega-Azurduy, Tan and Berger, 2009).

11.6   A web site for finding optimal designs

Optimal design ideas are increasingly used in various fields to provide more efficient estimates at the same or reduced cost. This is in part due to rising experimental cost and an increasing awareness of usefulness of optimal designs, see Atkinson (1996). To facilitate use of optimal design ideas in practice, the second author and his team of researchers have constructed a web site where users can freely generate a variety of optimal designs for a list of linear and nonlinear models available on the site.

The site contains introductory material on design issues for practitioners, background information on the construction of optimal designs and frequently asked questions and answers. We plan to continue to add programs and sometimes modify current programs when necessary on the web site. As such, there will be down time and when this happens, we ask the reader to return to the site at a later time. The reader should also be warned that the site is brand new and is not fully tested yet at this time when the book goes to the press. As always, we appreciate feedback and all comments be sent to the second author at email address wkwong@ucla.edu.

The web site address is at http://optimal-design.biostat.ucla.edu/optimal/. The main purpose of the site is to educate and inform the visitor of statistical design issues, including basic ideas of optimal designs and their construction. The other purposes of the site are to (i) make optimal designs in the literature easily accessible to practitioners and researchers and (ii) develop new programs for generating cost-effective designs in the biomedical sciences. The target audience is practitioners interested to incorporate optimal design ideas in their search for an efficient design for their problem. The user selects the model of interest and provides design parameters of their choice. The model can be linear or nonlinear and sometimes we permit the design problem to have multiple objectives with different degrees of interest. The site also evaluates efficiency of any user-specified design relative to the optimal design. Some algorithms are iterative in nature and for complicated models they are quite time consuming to run. In a few of these cases, we have also provided stand alone versions for the visitor to download and run the programs on their own computers. Doing so will free up running time on the server for other programs. We also expect this site to be a useful resource to experienced researchers in design because this site contains many optimal design results that are otherwise scattered all over the literature in statistical and non-statistical journals. This site also enables researchers to find the optimal design and investigate its sensitivities to changes in design parameters with a few keystrokes on the keyboard.

The most important page on the site is the one entitled 'Web Based Design Programs'. This is where all programs for generating optimal designs are kept. On this page, there are several programs for finding a variety of optimal designs for a list of models. Many of these models are widely used in the biomedical sciences. When the program is run, the site returns the optimal design, and also the efficiency of the user-supplied design if the 'additional options' box is checked. Specifically, the support points of the user-supplied design have to be entered in the box 'design points' with each point separated by a comma. Then the corresponding weight for each is entered in the same order and separated by a comma as well. Note that these weights have to sum to unity, and if they do not, the program automatically performs this task. For example, if '1, 2' is entered in the 'corresponding weights' box signifying that weight at the second point is twice the weight at the first point, the program automatically normalizes the weights to be 1/3 and 2/3.

Sometimes, when it is not too difficult to do so, we provide an accompanying plot to confirm the optimality of the generated design. In essence, this plot graphs the derivative of the optimality criterion. The main feature to examine in this plot is whether the graph peaks at the support points of the generated design and all peaks take on some common value. If the plot exhibits this feature, the generated design is optimal; otherwise, it is not. Therefore, when such a plot is provided, the visitor should use this plot to confirm the optimality of the generated design. In some of these programs, we also provide a plot of the mean response curve over the design space and 'nearby' models. References and related papers for the construction of each type of optimal designs are provided at the bottom of the webpage.

We now demonstrate how to use the site for finding optimal designs for a couple of popular models.

11.6.1   Optimal designs for the Michaelis–Menten and Emax models

On the webpage 'Web Based Design Programs' there is a list of optimal designs that can be generated for a model. For instance, we notice that there are a variety of optimal designs for the Michaelis–Menten model, including A-, D-, E-, D_S- or extrapolation optimal design. For illustrative purposes, let us suppose we wish to find a locally A-optimal design for the Michaelis–Menten model. We click on the A-optimal key and notice that there are default values in the boxes where input parameters are required. The default values are just for illustrative purposes. We also notice that there are three forms of parameterization for the Michaelis–Menten model. We provide three because some of the forms seem to be more popular than others, depending on the discipline. For instance, the third parameterization seems to be more frequently used in marine biology and in studying the health of the lakes and oceans. The practitioner first selects one of the forms and then inputs the design parameters in the given boxes; some are required and some are optional parameters. For example, the input parameter for a in the Michaelis–Menten model is needed only to graph its mean response curve and is not needed in the calculation of the optimal design. The site also provides the efficiency of a user-input design relative to the optimal design.

As an illustration, suppose we choose the second parameterization for the Michaelis–Menten model, which is most frequently used. Suppose further that we choose the design interval to be [0, 4] and input $a = b = 1$ to find the locally A-optimal design. We recall that an A-optimal design minimizes the sum of the variances of the estimated parameters. The results are

Design Results
Design Space $= [0, 4]$
YOUR DESIGN IS

N/A
THE OPTIMAL DESIGN IS
Design points = [0.504, 4.000]
Design weights = [0.670, 0.330]

There are two accompanying plots in the output. The first is the plot of the derivative function of the optimality criterion. The plot for this case shows that the plot takes on its maximum value at both the support points of the locally A-optimal design, thus confirming the generated design supported at 0.504 and 4.0 with weight at 4 equal to 0.330 is locally A optimal. The second plot simply shows the mean response function along with some neighbouring models.

Here is another simple illustration for finding the locally D-optimal designs for the Emax model commonly used in the pharmaceutical sciences. The visitor clicks on the Emax model key entitled 'D-optimal design' of the 'Web Based Design Programs' page. As before, there are default values on the page that we provided mainly for illustrative purposes. Suppose our design interval of interest is $[0, x_0] = [0, 10]$ and the set of nominal values for (a, b, h) is $(1, 2, 1)$. We input $x_0 = 10$ and also check the box that evaluates the efficiency of a user-selected design. What design is of interest? Well, uniform design is popular because of its simplicity. Let us evaluate the efficiency of a uniform design to see if it performs well and choose a three-point uniform design supported at three points and equally spread out at 0.5, 5 and 10. We type '1, 1, 1' in the 'corresponding weights' box signifying that weights at these points are equal. When we click 'okay', the program runs and returns the following:

Design Space = [0, 10]
YOUR DESIGN IS
Points = [0.500, 5.000, 10.000]
(normalized) Weights = [0.333, 0.333, 0.333]
Input weights = [1.000, 1.000, 1.000]
Efficiency = 0.816

THE OPTIMAL DESIGN IS
Optimal design points = [0.423, 2.597, 10.000]
Optimal design weights = [0.333, 0.333, 0.333]

The output prints some of the input parameters for the design problem including the user-selected design for evaluation. The locally D optimal design for estimating the parameters $(a, b, h) = (1, 2, 1)$ is equally supported at 0.423, 2.597 and 10.00, and the efficiency of the input design is 81.6%. The output includes the plot of the derivative function. This plot is bounded above by 3, which is the number of parameters in the model and the graph attains the maximal value 3 at the support points of the generated design. This confirms the D optimality of the generated design. The output also includes a plot of the mean response function for the input nominal parameters, along with a couple of other sets of parameters. A reference and related papers for the construction of the optimal design are given at the bottom of the output page.

The web site also has a program for finding the optimal design for estimating only the parameter h. We recall that this is a D_s-optimal design discussed earlier in Chapters 3 and 9. If one clicks on the Emax model key entitled 'D_1-optimal design', a similar page shows up and one can input parameters as we did for finding the locally D-optimal design. We use the same set of parameters we used for finding the locally D-optimal design, and now decide to evaluate how well the locally D-optimal design is for estimating only the parameter h. We can do this easily by checking the box and inputting the design points of the locally D-optimal design, that is 0.423, 2.597, 10 and followed by 1, 1, 1 in the 'corresponding weights' box. This is the locally D-optimal design we just found. The output from the site is

Design Space = [0, 10]

YOUR DESIGN IS
Points = [0.423, 2.597, 10.000]
(normalized) Weights = [0.333, 0.333, 0.333]
Input weights = [1.000, 1.000, 1.000]
Efficiency = 0.718

THE OPTIMAL DESIGN IS
Optimal design points = [0.231, 2.699, 10]
Optimal design weights = [0.576, 0.293, 0.130]

This shows that for the nominal values specified, the D_1-optimal design for estimating h is supported at 0.231, 2.699 and 10 and the weights at these points are 0.576, 0.293, and 0.130, respectively. Graphical plots and references are also provided as in the D-optimal page. We notice that the efficiency of the locally D-optimal design for estimating h is 71.8%.

11.6.2 Optimal designs for discriminating among toxicological models

Slob and Pieters (1998), Slob (2002) and Slob et al. (2005) proposed this nested class for studying continuous toxicological endpoints. In this series of papers, the authors gave cogent arguments that models in this class are sufficiently flexible to accommodate most continuous endpoints in the toxicological sciences. The assumption is that the mean of the continuous response outcome y_i at dose level x_i can be modelled by $\eta(x_i, \theta)$ where θ is a p-dimensional vector of parameters. The function $\eta(x_i, \theta)$ may take on one of the following forms:

Model 1: $\eta(x_i, \theta) = a \exp(-bx_i); \theta = (a, b)', \ a > 0$ and $b > 0$.

Model 2: $\eta(x_i, \theta) = a \exp(-bx_i^d)$; $\theta = (a, b, d)'$, $a > 0, b > 0$

and $d \geq 1$.

Model 3: $\eta(x_i, \theta) = a[c - (c - 1) \exp(-bx_i)]$; $\theta = (a, b, c)'$, $\quad$ (11.11)

$a > 0, b > 0$ and $0 \leq c \leq 1$.

Model 4: $\eta(x_i, \theta) = a[c - (c - 1) \exp(-bx_i^d)]$; $\theta = (a, b, c, d)'$,

$a > 0, b > 0, 0 \leq c \leq 1$ and $d \geq 1$.

The parameters in θ indicate different characteristics of these models. The parameter a represents the background level of the endpoint, b is the relative efficacy of dose, c is the maximum effect relative to a and d is the shape parameter. These models are hierarchically related to each other as follows: *Model* 4 reduces to *Model* 3 if $d = 1$; *Model* 4 reduces to *Model* 2 if $c = 0$; *Model* 3 reduces to *Model* 1 if $c = 0$ and *Model* 2 reduces to *Model* 1 if $d = 1$.

Figure 11.4 illustrates how these toxicological models differ in shape. It should be noted that depending on the values of the parameters in θ, these models can have various shapes other than the ones plotted here.

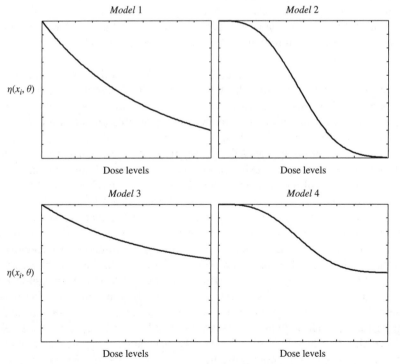

Figure 11.4 The shapes of the four toxicological models ($a = 1, b = 0.8, c = 0.5, d = 3$).

Slob and Pieters (1998) introduced the terms critical effect size (CES) and critical effect dose (CED) for continuous data. The CES reflects the quantitative change in a particular endpoint considered as non-adverse or acceptable at the individual level. The motivation for the CES comes from a biological point of view where it is felt that the most natural way of measuring an effect size is in terms of a percent change relative to the background value of the particular endpoint (Slob, 2002). The CES sought in a study with a continuous outcome is the minimal standardized change expected to be significant by the researcher (Moerbeek, Piersma and Slob, 2004; Slob, 2002; Woutersen *et al.* 2001).

More specifically, assume that the mean response $\eta(x_i, \theta)$ decreases as the doses increase, the CED is the dose that will result in observance of the CES. For a given CES and a given mean response function, the CED is estimated from the equation

$$CES = \frac{\eta(0, \theta) - \eta(CED, \theta)}{\eta(0, \theta)}. \tag{11.12}$$

Here, $\eta(0, \theta)$ is the mean response when the dose $x_i = 0$. Estimating the CED that results in a user-specified CES is an analogue for a continuous outcome similar in spirits to estimating a percentile or a threshold dose when the outcome is binary, which is discussed in Chapter 10.

Slob (2002) gave an example involving cholinesterase (ChE) activity, and ChE with CES = 0.20 when inhibition of ChE of less than 20% was postulated as non-adverse for an individual. Our interest here is to find an efficient design for estimating CED by minimizing the asymptotic variance of the estimated CED when $\eta(x_i, \theta)$ is known. This is a relatively straightforward problem similar to the design problem for estimating the turning point in a quadratic model discussed in Chapter 9. The main challenge here is that we do not know the model in advance; we are assuming that there are a few plausible models at the onset and we want to design the study with this in mind.

This is a rather complicated design problem and we discuss it only to show that there are very difficult design problems in practice and how one may design when there are multiple concerns. See Dette *et al.* (2008a,b) for the complicated theory behind the construction of such optimal designs. On the web site, we provide three types of optimal designs for this particular problem: maximin CED-optimal designs, maximin discriminating designs and maximin compound designs. The maximin here refers to maximizing the minimal efficiency of the optimal design regardless of which one of the models in the nested class is the 'true' model. This means that we construct an optimal design under a supposedly 'true' model and evaluate its efficiency under all the other models. The optimal design that maximizes the minimum of these efficiencies is the sought maximin optimal design. Throughout, errors are assumed to be independent, and either (i) normally distributed or (ii) log normally distributed. The latter assumption is especially popular in toxicology studies.

On the webpage entitled 'Optimal Designs for Toxicology Studies', we provide three types of optimal designs after the design parameters are supplied:

1. the optimal design that maximizes the minimal efficiency of the estimated CED from the different models;

2. the optimal design that maximizes the minimum of D- or D_l efficiencies taken over all models in the given class and under a set of user-specified range of plausible values for the nonlinear parameters;

3. the dual-objectives optimal design that selects the most appropriate model using a maximin approach and provides the best estimate for the CED of the selected model at the same time.

In addition, if the box for additional option is checked, the site reports the efficiencies of the user-supplied design for the four models in the above listed order. The input parameters are the type of optimal design desired, the design space, the error distribution and nominal values for the model parameters. Additionally, the researcher has to specify the number of support points desired for the optimal design and this number ranges from four to eight points. If this number is smaller than four, estimation problems may arise, and if it is larger than eight, the time required to generate the optimal design becomes quite long. For optimal designs described in 1 and 3 above, the user has to input the CES level as well. On this webpage, the format for inputting the design parameters is shown below, along with the default values:

Number of design points: 4.
Optimality criterion: maximin CED-optimal designs.
Assumption on errors: normality assumption.
Design interval: lower bound = 0, upper bound = 1.
CES = 0.05
a = 1, b = 1, c = 0.1, d = 1.1.

The output results are
YOUR DESIGN IS
Design points = [0.000, 0.250, 0.500, 1.000]
(normalized) design weights = [0.250, 0.250, 0.250, 0.250]
Input weights = [1.000, 1.000, 1.000, 1.000]
Design efficiencies = [0.553, 0.807, 0.760, 0.433]
THE OPTIMAL DESIGN IS
Design points = [0.000, 0.212, 0.626, 1.000]
Design weights = [0.273, 0.224, 0.267, 0.236]
Design efficiencies = [0.625, 0.640, 0.625, 0.625]

As a comparison, here are the corresponding results when we assumed errors are log normally distributed.

YOUR DESIGN IS
Design points = [0.000, 0.250, 0.500, 1.000]
(normalized) design weights = [0.250, 0.250, 0.250, 0.250]

Input weights = [1.000, 1.000, 1.000, 1.000]
Design efficiencies = [0.547, 0.774, 0.642, 0.478]

THE OPTIMAL DESIGN IS
Design points = [0.000, 0.218, 0.624, 1.000]
Design weights = [0.341, 0.210, 0.207, 0.243]
Design efficiencies = [0.636, 0.677, 0.635, 0.635]

We end this section with a note on minimally supported optimal designs available on the web site. These are optimal designs found only among the designs on the design space with number of design points equal to the number of parameters in the mean response function. As such, these designs may or may not be optimal among all designs on the design space. Analytical formulae for the minimally supported optimal design may exist when they are not available for the optimal design found among all designs on the design space. Minimally supported optimal designs are particularly appealing in situations where it is expensive to take observations at different sites. **Use the web site and have fun!**

11.7   Summary

For complicated models, it can still be a challenging task to find an optimal design theoretically or computationally. We believe this is a rapidly developing area of research as exemplified by the computer programs and the web site described in this chapter. A second point to be emphasized is that sometimes social and biomedical researchers may feel that it is not feasible or interesting to implement an optimal design in practice. However, this does not mean that it is useless for them to study or compute optimal designs. On the contrary, even if a theoretically established optimal design is not feasible in practice, it is still illuminating to study and understand characteristics of the optimal design in the best-case scenario where ideal model assumptions were made. Researchers should find optimal design techniques and the computational developments described in this chapter useful in their search for more efficient designs. At the very minimum, it is helpful to view optimal designs as calibration designs that measure trade-off between competing designs and at the same time also enable researchers to sensibly deviate from the optimum to meet their practical needs in the study.

References

Abdelbasit, K.M. and Plankett, R.L. (1983) Experimental design for binary data. *Journal of the American Statistical Association*, **78**, 90–98.

Abt, M., Gaffke, N., Liski, E.P. and Sinha, B.K. (1998) Optimal designs for growth curve models: Part II. Correlated model for quadratic growth: optimal designs for slope parameters and growth prediction. *Journal of Statistical Planning and Inference*, **67**, 287–96.

Abt, M., Liski, E.P., Mandal, N.K. and Sinha, B.K. (1997) Optimal design in growth curve models: Part I. Correlated model for linear growth: optimal designs for slope parameter estimation and growth prediction. *Journal of Statistical Planning and Inference*, **64**, 141–50.

Afsarinejad, K. (1990) Repeated measurements designs – a review. *Communications in Statistics: Theory and Methods*, **19**, 3985–4028.

Akaike, H. (1974) A new look at the statistical model identification. *IEEE Transactions on Automatic Control*, **19**, 716–23.

Allison, D.B., Allison, R.L., Faith, M.S. *et al.* (1997) Power and money: designing statistically powerful studies while minimizing financial costs. *Psychological Methods*, **2**, 20–33.

American Psychological Association (2002) *Ethical Principles of Psychologists and Code of Conduct*, Author, Washington, DC.

Ankerman, B.E., Aviles, A.I. and Pinheiro, J.C. (2003) Optimal designs for mixed-effects models with two random nested factors. *Statistica Sinica*, **13**, 385–401.

Atkins, J.E. and Cheng, C.S. (1999) Optimal regression designs in the presence of random block effects. *Journal of Statistical Planning and Inference*, **77**, 321–35.

Atkinson, A.C. (1982) Optimum biased coin designs for sequential clinical trials with prognostic factors. *Biometrika*, **69**, 61–67.

Atkinson, A.C. (1992) A segmented algorithm for simulated annealing. *Statistics and Computing*, **2**, 221–30.

Atkinson, A.C. (1996) The usefulness of optimum experimental designs. *Journal of the Royal Statistical Society. Series B*, **58**, 59–76.

Atkinson, A.C. and Cook, R.D. (1995) D-optimal designs for heteroscedastic linear models. *Journal of the American Statistical Association*, **90**, 204–12.

Atkinson, A.C. and Donev, A.N. (1992) *Optimum Experimental Designs*, Clarendon Press, Oxford.

Atkinson, A.C., Donev, A.N. and Tobias, R.D. (2007) *Optimum Experimental Designs, with SAS*, Oxford University Press.

Atkinson, A.C. and Fedorov, V.V. (1975a) The design of experiments for discriminating between two rival models. *Biometrika*, **62**, 57–70.

Atkinson, A.C. and Fedorov, V.V. (1975b) Optimal design: experiments for discriminating between several models. *Biometrika*, **62**, 289–303.

Atwood, C.L. (1976) Sequences converging to D-optimal designs of experiments. *Annals of Statistics*, **1**, 342–52.

Baba, R. (2002) Simple exponential regression model to describe the relation between minute and oxygen uptake during incremental exercise. *Nagoya Journal of Medical Science*, **65**, 95–102.

Bailey, R.A. (2008) *Design of Comparative Experiments*, Cambridge University Press, Cambridge.

Balaam, L.N. (1968) A two period design with t^2 experimental units. *Biometrics*, **24**, 61–73.

Begg, C.C. and Kalish, L.A. (1984) Treatment allocation for nonlinear models in clinical trials: the logistic model. *Biometrics*, **40**, 409–20.

Benson, K. and Hartz, A.J. (2000) A comparison of observational studies and randomized, controlled trials. *New England Journal of Medicine*, **342**, 1878–86.

Berger, M.P.F. (1992) Sequential sampling designs for the two-parameter item response theory model. *Psychometrika*, **57**, 521–38.

Berger, M.P.F. (1994) D-optimal sequential sampling designs for item response theory models. *Journal of Educational Statistics*, **19**, 43–56.

Berger, M.P.F. (2005) Optimal designs for categorical variables, in *Encyclopaedia of Statistics in Behavioural Science*, (eds B. Everitt and D. Howell), John Wiley & Sons, Inc., New York.

Berger, M.P.F. and Tan, F.E.S. (2004) Robust designs for linear mixed effects models. *Journal of the Royal Statistical Society. Series C*, **53**, 569–81.

Berger, M.P.F. and Wong, W.K. (eds) (2005) *Applied Optimal Designs*, John Wiley & Sons, Inc.

Bischoff, W. (1993) On D-optimal designs for linear models under correlated observations with an application to a linear model with multiple response. *Journal of Statistical Planning and Inference*, **37**, 69–80.

Bock, R.D. (1975) *Multivariate Statistical Methods in Behavioral Research*, McGraw-Hill, New York.

Box, G.E.P. (1982) Choice of response surface design and alphabetic optimality. *Utilitas Mathematica*, **B21**, 11–51.

Box, G.E.P. (1990) Must we randomize our experiment? *Quality Engineering*, **2**, 497–502.

Box, G.E.P. and Cox, D.R. (1984) An analysis of transformations revisited, rebuttal. *Journal of the American Statistical Association*, **79**, 209–10.

Box, G.E.P. and Draper, N.R. (1987) *Empirical Model Building and Response Surfaces*, John Wiley & Sons, Inc., New York.

Box, G.E.P., Hunter, W.G. and Hunter, J.S. (1978) *Statistics for Experimenters*, John Wiley & Sons, Inc., New York.

Box, G.E.P. and Lucas, H.L. (1959) Design of experiments in nonlinear situations. *Biometrika*, **46**, 77–90.

Bozdogan, H. (1987) Model selection and Akaike's information criterion (AIC): the general theory and its analytical extensions. *Psychometrika*, **52**, 345–70.

Brown, B.W. (1980) The crossover experiment for clinical trials. *Biometrics*, **36**, 69–79.

Brown, M.B. and Forsythe, A.B. (1974) The small sample behavior of some statistics which test the equality of several means. *Technometrics*, **16**, 129–32.

Brown, L.D. and Wong, W.K. (2000) An algorithmic construction of optimal minimax designs for heteroscedastic linear models. *Journal of Statistical Planning and Inference*, **85**, 103–14.

Campbell, D.T. and Stanley, J.C. (1963) *Experimental and Quasi-Experimental designs for Research*, Rand McNally, Chicago.

Candel, M.J.J.M., van Breukelen, G.J.P., Kotova, L. and Berger, M.P.F. (2008) Optimality of equal versus unequal cluster sizes in multilevel intervention studies: A Monte Carlo study for small sample sizes. *Communications in Statistics: Simulation and Computation*, **37**, 22–239.

Carriere, K.C. (1994) Crossover designs for clinical trials. *Statistics in Medicine*, **13**, 1063–69.

Carriere, K.C. (1995) Optimal two-period crossover designs: a practical solution. *Clinical Research and Regulatory Affairs*, **12**, 183–91.

Carriere, K.C. and Huang, R. (2000) Crossover designs for two-treatment clinical trials. *Journal of Statistical Planning and Inference*, **87**, 125–34.

Carriere, K.C. and Reinsel, G.C. (1992) Investigation of dual-balanced cross-over designs for two treatments. *Biometrics*, **48**, 1157–64.

Carriere, K.C. and Reinsel, G.C. (1993) Optimal two-period repeated measurement design with two or more treatments. *Biometrika*, **80**, 924–29.

Carroll, R.J. and Ruppert, D. (1984) Power transformations when fitting theoretical models to data. *Journal of the American Statistical Association*, **79**, 321–28.

Chaloner, K. (1984) Optimal Bayesian experimental designs for linear models. *Annals of Statistics*, **12**, 283–300.

Chaloner, K. (1989) Bayesian design for estimating the turning point of a quadratic regression. *Communications in Statistics: Theory and Methods*, **18**, 1385–400.

Chaloner, K., Church, T., Louis, T.A. and Matts, J.P. (1993) Graphical elicitation of a prior distribution for a clinical trial. Special issue: conference on practical Bayesian statistics. *The Statistician*, **42**, 341–53.

Chaloner, K. and Duncan, G.T. (1983) Assessment of a beta prior distribution: PM elicitation. Proceedings of the 1982 I.O.S. annual conference on practical Bayesian statistics. *The Statistician*, **32**, 174–80.

Chaloner, K. and Duncan, G.T. (1987) Some properties of the dirichlet-multinomial distribution and its use in prior elicitation. *Communications in Statistics: Theory and Methods*, **16**, 511–23.

Chaloner, K. and Larntz, K. (1989) Optimal Bayesian design applied to logistic regression experiments. *Journal of Statistical Planning and Inference*, **21**, 191–208.

Chaloner, K. and Verdinelli, I. (1995) Bayesian experimental design: a review. *Statistical Science*, **10**, 273–304.

Chassan, J.B. (1970) A note on relative efficiency in clinical trials. *Journal of Clinical Pharmacology*, **10**, 359–60.

Chatterjee, S. and Price, B. (1991) *Regression Analysis by Example*, John Wiley & Sons, Inc., New York.

Cheng, C.S. (1995) Optimal regression designs under random block effects models. *Statistica Sinica*, **5**, 485–97.

Cheng, C. and Wu, C. (1980) Balanced repeated measurement designs. *Annals of Statistics*, **8**, 1272–83.

Chernoff, H. (1953) Locally optimal designs for estimating parameters. *The Annals of Mathematical Statistics*, **24**, 586–602.

Chi, E.M. and Reinsel, G.C. (1989) Models for longitudinal data with random effects and AR(1) errors. *Journal of the American Statistical Association*, **84**, 452–59.

Chinchilli, V.M. and Esinhart, J.D. (1996) Design and analysis of intra-subject variability in cross-over experiments. *Statistics in Medicine*, **15**, 1619–34.

Chow, S.C., Shao, J. and Wang, H.S. (2007) *Sample Size Calculations in Clinical Research*, 2nd edn, Chapman & Hall/CRC, Boca Raton.

Cohen, M.P. (1998) Determining sample sizes for surveys with data analyzed by hierarchical linear models. *Journal of Official Statistics*, **14**, 267–75.

Cohen, M.P. (2005) Sample size considerations for multilevel surveys. *International Statistical Review*, **73**, 279–87.

Concato, J., Shah, N. and Horwitz, R.I. (2000) Randomized, controlled trials, observational studies and the hierarchy of research designs. *The New England Journal of Medicine*, **342**, 1887–92.

Cook, T.D. and Campbell, D.T. (1979) *Quasi-Experimentation: Design & Analysis Issues for Field Settings*, Rand McNally College Publishing Company, Chicago.

Cook, R.D. and Nachtsheim, C.J. (1980) A comparison of algorithms for constructing exact D-optimal designs. *Technometrics*, **22**, 315–24.

Cook, R.D. and Wong, W.K. (1994) On the equivalence of constrained and compound optimal designs. *Journal of the American Statistical Association*, **89**, 687–92.

Cox, D.R. (1958) *Planning of Experiments*, John Wiley & Sons, Inc., New York.

Cox, D.R. and Reid, N. (2000) *The Theory of the Design of Experiments*, Chapman & Hall/CRC, Boca Raton.

Crainiceanu, C.M., Ruppert, D. and Vogelsang, T.J. (2003). Some Properties of Likelihood ratio Tests in Linear Mixed Models. To be published. (available from www.orie.cornell.edu/~davidr/papers).

Davidson, M.L. (1972) Univariate versus multivariate tests in repeated measures experiments. *Psychological Bulletin*, **77**, 446–52.

Declaration of Helsinki (2000) *Ethical Principles for Medical Research Involving Human Subjects*, World Medical Association, Edinburgh.

Derendorf, H. and Meibohm, B. (1999) Modeling of phamacokinetic / phamacodynamic (PK/PD) relationships: concepts and perspectives. *Pharmaceutical Research*, **16**, 176–85.

Dette, H. (1995) Discussion paper of constrained optimization of experimental design by Cook, R.D. and Fedorov, V.V. *Statistics*, **26**, 129–78.

Dette, H. (1997) Designing experiments with respect to 'standardized' optimality criteria. *Journal of the Royal Statistical Society. Series B*, **59**, 97–110.

Dette, H. (2004) On robust and efficient designs for risk estimation in epidemiological studies. *Scandinavian Journal of Statistics*, **31**, 319–31.

Dette, H., Kiss, C. and Wong, W.K. (2008). Robustness of Optimal Designs for the Michaelis-Menten model under a Variation of Criteria. To be published in: Statistics for Biopharmaceutical Research.

Dette, H. and Pepelysheve, A. (2008) Efficient experimental designs for sigmoidal growth models. *Journal of Statistical Planning and Inference*, **138**, 2–17.

Dette, H., Pepelysheve, A., Shpilev, P. and Wong, W.K. (2008a). Optimal Designs for Discriminating dose Response Models in Toxicology Studies. Under Review.

Dette, H., Pepelysheve, A., Shpilev, P. and Wong, W.K. (2008b). Optimal Designs for Estimating Critical Effective dose under Model Uncertainty in a dose Response Study. To be published in Statistics and its Interface.

Dette, H., Melas, V.B. and Wong, W.K. (2005) Optimal designs for goodness of fit of the Michaelis-Menten enzyme kinetic function. *Journal of the American Statistical Association*, **100**, 1370–81.

Dette, H. and Wong, W.K. (1996) Bayesian optimal designs for models with partially specified heteroscedastic structure. *Annals of Statistics*, **24**, 2108–127.

Diggle, P.J. (1988) An approach to the analysis of repeated measures. *Biometrics*, **44**, 959–71.

Diggle, P.J., Heagerty, P.J., Liang, K.-Y. and Zeger, S.L. (2002) *Analysis of Longitudinal Data*, Oxford University Press, New York.

Diggle, P.J., Liang, K.-Y. and Zeger, S.L. (1994) *Analysis of Longitudinal Data*, Oxford University Press, New York.

Donner, A., Birkett, N. and Buck, C. (1981) Randomization by cluster, sample size requirements and analysis. *American Journal of Epidemiology*, **114**, 906–14.

Dror, H.A. and Steinberg, D.M. (2006) Robust experimental design for multivariate generalized linear models. *Technometrics*, **48**, 520–29.

Duimel-Peeters, I.G.P., Halfens, R.J.G., Ambergen, A.W. *et al.* (2007) The effectiveness of massage with and without dimethyl sulfoxide in preventing pressure ulcers: a randomized, double-blind cross-over trial in patients prone to pressure ulcers. *International Journal of Nursing Studies*, **8**, 1285–95.

Duimel-Peeters, I.G.P., Halfens, R.J.G., Berger, M.P.F. and Snoeckx, L.H.E.H. (2004) The effects of massage as a method to prevent pressure sores. *Ostomy Wound Management*, **51**, 70–80.

Dunn, G. (1988) Optimal designs for drug, neurotransmitter and hormone receptor assays. *Statistics in Medicine*, **7**, 805–15.

Dykstra, O. (1971) The augmentation of experimental data to maximize $|X^T X|$. *Technometrics*, **13**, 682–88.

Ebbutt, A.F. (1984) Three-period crossover designs for two treatments. *Biometrics*, **40**, 219–24.

Emanuel, E.J., Wendler, D. and Grady, C. (2000) What makes clinical research ethical? *Journal of the American Medical Association*, **283**, 2701–10.

Everitt, B.S. (1995) The analysis of repeated measures: a practical review with examples. *Journal of the Royal Statistical Society. Series D*, **44**, 113–35.

Faddy, M.J. (1993) A structured compartmental model for drug kinetics. *Biometrics*, **49**, 243–48.

Fedorov, V.V. (1972). *Theory of Optimal Experiments*. Translated and edited by (eds W.J. Studden and E.M. Klimko), Academic Press, New York.

Fedorov, V.V., Gagnon, R.C. and Leonov, S.L. (2002) Design of experiments with unknown parameters in variance. *Applied Stochastic Models in Business and Industry*, **18**, 207–18.

Fedorov, V.V. and Leonov, S.L. (2005) Response driven designs in drug development, in *Applied Optimal Designs*, (eds M.P.F. Berger and W.K. Wong), John Wiley & Sons, Inc., New York, pp. 103–36.

Fedorov, V.V. and Mueller, W.G. (1997) Another view on optimal design for estimating the point of extremum in quadratic regression. *Metrika*, **46**, 147–57.

Fisher, R.A. (1925) *Statistical Methods for Research Workers*, Oliver & Boyd, Edinburgh.

Fisher, R.A. (1926) The arrangement of field experiments. *Journal of the Ministry of Agriculture of Great Britain*, **33**, 503–13.

Fisher, R.A. (1935) *The Design of Experiments*, Oliver & Boyd, Edinburgh.

Fisher, R.A. and Yates, F. (1963) *Statistical Tables for Biological, Agricultural and Medical Research*, Oliver & Boyd, Edinburgh.

Fleiss, J.L. (1989) A critique of recent research on the two-treatment crossover design. *Controlled Clinical Trials*, **10**, 237–43.

Frison, L. and Pocock, S.J. (1992) Repeated measures in clinical trials: analysis using mean, summary statistics and its implications for design. *Statistics in Medicine*, **11**, 1685–1704.

Flynn, T.N., Whitley, E. and Peters, T.J. (2002) Recruitment strategies in a cluster randomized trial - cost implications. *Statistics in Medicine*, **21**, 397–405.

Ford, I., Kitsos, C.P. and Titterington, D.M. (1989) Recent advances in nonlinear experimental design. *Technometrics*, **31**, 49–60.

Fornius, E.F. (2008) Sequential designs for binary data with the purpose to maximize the probability of response. *Communications in Statistics: Simulation and Computation*, **37**, 1219–38.

Freeman, P. (1989) The performance of the two-stage analysis of two-treatment, two period cross-over trials. *Statistics in Medicine*, **8**, 1421–32.

Gaffke, N. and Krafft, O. (1982) Exact D-optimum designs for quadratic regression. *Journal of the Royal Statistical Society. Series B*, **44**, 394–97.

Gagnon, R. and Leonov, S.L. (2005) Optimal population designs for PK models with serial sampling. *Journal of Biopharmaceutical Statistics*, **15**, 143–63.

Galbraith, S. and Marschner, I.C. (2002) Guidelines for the design of clinical trials with longitudinal outcomes. *Controlled Clinical Trials*, **23**, 257–73.

Galil, Z. and Kiefer, J. (1980) Time- and space-saving computer methods, related to Mitchell's DETMAX, for finding D-optimum designs. *Technometrics*, **21**, 301–13.

Galton, F. (1878) Composite portraits. *Journal of the Anthropological Institute of Great Britain and Ireland*, **8**, 132–42.

Gaudard, M.A., Karson, M.J., Linder, E. and Tse, S.K. (1993) Efficient designs for estimation in the power logistic quantal response model. *Statistica Sinica*, **3**, 233–43.

Gavasakar, U. (1988) A comparison of two elicitation methods for a prior distribution for a binomial parameter. *Management Science*, **34**, 784–90.

Gaylor, D.W., Chen, J.J. and Kodell, R.L. (1985) Experimental design of bioassays for screening and low dose extrapolation. *Risk Analysis*, **5**, 9–16.

Goldstein, H. (1995) *Multilevel Statistical Models*, Edward Arnold, London.

Goos, P. (2002). *The Optimal Design of Blocked and Split-Plot Experiments*, Lecture Notes in Statistics, Springer, New York.

Gottschalk, P.G. and Dunn, J.R. (2005) The five parameter logistic: a characterization and comparison with the four-parameter logistic. *Analytical Biochemistry*, **343**, 54–65.

Greenhouse, S. and Geisser, S. (1959) On the methods in the analysis of profile data. *Psychometrika*, **24**, 95–112.

Grizzle, J.E. (1965) The two-period change over design and its use in clinical trials. *Biometrics*, **21**, 467–80.

Grizzle, J.E. (1974) Correction to Grizzle (1965). *Biometrics*, **30**, 727.

Guest, P. (1958) The spacing of observations in polynomial regression. *The Annals of Mathematical Statistics*, **29**, 294–99.

Hahn, E.D. (2005) Re-examining informative prior elicitation through the lens of Markov Chain Monte Carlo methods. *Journal of the Royal Statistical Society. Series A*, **169**, 37–48.

Haines, L.M. (1987) The application of the simulated annealing algorithm to the construction of exact optimal designs for linear regression models. *Technometrics*, **29**, 439–47.

Hand, D.J. and Crowder, M.J. (1996) *Practical Longitudinal Data Analysis*, Chapman & Hall, London.

Hansen, M.H., Hurwitz, W.N. and Madow, W.G. (1953) *Sample Survey Methods and Theory*, Vol. II (Theory), John Wiley & Sons, Inc., New York.

Harman, R. and Pronzato, L. (2007) Improvements on removing support points in D-optimum design algorithms. *Statistics and Probability Letters*, **77**, 90–94.

Harville, D.A. (1974) Bayesian inference for variance components using only error contrasts. *Biometrika*, **61**, 383–85.

Harville, D.A. (1977) Maximum likelihood approaches to variance component estimation and to related problems. *Journal of the American Statistical Association*, **72**, 320–40.

Harville, D.A. (1997) *Matrix Algebra from a Statistician's Perspective*, Springer, New York.

Headrick, T.C. and Zumbo, B.D. (2005) On optimizing multi-level designs: power under budget constraints. *Australian and New Zealand Journal of Statistics*, **13**, 219–29.

Hedayat, A. and Afsarinejad, K. (1975) Repeated measurements designs I, in *A Survey of Statistical Design and Linear Models*, (ed. J.N. Srivastava), North-Holland, Amsterdam, pp. 229–42.

Hedayat, A. and Afsarinejad, K. (1978) Repeated measurements designs II. *Annals of Statistics*, **6**, 619–28.

Hedayat, A., Stufken, J. and Yang, M. (2006) Otpimal and efficient crossover designs when subject effects are random. *Journal of the American Statistical Association*, **101**, 1031–38.

Heredia-Langner, A., Carlyle, W.M., Montgomery, D.C. *et al.* (2003) Generic algorithms for the construction of D-optimal designs. *Journal of Quality Technology*, **35** (1), 28–46.

Hill, D.H. (1978a) A note on the equivalence of D-optimal design measures for three rival linear models. *Biometrika*, **68**, 666–67.

Hill, D.H. (1978b) A review of experimental design procedures for regression model discrimination. *Technometrics*, **20**, 15–21.

Hill, D.H. (1980) D-optimal designs for partially nonlinear regression. *Technometrics*, **22**, 275–76.

Hill, W.J., Hunter, W.G. and Wichern, D.W. (1968) A joint design criterion for the dual problem of model discrimination and parameter estimation. *Technometrics*, **10**, 145–60.

Holmberg, A. (1982) On the practical identifiability of microbial growth models incorporating Michaelis-Menten type nonlinearities. *Mathematical Biosciences*, **62**, 23–43.

Horan, M.J., Chair (1987) Task Force on Blood Pressure Control in Children. Report of the second task force on blood pressure control in children. *Pediatrics*, **79** (1), 1–25.

Hosmer, D.W. and Lemeshow, S. (1989) *Applied Logistic Regression*, John Wiley & Sons, Inc., New York.

Houston, J.B. and Kenworthy, K.E. (2000) In vitro-in vivo scaling of cyp kinetic data not consistent with the classical Michaelis-Menten model. *Drug Metabolism and Disposition*, **28**, 246–54.

Hox, J.J. (2002) *Multilevel Analysis: Techniques and Applications*, Lawrence Erlbaum.

Hsieh, F.Y. (1988) Sample size formulae for intervention studies with the cluster as unit of randomization. *Statistics in Medicine*, **8**, 1195–201.

Huang, Y.C. and Wong, W.K. (1998) Multiple-objective optimal designs. *Journal of Biopharmaceutical Statistics*, **8**, 635–43.

Hungerford, T.W. (1990) *Abstract Algebra*, Holt, Rinehart and Winston, Orlando.

Huynh, H. and Feldt, L.S. (1976) Estimates of the box correction for degrees of freedom for sample data in randomized block and split-plot designs. *Journal of Educational Statistics*, **1**, 69–82.

Imhof, L.A., Song, D. and Wong, W.K. (2002) Optimal design of experiments with possibly failing trials. *Statistica Sinica*, **12**, 1145–55.

Imhof, L.A., Song, D. and Wong, W.K. (2004) Optimal design of experiments with anticipated pattern of missing observations. *Journal of Theoretical Biology*, **228**, 251–60.

Jennrich, R.I. and Schluchter, M.D. (1986) Unbalanced repeated-measures models with structured covariance matrices. *Biometrics*, **42**, 805–20.

Jensen, P.S., Watanabe, H.K. and Richters, J.E. (1999) Who's up first? Testing for order effects in structured interviews using a counterbalanced experimental design. *Journal of Abnormal Child Psychology*, **27**, 439–45.

Jin, R., Chen, W. and Sudjianto, A. (2005) An efficient algorithm for constructing optimal design of computer experiments. *Journal of Statistical Planning and Inference*, **134**, 268–87.

Johnson, M.E. and Nachtsheim, C.J. (1983) Some guidelines for constructing exact D-optimal designs on convex design spaces. *Technometrics*, **25**, 271–77.

Jones, R.H. (1990) Serial correlation or random subjects effects? *Communications in Statistics, Part A: Theory and Methods*, **19**, 1105–23.

Jones, R.H. and Boardi-Boteng, F. (1991) Unequally spaced longitudinal data with serial correlation. *Biometrics*, **47**, 161–75.

Jones, B. and Kenward, M.G. (2003) *Design and Analysis of Cross-Over Trials*, Chapman & Hall/CRC, Florida.

Kalish, L.A. (1990) Efficient design for estimation of median lethal dose and quantal dose-response curves. *Biometrics*, **46**, 737–48.

Kenward, M.G. and Rogers, J.H. (1997) Small sample inference for fixed effects from restricted maximum likelihood. *Biometrics*, **53**, 983–97.

Kershner, R.P. and Federer, W.T. (1981) Two-treatment crossover designs for estimating a variety of effects. *Journal of the American Statistical Association*, **76**, 612–19.

Khan, M.K. and Yazdi, A.A. (1988) On D-optimal designs for binary data. *Journal of Statistical Planning and Inference*, **18**, 83–91.

Khuri, A.I. (1984) A note on D-optimal designs for partially nonlinear regression models. *Technometrics*, **26**, 59–61.

Khuri, A., Mukherjee, B., Sinha, K. and Ghosh, M. (2006) Design issues for generalized linear models: a review. *Statistical Science*, **21**, 376–910.

Kiefer, J. (1959) Optimum experimental designs. *Journal of the Royal Statistical Society. Series B*, **21**, 272–319.

Kiefer, J. (1974) General equivalence theory for optimum designs (Approximate theory). *Annals of Statistics*, **2**, 849–79.

Kiefer, J. (1985) *Jack Carl Kiefer Collected Papers III: Design of Experiments*, Springer Verlag, New York.

Kiefer, J. and Wolfowitz, J. (1959) Optimum designs in regression problems. *The Annals of Mathematical Statistics*, **30**, 271–94.

Kiefer, J. and Wolfowitz, J. (1960) The equivalence of two extremum problems. *Canadian Journal of Mathematics*, **12**, 363–66.

King, J. and Wong, W.K. (2000) Minimax D-optimal designs for the logistic model. *Biometrics*, **56**, 1263–67.

Kirk, R.E. (1995) *Experimental Design: Procedures for the Behavioral Sciences*, 3rd edn, Brooks/Cole.

Kirkpatrick, S., Gellat, C.D. and Vecchi, M.P. (1983) Optimization by simulated annealing. *Science*, **220**, 671–80.

Kish, L. (1965) *Survey Sampling*, John Wiley & Sons, Inc., New York.

Kish, L. (1987) *Statistical Design for Research*, John Wiley & Sons, Inc., New York.

Kleinbaum, D.G. and Klein, M. (2002) *Logistic Regression: A Self-Learning Text Series: Statistics for Biology and Health*, Springer Verlag, New York.

Kleinbaum, D.G., Kupper, L.L., Muller, K.E. and Nizam, A. (1998) *Applied Regression Analysis and Multivariable Methods*, Duxbury Press, Belmont, California.

Krewski, D. and Kovar, J. (1982) Low-dose extrapolation under single parameter dose response models. *Communications in Statistics: Simulation and Computation*, **11**, 27–46.

Kshirsagar, A.M. and Smith, W.B. (1995) *Growth Curves*, Marcel Dekker, New York.

Kuhfeld, W.F. and Tobias, R.D. (2005) Large factorial designs for product engineering and marketing research applications. *Technometrics*, **47**, 132–41.

Kunert, J. (1991) Cross-over designs for two treatments and correlated errors. *Biometrika*, **78**, 315–24.

Kushner, H.B. (1997) Optimal repeated measurements designs: the linear optimality equations. *Annals of Statistics*, **25**, 2328–44.

Kushner, H.B. (1998) Optimal and efficient repeated measurements designs for uncorrelated observations. *Journal of the American Statistical Association*, **93**, 1176–87.

Lagarias, J.C., Reeds, J.A., Wright, M.H. and Wright, P.E. (1998) Convergence properties of the nelder-mead simplex method in low dimensions. *SIAM Journal of Optimization*, **9**, 112–47.

Laird, N.M. and Ware, J.H. (1982) Random-effects models for longitudinal data. *Biometrics*, **38**, 963–74.

Landaw, E. (1980). Optimal experimental design for biologic compartmental systems. Department of Biomathematics. PhD Thesis, UCLA.

Lange, K. (1999) *Numerical Analysis for Statisticians*, Springer Verlag, New York.

Langlois, J.H. and Roggeman, L.A. (1990) Attractive faces are only average. *Psychological Science*, **1**, 115–21.

Laska, E.M. and Meisner, M. (1985) A variational approach to optimal two-treatment crossover designs: applications of carryover effect models. *Journal of the American Statistical Association*, **80**, 704–10.

Laska, E.M., Meisner, M. and Kushner, H.B. (1983) Optimal crossover designs in the presence of carryover effects. *Biometrics*, **39**, 1087–91.

Läuter, E. (1974) Experimental design planning in a class of models. *Mathematische Operationsforschung und Statistik*, **5**, 379–98.

Läuter, E. (1976) Optimal multipurpose design for regression model. *Mathematische Operationsforschung und Statistik*, **7**, 51–68.

Lee, C.M.S. (1987) Constrained optimal designs for regression models. *Communications in Statistics, Part A: Theory and Methods*, **16**, 765–83.

Lee, C.M.S. (1988) Constrained optimal designs. *Journal of Statistical Planning and Inference*, **18**, 377–89.

Leibowitz, H.W. and Gwozdicki, J. (1967) The magnitude of the poggendorff illusion as a function of age. *Child Development*, **38**, 573–80.

Leibowitz, H.W. and Judisch, J.M. (1967) The relation between age and the magnitude of the Ponzo illusion. *The American Journal of Psychology*, **80**, 105–9.

Lejeune, M.A. (2003) Heuristic optimization of experimental designs. *European Journal of Operational Research*, **147**, 484–98.

Lima Passos, V. and Berger, M.P.F. (2004) Maximin calibration designs for the nominal response model: an empirical evaluation. *Applied Psychological Measurement*, **28**, 72–87.

Lima Passos, V., Berger, M.P.F. and Tan, F.E.S. (2007) Test design optimization in CAT early stage with the nominal response model. *Applied Psychological Measurement*, **31**, 213–32.

Lima Passos, V., Berger, M.P.F. and Tan, F.E.S. (2008) The D-optimality item selection criterion in the early stage of CAT: A study with the graded response model. *Journal of Educational and Behavioral Statistics*, **33**, 88–110.

Lindsey, J.K. (1993) *Models for Repeated Measurements*, Clarendon Press, Oxford.

Lindsey, J.K. (2001). *Nonlinear Models in Medical Statistics*, *Oxford Statistical Science Series #24*, Oxford University Press.

Little, R.J.A. and Rubin, D.B. (1987) *Statistical Analysis with Missing Data*, John Wiley & Sons, Inc., New York.

Liu, X. (2003) Statistical power and optimum allocation ratio for treatment and control having unequal costs per unit of randomization. *Journal of Educational and Behavioral Statistics*, **28**, 231–48.

Lloyd, T., Andon, M.B., Rollings, N. *et al.* (1993) Calcium supplementation and bone mineral density in adolescent girls. *Journal of the American Medical Association*, **270**, 841–44.

Lopez-Fidalgo, J., Tommasi, C. and Trandafir, P.C. (2008) Optimal designs for discriminating between some extensions of the Michaelis-Menten model. *Journal of Statistical Planning and Inference*, **138**, 3797–804.

Mandal, N.K. and Heiligers, B. (1992) Minimax designs for estimating the optimum point in a quadratic response surface. *Journal of Statistical Planning and Inference*, **31**, 235–44.

Mathew, Th. and Sinha, B.K. (2001) Optimal designs for binary data under logistic regression. *Journal of Statistical Planning and Inference*, **93**, 295–307.

Matlab (2004) *Handbook Matlab*, Version 7.0.1.(R14), Mathworks Inc., http://www.mathworks.com/products/matlab.

Matthews, J.N.S. (1987) Optimal crossover designs for comparison of two treatments in the presence of carryover effects and autocorrelated errors. *Biometrika*, **74** (2), 311–20.

Matthews, J.N.S. (1988) Recent developments in crossover designs. *International Statistical Review*, **56**, 117–27.

Matthews, J.N.S. (1990) Optimal dual-balanced two treatment crossover designs. *Sankhya*, **52**, 332–37.

Matthews, J.N.S., Altman, D.G., Campbell, M.J. and Royston, P. (1990) Analysis of serial measurements in medical research. *British Medical Journal*, **300**, 230–35.

Mauchly, J.W. (1940) Significance test for sphericity of a normal n-variate distribution. *The Annals of Mathematical Statistics*, **29**, 204–9.

McClelland, G.H. (1997) Optimal design in psychological research. *Psychological Methods*, **2**, 3–19.

McCulloch, C.E. and Searle, S.R. (2001) *Generalized, Linear, and Mixed Models*, John Wiley & Sons, Inc., New York.

Mentré, F., Mallet, A. and Baccar, D. (1997) Optimal design in random effect regression models. *Biometrika*, **84**, 429–42.

Meyer, R.K. and Nachtsheim, C.J. (1988) Simulated annealing in the construction of exact optimal design of experiments. *American Journal of Mathematical and Management Sciences*, **8**, 329–59.

Michaelis, L. and Menton, M.L. (1913) Die Kinetik der invertinwirkung. *Biochemische Zeitschrift*, **4**, 334–69.

Miller, R.G. (1966) *Simultaneous Statistical Inference*, McGraw-Hill, New York.

Miller, A.J. and Nguyen, N.-K. (1994) A Fedorov exchange algorithm for D-optimal design. *Applied Statistics*, **43**, 669–78.

MINITAB® Statistical Software (2008) Release 15. 2008, Minitab Inc.

Minkin, S. (1987) Optimal designs for binary data. *Journal of the American Statistical Association*, **82**, 1098–103.

Mitchell, T.J. (1974) An algorithm for the construction of 'D-optimum' experimental designs. *Technometrics*, **16**, 203–10.

Moerbeek, M. (2005a) Robustness properties of A-, D-, and E-optimal designs for polynomial growth models with auto correlated errors. *Computational Statistics and Data Analysis*, **48**, 765–78.

Moerbeek, M. (2005b) Randomization of clusters versus randomization of persons within clusters: which is preferable? *The American Statistician*, **59**, 72–78.

Moerbeek, M. (2006) Power and money in cluster randomized trials: when is it worth measuring a covariate? *Statistics in Medicine*, **25**, 2607–17.

Moerbeek, M. (2008) Powerful and cost-efficient designs for longitudinal intervention studies with two treatment groups. *Journal of Educational and Behavioral Statistics*, **33**, 41–61.

Moerbeek, M., Piersma, A.H. and Slob, W. (2004) A comparison of three methods for calculating confidence intervals for benckmark dose. *Risk Analysis*, **24**, 31–40.

Moerbeek, M., van Breukelen, G.J.P., Ausems, M. and Berger, M.P.F. (2003) Optimal sample sizes in experimental designs with individuals nested within clusters. *Understanding Statistics*, **2**, 151–75.

Moerbeek, M., van Breukelen, G.J.P. and Berger, M.P.F. (2000) Design issues for experiments in multilevel populations. *Journal of Educational and Behavioral Statistics*, **25** (2), 271–84.

Moerbeek, M., van Breukelen, G.J.P. and Berger, M.P.F. (2001) Optimal experimental designs for multilevel models with covariates. *Communications in Statistics, Part A: Theory and Methods*, **30**, 2683–97.

Moerbeek, M., van Breukelen, G.J.P. and Berger, M.P.F. (2008) Optimal designs for multilevel studies, in *Handbook of Multilevel Analysis*, (eds J. De.Leeuw and E. Meijer), Springer, New York.

Mok, M. (1995) Sample size requirements for 2-level designs in educational research. *Multilevel Newsletter*, **7**, 11–15.

Montepiedra, G. and Wong, W.K. (2001) A new criterion when heteroscedasticity is ignored. *Annals of the Institute of Statistical Mathematics*, **53** (2), 418–26.

Montgomery, D.C. (2000) *Design and Analysis of Experiments*, 5th edn, John Wiley & Sons, Inc., New York.

Montgomery, D.C. and Peck, E.A. (1992) *Introduction to Linear Regression Analysis*, John Wiley & Sons, Inc., New York.

Motulsky, H. and Christopoulos, A. (2004) *Fitting Models to Biological Data using Linear and Nonlinear Regression: A Practical Guide to Curve Fitting*, Oxford University Press, London.

Murray, D.M. (1998) *Design and Analysis of Group-Randomized Trials*, Oxford University Press, New York.

Murray, D.M., Varnell, S.P. and Blitstein, J.L. (2004) Design and analysis of group-randomized trials: a review of recent methodological developments. *American Journal of Public Health*, **94**, 423–32.

Neter, J., Kutner, M.H. and Wasserman, W. (1985) *Applied Linear Statistical Models*, 2nd edn, Richard Irwin, Illinois.

Neter, J., Wasserman, W. and Kutner, M.H. (1983) *Applied Linear Regression Models*, Richard D. Irwin, Homewood.

Nguyen, N.-K. and Miller, A.J. (1992) A review of some exchange algorithms for constructing discrete d-optimal designs. *Computational Statistics and Data Analysis*, **14**, 489–98.

Nuremberg Code (1947) Mitscherlich, A. and Mielke, F. (eds) (1949) *Doctors of Infamy: The Story of the Nazi Medical Crimes*, Schuman, New York, pp. xxiii–xxv.

Ortega-Azurduy, S.A., Tan, F.E.S. and Berger, M.P.F. (2008a) On the Incorrect Specification of Optimal Designs for Heterogeneous Mixed Models, Research Report, University of Maastricht.

Ortega-Azurduy, S.A., Tan, F.E.S. and Berger, M.P.F. (2008b) The effect of drop-out on the efficiency of D-optimal designs of linear mixed models. *Statistics in Medicine*, **27**, 2601–17.

Ortega-Azurduy, S.A., Tan, F.E.S. and Berger, M.P.F. (2009) Highly efficient designs to handle the incorrect specification of linear mixed models. *Communications in Statistics: Simulation and Computation*, **38**, 14–30.

Ouwens, M.J.N.M., Tan, F.E.S. and Berger, M.P.F. (2002) Maximin D-optimal designs for longitudinal mixed effects models. *Biometrics*, **58**, 735–41.

Ouwens, M.J.N.M., Tan, F.E.S. and Berger, M.P.F. (2006) A maximin criterion for the logistic random intercept model with covariates. *Journal of Statistical Planning and Inference*, **136**, 962–81.

Palmer, C.R. (2002) Ethics, data-dependent designs and the strategy of clinical trials: time to start learning-as-we-go. *Statistical Methods in Medical Research*, **11**, 381–402.

Parkes, K.R. (1982) Occupational stress among student nurses: a natural experiment. *The Journal of Applied Psychology*, **67**, 784–96.

Patel, H. (1983) Use of baseline measurements in the two-period crossover design. *Communications in Statistics: Theory and Methods*, **12**, 2693–712.

Pázman, A. (1993) Higher dimensional nonlinear regression – a statistical use of the Riemannian curvature tensor. *Statistics*, **25**, 17–25.

Pocock, S.J. (1983) *Clinical Trials: A Practical Approach*, John Wiley & Sons, Inc., New York.

Pocock, S.J. and Elbourne, D.R. (2000) Randomized trials or observational tribulations. *The New England Journal of Medicine*, **342**, 1907–9.

Poland, J., Mitterer, A., Knödler, K. and Zell, A. (2001) Genetic algorithms can improve the construction of D-optimal experimental designs, in *Advances in Fuzzy Systems and Evolutionary Computation (Proceedings of WSES Conference on Evolutionary Computation EC 2001)*, (ed. N. Mastorakis), WSES Press International, pp. 227–31.

Porter, A.C. and Raudenbush, S.W. (1987) Analysis of covariance: its model and use in psychological research. *Journal of Counseling Psychology*, **34**, 383–92.

Preitschopf, F. and Pukelsheim, F. (1987) Optimal designs for quadratic regression. *Journal of Statistical Planning and Inference*, **16**, 213–18.

Prentice, R.L. (1976) A generalization of the probit and logit methods for dose response curves. *Biometrics*, **32**, 761–68.

Pronzato, L. (2003) Removing non-optimal support points in D-optimum design algorithms. *Statistics and Probability Letters*, **63**, 223–28.

Pukelsheim, F. (1993) *Optimal Design of Experiments*, John Wiley & Sons, Inc., New York.

Pukelsheim, F. and Rieder, S. (1992) Efficient rounding of approximate designs. *Biometrika*, **79**, 763–70.

Raaijmakers, J.G.W. (1987) Statistical analysis of the Michaelis-Menten equation. *Biometrics*, **43**, 793–80.

Rao, C.R. (1973) *Linear Statistical Inference and its Applications*, 2nd edn, John Wiley & Sons, Inc., New York.

Ratkowsky, D.A., Evans, M.A. and Alldredge, J.R. (1993) *Cross-Over Experiments: Design, Analysis and application*, Marcel Dekker, New York.

Raudenbush, S.W. (1997) Statistical analysis and optimal design for cluster randomized trials. *Psychological Methods*, **2** (2), 173–85.

Raudenbush, S.W. and Bryk, A.S. (2002) *Hierarchical Linear Models: Applications and Data Analysis Methods*, 2nd edn, Sage Publications, Thousand Oaks.

Raudenbush, S.W. and Liu, X. (2000) Statistical power and optimal design for multi site trials. *Psychological Methods*, **5**, 199–213.

Raudenbush, S.W., Spybrook, J., Liu, X. and Congdon, R. (2004) *Optimal Design for Longitudinal and Multilevel Research: Documentation for the Optimal Design Software, Version 1.76*, University of Michigan, Ann Arbor.

Rochon, J. (1992) ARMA covariance structures with time heteroscedasticity for repeated measures experiments. *Journal of the American Statistical Association*, **87**, 777–84.

Rosenberger, W.F. and Grill, S.E. (1997) A sequential design for psychophysical experiments: an application to estimating timing of sensory events. *Statistics in Medicine*, **16**, 2245–60.

Rothman, K.J. and Greenland, S. (1998) *Modern Epidemiology*, 2nd edn, Lippincot-Raven, Philadelphia.

Rouanet, H. and Lepine, D. (1970) Comparison between treatments in a repeated measures design: ANOVA and multivariate methods. *The British Journal of Mathematical and Statistical Psychology*, **23**, 147–63.

SAS (1999) *SAS/QC User's Guide*, Version 8, SAS Institute, Cary.

Sastry, N., Ghosh-Dastidar, B., Adams, J. and Pebley, A.R. (2006) The design of a multilevel survey of children, families and communities: the Los Angeles Family and Neighborhood Survey. *Social Science Research*, **35**, 1000–24.

Searle, S., Casella, G. and McCulloch, C. (1992) *Variance Components*, John Wiley & Sons, Inc., New York.

Sebastiani, P. and Settimi, R. (1997) A note on D-optimal designs for a logistic regression model. *Journal of Statistical Planning and Inference*, **59**, 359–68.

Sebastiani, P. and Settimi, R. (1998) First-order optimal designs for non-linear models. *Journal of Statistical Planning and Inference*, **74**, 177–92.

Senn, S. (2002) *Cross-Over Trials in Clinical Research*, John Wiley & Sons, Inc., Chichester, England.

Senn, S. and Hildebrand, H. (1991) Cross-over trials, degrees of freedom, the carryover problem and its dual. *Statistics in Medicine*, **10**, 1361–74.

Senn, S. and Lee, S. (2004) The analysis of the AB/BA cross-over trial in the medical literature. *Pharmaceutical Statistics*, **3**, 123–31.

Sexton, C.J., Anthony, D.K., Lewis, S.M. *et al.* (2006) Design of experiment algorithms for assembled products. *Journal of Quality Technology*, **38**, 298–308.

Shadish, W.R., Cook, T.D. and Campbell, D.T. (2002) *Experimental and Quasi-Experimental Designs for Generalized Causal Inference*, Houghton-Mifflin, Boston.

Silvey, S.D. (1980) *Optimal Design*, Chapman and Hall, London.

Sitter, R. (1992) Robust designs for binary data. *Biometrics*, **48**, 1145–55.

Sitter, R.R. and Torsney, B. (1995) Optimal designs for binary response experiments with two design variables. *Statistica Sinica*, **5**, 405–19.

Sitter, R.R. and Wu, C.-F.J. (1999) Two-stage design of quantal response studies. *Biometrics*, **55**, 396–402.

Sloane, N.J.A. and Hardin, R.H. (2003). GOSSET: A General-purpose Program for Designing Experiments, www.research.att.com/~njas/gosset/.

Slob, W. (2002) Dose–response modeling of continuous endpoints. *Toxicological Sciences*, **66**, 298–312.

Slob, W. and Pieters, M.N. (1998) A probabilistic approach for deriving acceptable human intake limits and human health risks form toxicological studies: general framework. *Risk Analysis*, **18**, 787–98.

Slob, W., Moerbeek, M., Rauniomaa, E. and Piersma, A.H. (2005) A statistical evaluation of toxicity study designs for the estimation of the benchmark dose in continuous endpoints. *Toxicological Sciences*, **84**, 167–85.

Smith, K. (1918) On the standard deviations of adjusted and interpolated values of an observed polynomial function and its constants and the guidance they give towards a proper choice of the distribution of observations. *Biometrika*, **12**, 1–85.

Snijders, T.A.B. and Bosker, R.J. (1993) Standard errors and sample sizes for two-level research. *Journal of Educational Statistics*, **18**, 237–59.

Snijders, T.A.B. and Bosker, R.J. (1999) *Multilevel Analysis: An Introduction to Basic and Advanced Multilevel Modeling*, Sage Publications, London.

Soria, I., Harrison, L., Myhre, P. *et al.* (2002) Bioequivalence of press-and-breathe and breath-actuated inhalers of beclomethasone dipropinate extrafine aerosol. *Clinical Drug Investigation*, **22**, 523–31.

Stehlik, M. (2006) *Some Properties of Exchange Design Algorithms Under Correlation*, Research Report Series No. 28, Wirtschafts Universität, Wien, http://statmath.wu-wien.ac.at/.

Stevens, S.S. (1951) Mathematics, measurement and psychophysics, in *Handbook of Eperimental Psychology*, (ed. S.S. Stevens), John Wiley & Sons, Inc., New York, pp. 1–49.

Stocking, M.L. (1990) Specifying optimum examinees for item parameter estimation in item response theory. *Psychometrika*, **55**, 461–75.

Stoddard, J.T. (1886) Composite portraiture. *Science*, **8** (182), 89–91.

Stufken, J. (1996) Optimal cross over designs, in *Handbook of Statistics, 13*, (eds S. Ghosh and C.R. Rao), North Holland, Amsterdam, pp. 63–90.

Tan, F.E.S. and Berger, M.P.F. (1999) Optimal allocation of time points for random effects models. *Communications in Statistics: Simulations and Computations*, **28**, 517–40.

Tekle, F.B., Tan, F.E.S. and Berger, M.P.F. (2008a) Highly efficient designs for logistic models with categorical variables. *Communications in Statistics: Theory and Methods*, **37**, 746–59.

Tekle, F.B., Tan, F.E.S. and Berger, M.P.F. (2008b) D-optimal cohort designs for linear mixed effects models. *Statistics in Medicine*, **27**, 2586–600.

Tekle, F.B., Tan, F.E.S. and Berger, M.P.F. (2009) Interactive computer program for optimal designs of longitudinal studies. *Computer Methods and Programs in Biomedicine*, Doi:10.1016/j.cmpb.2008.11.002.

Tsutakawa, R.K. (1972) Design of experiment for bioassay. *Journal of the American Statistical Association*, **67**, 584–90.

Ucinski, D. and Atkinson, A.C. (2004). Experimental design for time-dependent models with correlated observations. *Studies in Nonlinear Dynamics and Econometrics*, **8**, 2, *Article 13*. http:/www.bepress.com/snde/vol8/iss2/art13.

US Government Printing Office (1949) *Trials of War Criminals before the Nuremberg Military Tribunals under Control Council Law*, **2** (10), Author, Washington, DC, pp. 181–82.

Vallejo, G. and Livacic-Rojas, P. (2005) Comparison of two procedures for analyzing small sets of repeated measures data. *Multivariate Behavioral Research*, **40** (2), 179–205.

Van Breukelen, G.J.P. (2006) ANOVA versus change from baseline had more power in randomized studies and more bias in nonrandomized studies. *Journal of Clinical Epidemiology*, **59**, 920–25.

Van Breukelen, G.J.P., Candel, M.J.J.M. and Berger, M.P.F. (2007) Relative efficiency of unequal versus equal cluster sizes in cluster randomized and multicentre trials. *Statistics in Medicine*, **26**, 2589–603.

Van Breukelen, G.J.P., Candel, M.J.J.M. and Berger, M.P.F. (2008) Relative efficiency of unequal cluster sizes for variance component estimation in cluster randomized and multicentre trials. *Statistical Methods in Medical Research*, **17**, 439–58.

Van Dongen, S. (2006) Prior specification in Bayesian statistics: three cautionary tales. *Journal of Theoretical Biology*, **242**, 90–100.

Van der Linden, W.J. (2005) *Linear Models for Optimal Test Design*, Springer, New York.

Van der Linden, W.J. and Hambleton, R.K. (1997) *Handbook of Modern Test Theory*, Springer, New York.

Vander Linden, W.J. and Glas, C.A.W. (eds) (2000) *Computerized Adaptive Testing: Theory and Practice*, Kluwer, Dordrecht.

Verbeke, G. and Molenberghs, G. (2000) *Linear Mixed Models for Longitudinal Data*, Springer Verlag, New York.

Vickers, A.J. (2003) How many repeated measures in repeated measures designs? Statistical issues for comparative trials. *BMC Medical Research Methodology*, **3**, 1–9.

Volund, A. (1978) Applications of the four-parameter logistic model to bioassay: comparison with slope ratio and parallel line models. *Biometrics*, **34**, 357–65.

Vonesh, E.F. and Chinchilli, V.M. (1997) *Linear and Nonlinear Models for the Analysis of Repeated measurements*, Marcel Dekker, New York.

Wainer, H. (2000) *Computerized Adaptive Testing: a Primer*, Lawrence Erlbaum Associates.

Wallenstein, S. (1979) Inclusion of baseline values in the analysis of crossover designs (abstract). *Biometrics*, **35**, 894.

Walter, S.D. (1976) The estimation and interpretation of attributable risk in health research. *Biometrics*, **32**, 829–49.

Walter, S.D. (1977) Determination of significant relative risks and optimal samplingprocedures in prospective and retrospective comparative studies of various sizes. *American Journal of Epidemiology*, **106**, 387–97.

Wheeler, R.E. (2004) AlgDesign. The R Project for Statistical Computing, http://www.r-project.org/.

Willett, J.B., Singer, J.D. and Martin, N.C. (1998) The design and analysis of longitudinal studies of development and psychopathology in context: Statistical models and methodological recommendations. *Development and Psychopathology*, **10**, 395–426.

Winer, B.J., Brown, D.R. and Michels, K.M. (1991). *Statistical Principles in Experimental Design*, 3rd edn, McGraw-Hill.

Winkens, B., Schouten, H.J.A., van Breukelen, G.J.P. and Berger, M.P.F. (2005) Optimal time-points in clinical trials with linearly divergent treatment effects. *Statistics in Medicine*, **24**, 3743–56.

Winkens, B., Schouten, H.J.A., van Breukelen, G.J.P. and Berger, M.P.F. (2006) Optimal number of repeated measures and group sizes in clinical trials with linear divergent treatment effects. *Contemporary Clinical Trials*, **27**, 57–69.

Wise, M.E. (1985) Negative power functions of time in pharmacokinetics and their implications. *Journal of Pharmacokineticsand Biopharmaceutics*, **13**, 309–46.

Wong, W.K. (1990). Heteroscedastic G-optimal designs. PhD Thesis, School of Statistics, University of Minnesota.

Wong, W.K. (1992) A unified approach to the construction of minimax designs. *Biometrika*, **79**, 611–20.

Wong, W.K. (1994) Comparing robustness properties of A, D, E and G-optimal designs. *Computational Statistics and Data Analysis*, **18**, 441–48.

Wong, W.K. (1995a) On the equivalence of D- and G-optimal designs in heteroscedastic models. *Statistics and Probability Letters*, **25**, 317–21.

Wong, W.K. (1995b) A graphical approach to constructing constrained D- and L-optimal designs using efficiency plots. *Journal of Statistical Simulation and Computation*, **53**, 143–52.

Wong, W.K. (1999) Recent advances in constrained optimal design strategies. *Statistica Neerlandica*, **53**, 257–76.

Wong, W.K. and Cook, R.D. (1993) Heteroscedastic G-optimal designs. *Journal of the Royal Statistical Society. Series B*, **55**, 871–80.

Wong, W.K. and Lachenbruch, P.A. (1996) Designing studies for dose response. *Statistics in Medicine*, **15**, 343–60.

Worchel, S. and Shakelford, S.L. (1991) Groups under stress: the influence of group structure and environment on process and performance. *Personality and Social Psychology Bulletin*, **17**, 640–47.

Woutersen, R.A., Jonker, D., Sevenson, H. *et al.* (2001) The bench approach applied to a 28-day toxicity study with Rhodorsil silane in rats: the impact of increasing the number of dose groups. *Food and Chemical Toxicology*, **39**, 697–707.

Wu, C.F.J. (1985) Efficient sequential designs with binary data. *Journal of the American Statistical Association*, **80**, 974–84.

Wu, C.F.J. (1988) Optimal design for percentile estimation of a quantal response curve, in *Optimal Design and Analysis of Experiments*, (eds Y. Dodge, V., Federov and H.P. Wynn), Elsevier Science Publishers, North-Holland, pp. 213–24.

Wu, C.F.J. and Wynn, H.P. (1978a) The convergence of general step length algorithm for regular optimum design criteria. *Annals of Statistics*, **6**, 1286–301.

Wu, C.F.J. and Wynn, H.P. (1978b) General-step-length algorithms for optimal design criteria. *Annals of Statistics*, **6**, 1273–85.

Wynn, H.P. (1970) The sequential generation of D-optimal experimental designs. *The Annals of Mathematical Statistics*, **41**, 1055–64.

Wynn, H.P. (1972) Results in the theory and construction of D-optimal experimental designs. *Journal of the Royal Statistical Society. Series B*, **34**, 133–47.

Yu, R.C. and Rappaport, S.M. (1996) Relation between pulmonary clearance and particle burden: a Michaelis-Menten-like kinetic model. *Occupational and Environmental Medicine*, **53**, 567–72.

Yuan, Y. and Zhou, J. (2005). Cost-efficient higher-order crossover designs for two treatment clinical trials. *Pharmaceutical Statistics*, **4**, 245–52.

Zhu, W., Ahn, H. and Wong, W.K. (1998) Multiple-objective optimal designs for the logit model. *Communications in Statistics: Theory and Methods*, **27**, 1581–92.

Zhu, W. and Wong, W.K. (2000a) Multiple-objective designs in a dose response experiment. *Journal of Biopharmaceutical Statistics*, **10**, 1–14.

Zhu, W. and Wong, W.K. (2000b) Bayesian optimal designs for estimating a set of symmetric quantiles. *Statistics in Medicine*, **20**, 123–37.

Author Index

Abdelbasit, K. M., 126
Abt, M., 200
Afsarinejad, K., 216, 221, 229
Ahn, H., 268, 269
Akaike, H., 194
Alldredge, J. R., 213
Allison, D. B., 165
American Psychological
    Association, 16
Ankerman, B. E., 165, 188
Atkins, J. E., 200
Atkinson, A. C., vii, 2, 3, 9, 30, 40,
    43, 62, 63, 79, 200, 207, 240,
    241, 255, 265, 284, 286
Atwood, C. L., 245
Aviles, A. I., 165, 188

Baba, R., 274
Baccar, D., 200
Bailey, R. A., 213
Balaam, L. N., 217, 231
Begg, C. C., 271
Benson, K., 9
Berger, M. P. F., 3, 26, 124, 137,
    148, 150, 151, 155, 165, 171,
    179, 197, 200, 201, 207, 208,
    271, 286
Birkett, N., 155
Bischoff, W., 200
Blitstein, J. L., 147
Boardi-Boteng, F., 186
Bock, R. D., 21, 57, 71, 107
Bosker, R. J., 143, 148, 155, 166,
    176

Box, G. E. P., 2, 9, 15, 40, 66,
    257
Bozdogan, H., 194
Brown, B. W., 225, 227
Brown, D. R., 8, 109
Brown, L. D., 247
Brown, M. B., 189
Bryk, A. S., 143
Buck, C., 155

Campbell, D. T., vii, 1, 9, 10, 12,
    100, 147, 215
Candel, M. J. J. M., 148, 150, 151,
    166, 171
Carriere, K. C., 229, 231
Carrol, R. J., 66
Casella, G., 148, 150, 171, 188,
    191, 223
Chaloner, K., 134, 243, 246, 247,
    267, 268, 270, 271
Chassan, J. B., 225
Chatterjee, S., 130
Chen, J. J., 241
Chen, W., 285
Cheng, C. S., 200, 229, 230, 231
Chernoff, H., 262
Chi, E. M., 184, 186
Chinchilli, V. M., 176, 189, 221,
    223
Chow, S. C., 11
Christopoulos, A., 257, 267
Cohen, M. P., 155, 169
Concato, J., 9
Cook, R. D., 207, 248, 253, 283

An Introduction to Optimal Designs for Social and Biomedical Research M. P. F. Berger & W. K. Wong
© 2009, John Wiley & Sons, Ltd

Cook, T. D., vii, 1, 9, 10, 12, 100,
    147, 215
Cox, D. R., vii, 1, 2, 9, 66, 215
Crainiceanu, C. M., 188
Crowder, M. J., 176

Davidson, M. L, ., 181
Declaration of Helsinki, 16
Derendorf, H., 261
Dette, H., 71, 118, 120, 121, 249,
    250, 266, 273, 274, 292
Diggle, P. J., 176, 181, 184, 223
Donev, A. N., vii, 2, 9, 30, 40, 43,
    62, 63, 79, 200, 240, 241,
    265, 284
Donner, A., 155
Draper, N. R., 9
Dror, H. A., 285
Duimel-Peeters, I. G. P., 155
Duncan, G. T., 247
Dunn, G., 260
Dunn, J. R., 275
Dykstra, O., 279

Ebbutt, A. F., 215
Elbourne, D. R., 9
Emanuel, E. J., 16
Esinhart, J. D., 223
Evans, M. A., 213
Everitt, B. S., 180

Faddy, M. J., 259
Federer, W. T., 215
Fedorov, V. V., vii, 2, 197, 245,
    249, 255, 272, 278, 283
Feldt, L. S., 181, 185
Fisher, R. A., 2, 15, 107
Fleiss, J. L., 215, 232
Flynn, T. N., 163
Ford, I., 257
Fornius, E. F., 243
Forsythe, A. B., 189
Freeman, P., 216
Frison, L., 180

Gaffke, N., 70
Gagnon, R. C., 197
Galbraith, S., 208
Galil, Z., 283
Galton, F., 23
Gaudard, M. A., 273
Gavasakar, U., 247
Gaylor, D. W., 241
Geisser, S., 181, 185
Gellat, C. D., 284
Glas, C. A. W., 135
Goldstein, H., 143, 176
Goos, P., 40, 63, 79, 124, 189
Gottschalk, P. G., 275
Grady, C., 16
Greenhouse, S., 181, 185
Greenland, S., 8
Grill, S. E., 272
Grizzle, J. E., 216
Guest, P., 66, 200
Gwozdicki, J., 21

Hahn, E. D., 247
Haines, L. M., 284
Hambleton, R. K., 24, 25
Hand, D. J., 176
Hansen, M. H., 169
Hardin, R. H., 285
Harman, R., 284
Hartz, A. J., 9
Harville, D. A., 188, 191, 223, 283
Headrick, T. C., 165
Hedayat, A., 216, 229, 230
Heiligers, B., 245
Heredia-Lagner, A., 285
Hildebrand, H., 232
Hill, D. H., 255, 263
Hill, W. J., 255
Holmberg, A., 260
Horwitz, R. I., 9
Hosmer, D. W., 116
Houston, J. B., 260
Hox, J. J., 143, 148, 166
Hsieh, F. Y., 155

Subject Index

STATISTICS IN PRACTICE

Human and Biological Sciences

Berger – Selection Bias and Covariate Imbalances in Randomized Clinical Trials

Brown and Prescott – Applied Mixed Models in Medicine, Second Edition

Chevret (Ed) – Statistical Methods for Dose-Finding Experiments

Ellenberg, Fleming and DeMets – Data Monitoring Committees in Clinical Trials: A Practical Perspective

Hauschke, Steinijans & Pigeot – Bioequivalence Studies in Drug Development: Methods and Applications

Lawson, Browne and Vidal Rodeiro – Disease Mapping with WinBUGS and MLwiN

Lesaffre, Feine, Leroux & Declerck – Statistical and Methodological Aspects of Oral Health Research

Lui – Statistical Estimation of Epidemiological Risk

Marubini and Valsecchi – Analysing Survival Data from Clinical Trials and Observation Studies

Molenberghs and Kenward – Missing Data in Clinical Studies

O'Hagan, Buck, Daneshkhah, Eiser, Garthwaite, Jenkinson, Oakley & Rakow – Uncertain Judgements: Eliciting Expert's Probabilities

Parmigiani – Modeling in Medical Decision Making: A Bayesian Approach

Pintilie – Competing Risks: A Practical Perspective

Senn – Cross-over Trials in Clinical Research, Second Edition

Senn – Statistical Issues in Drug Development, Second Edition

Spiegelhalter, Abrams and Myles – Bayesian Approaches to Clinical Trials and Health-Care Evaluation

Whitehead – Design and Analysis of Sequential Clinical Trials, Revised Second Edition

Whitehead – Meta-Analysis of Controlled Clinical Trials

Willan and Briggs – Statistical Analysis of Cost Effectiveness Data

Winkel and Zhang – Statistical Development of Quality in Medicine

Earth and Environmental Sciences

Buck, Cavanagh and Litton – Bayesian Approach to Interpreting Archaeological Data
Glasbey and Horgan – Image Analysis in the Biological Sciences
Helsel – Nondetects and Data Analysis: Statistics for Censored Environmental Data
Illian, Penttinen, Stoyan, H and Stoyan D-Statistical Analysis and Modelling of Spatial Point Patterns
McBride – Using Statistical Methods for Water Quality Management
Webster and Oliver – Geostatistics for Environmental Scientists, Second Edition
Wymer (Ed) – Statistical Framework for Recreational Water Quality Criteria and Monitoring

Industry, Commerce and Finance

Aitken – Statistics and the Evaluation of Evidence for Forensic Scientists, Second Edition
Agung – Time Series Data Analysis Using EViews
Balding – Weight-of-evidence for Forensic DNA Profiles
Brandimarte – Numerical Methods in Finance and Economics: A MATLAB-Based Introduction, Second Edition
Brandimarte and Zotteri – Introduction to Distribution Logistics
Chan – Simulation Techniques in Financial Risk Management
Coleman, Greenfield, Stewardson and Montgomery (Eds) – Statistical Practice in Business and Industry
Frisen (Ed) – Financial Surveillance
Fung and Hu – Statistical DNA Forensics
Lehtonen and Pahkinen – Practical Methods for Design and Analysis of Complex Surveys, Second Edition
Ohser and Mücklich – Statistical Analysis of Microstructures in Materials Science
Pourret, Naim & Marcot (Eds) – Bayesian Networks: A Practical Guide to Applications
Taroni, Aitken, Garbolino and Biedermann – Bayesian Networks and Probabilistic Inference in Forensic Science